Probability and Mathematical Statistics (Con

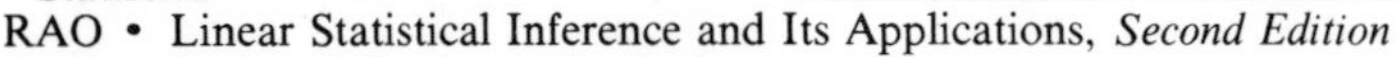

PURI and SEN • Nonparametric Meth
RANDLES and WOLFE • Introductic
Statistics
RAO • Linear Statistical Inference and Its Applications, *Second Edition*
RAO • Real and Stochastic Analysis
RAO and SEDRANSK • W.G. Cochran's Impact on Statistics
RAO • Asymptotic Theory of Statistical Inference
ROHATGI • An Introduction to Probability Theory and Mathematical Statistics
ROHATGI • Statistical Inference
ROSS • Stochastic Processes
RUBINSTEIN • Simulation and The Monte Carlo Method
SCHEFFE • The Analysis of Variance
SEBER • Linear Regression Analysis
SEBER • Multivariate Observations
SEN • Sequential Nonparametrics: Invariance Principles and Statistical Inference
SERFLING • Approximation Theorems of Mathematical Statistics
SHORACK and WELLNER • Empirical Processes with Applications to Statistics
TJUR • Probability Based on Radon Measures

Applied Probability and Statistics

ABRAHAM and LEDOLTER • Statistical Methods for Forecasting
AGRESTI • Analysis of Ordinal Categorical Data
AICKIN • Linear Statistical Analysis of Discrete Data
ANDERSON, AUQUIER, HAUCK, OAKES, VANDAELE, and WEISBERG • Statistical Methods for Comparative Studies
ARTHANARI and DODGE • Mathematical Programming in Statistics
BAILEY • The Elements of Stochastic Processes with Applications to the Natural Sciences
BAILEY • Mathematics, Statistics and Systems for Health
BARNETT • Interpreting Multivariate Data
BARNETT and LEWIS • Outliers in Statistical Data, *Second Edition*
BARTHOLOMEW • Stochastic Models for Social Processes, *Third Edition*
BARTHOLOMEW and FORBES • Statistical Techniques for Manpower Planning
BECK and ARNOLD • Parameter Estimation in Engineering and Science
BELSLEY, KUH, and WELSCH • Regression Diagnostics: Identifying Influential Data and Sources of Collinearity
BHAT • Elements of Applied Stochastic Processes, *Second Edition*
BLOOMFIELD • Fourier Analysis of Time Series: An Introduction
BOX • R. A. Fisher, The Life of a Scientist
BOX and DRAPER • Empirical Model-Building and Response Surfaces
BOX and DRAPER • Evolutionary Operation: A Statistical Method for Process Improvement
BOX, HUNTER, and HUNTER • Statistics for Experimenters: An Introduction to Design, Data Analysis, and Model Building
BROWN and HOLLANDER • Statistics: A Biomedical Introduction
BUNKE and BUNKE • Statistical Inference in Linear Models, Volume I
CHAMBERS • Computational Methods for Data Analysis
CHATTERJEE and PRICE • Regression Analysis by Example
CHOW • Econometric Analysis by Control Methods
CLARKE and DISNEY • Probability and Random Processes: A First Course with Applications, *Second Edition*
COCHRAN • Sampling Techniques, *Third Edition*
COCHRAN and COX • Experimental Designs, *Second Edition*
CONOVER • Practical Nonparametric Statistics, *Second Edition*
CONOVER and IMAN • Introduction to Modern Business Statistics
CORNELL • Experiments with Mixtures: Designs, Models and The Analysis of Mixture Data

Applied Probability and Statistics (Continued)

COX • Planning of Experiments
COX • A Handbook of Introductory Statistical Methods
DANIEL • Biostatistics: A Foundation for Analysis in the Health Sciences, *Third Edition*
DANIEL • Applications of Statistics to Industrial Experimentation
DANIEL and WOOD • Fitting Equations to Data: Computer Analysis of Multifactor Data, *Second Edition*
DAVID • Order Statistics, *Second Edition*
DAVISON • Multidimensional Scaling
DEGROOT, FIENBERG and KADANE • Statistics and the Law
DEMING • Sample Design in Business Research
DILLON and GOLDSTEIN • Multivariate Analysis: Methods and Applications
DODGE • Analysis of Experiments with Missing Data
DODGE and ROMIG • Sampling Inspection Tables, *Second Edition*
DOWDY and WEARDEN • Statistics for Research
DRAPER and SMITH • Applied Regression Analysis, *Second Edition*
DUNN • Basic Statistics: A Primer for the Biomedical Sciences, *Second Edition*
DUNN and CLARK • Applied Statistics: Analysis of Variance and Regression
ELANDT-JOHNSON and JOHNSON • Survival Models and Data Analysis
FLEISS • Statistical Methods for Rates and Proportions, *Second Edition*
FLEISS • The Design and Analysis of Clinical Experiments
FOX • Linear Statistical Models and Related Methods
FRANKEN, KÖNIG, ARNDT, and SCHMIDT • Queues and Point Processes
GIBBONS, OLKIN, and SOBEL • Selecting and Ordering Populations: A New Statistical Methodology
GNANADESIKAN • Methods for Statistical Data Analysis of Multivariate Observations
GREENBERG and WEBSTER • Advanced Econometrics: A Bridge to the Literature
GROSS and HARRIS • Fundamentals of Queueing Theory, *Second Edition*
GUPTA and PANCHAPAKESAN • Multiple Decision Procedures: Theory and Methodology of Selecting and Ranking Populations
GUTTMAN, WILKS, and HUNTER • Introductory Engineering Statistics, *Third Edition*
HAHN and SHAPIRO • Statistical Models in Engineering
HALD • Statistical Tables and Formulas
HALD • Statistical Theory with Engineering Applications
HAND • Discrimination and Classification
HOAGLIN, MOSTELLER and TUKEY • Exploring Data Tables, Trends and Shapes
HOAGLIN, MOSTELLER, and TUKEY • Understanding Robust and Exploratory Data Analysis
HOEL • Elementary Statistics, *Fourth Edition*
HOEL and JESSEN • Basic Statistics for Business and Economics, *Third Edition*
HOGG and KLUGMAN • Loss Distributions
HOLLANDER and WOLFE • Nonparametric Statistical Methods
IMAN and CONOVER • Modern Business Statistics
JAGERS • Branching Processes with Biological Applications
JESSEN • Statistical Survey Techniques
JOHNSON and KOTZ • Distributions in Statistics
Discrete Distributions
Continuous Univariate Distributions—1
Continuous Univariate Distributions—2
Continuous Multivariate Distributions

(*continued on back*)

Statistics and the Law

Statistics and the Law

EDITED BY

Morris H. DeGroot
Stephen E. Fienberg
Joseph B. Kadane

Department of Statistics
Carnegie-Mellon University
Pittsburgh, Pennsylvania

JOHN WILEY & SONS

New York Chichester Brisbane Toronto Singapore

Library of Congress Cataloging in Publication Data:

Statistics and the law.
(Wiley series in probability and mathematical statistics. Applied probability and statistics, ISSN 0271-6356)
Includes bibliographies and indexes.
1. Evidence (Law)—United States. 2. Evidence (Law)—United States—Statistical methods. I. DeGroot, Morris H., 1931– . II. Fienberg, Stephen E. III. Kadane, Joseph B. IV. Series.
KF8936.S7 1986 347.73′64 86-5637
ISBN 0-471-09435-8 347.30764

Printed in the United States of America

10 9 8 7 6 5 4 3 2 1

Contributors

Gordon J. Apple
Maun, Green, Hayes, Simon, Johanneson & Brehl
St. Paul, Minnesota

Donald A. Berry
School of Statistics
University of Minnesota
Minneapolis, Minnesota

Soren Bisgaard
Department of Statistics
University of Wisconsin
Madison, Wisconsin

Delores A. Conway
Graduate School of Business
University of Chicago
Chicago, Illinois

Robert F. Coulam
Deputy Budget Director
Budget Office
Massachusetts Department of Public Works
Boston, Massachusetts

William B. Fairley
Analysis and Inference, Inc.
Boston, Massachusetts

Stephen E. Fienberg
Departments of Statistics and of Social Sciences
Carnegie-Mellon University
Pittsburgh, Pennsylvania

Michael O. Finkelstein
Barret, Smith, Schapiro, Simon & Armstrong
New York, New York

Martin S. Geisel
Graduate School of Management
University of Rochester
Rochester, New York

Seymour Geisser
School of Statistics
University of Minnesota
Minneapolis, Minnesota

Dennis C. Gilliland
Department of Statistics and Probability
Michigan State University
East Lansing, Michigan

Jeffrey E. Glen
Special Assistant Corporate Counsel
New York City Law Department
New York, New York

Arthur S. Goldberger
Department of Economics
University of Wisconsin
Madison, Wisconsin

William G. Hunter
Department of Statistics
University of Wisconsin
Madison, Wisconsin

Joseph B. Kadane
Departments of Statistics and of Social Sciences
Carnegie-Mellon University
Pittsburgh, Pennsylvania

Benjamin F. King
Educational Testing Service
Princeton, New Jersey

John Lehoczky
Department of Statistics
Carnegie-Mellon University
Pittsburgh, Pennsylvania

Hans Levenbach
Core Analytic, Inc.
Short Hills, New Jersey

Bruce Levin
Division of Biostatistics
Columbia University School of Public Health
New York, New York

Jean L. Masson
Graduate School of Management
University of Rochester
Rochester, New York

Paul Meier
Department of Statistics
University of Chicago
Chicago, Illinois

Stephan Michelson
Econometric Research, Inc.
Washington, D.C.

John Pincus
The Rand Corporation
Santa Monica, California

Herbert Robbins
Columbia University
New York, New York

Harry V. Roberts
Graduate School of Business
University of Chicago
Chicago, Illinois

John E. Rolph
The Rand Corporation
Santa Monica, California

Jerome Sacks
Department of Statistics
University of Illinois
Urbana, Illinois

Herbert Solomon
Department of Statistics
Stanford University
Stanford, California

D. E. Splitstone
IT Corporation
Pittsburgh, Pennsylvania

G. A. Whitmore
Faculty of Management
McGill University
Montreal, Canada

Donald Ylvisaker
Department of Mathematics
University of California
Los Angeles, California

Sandy L. Zabell
Department of Mathematics
Northwestern University
Evanston, Illinois

Preface

During the past decade, the use of statistical methods in legal settings has undergone a rapid and broad development that continues unabated today. Statisticians are now regularly hired as consultants or expert witnesses by one side or the other in connection with a wide variety of criminal cases and administrative proceedings conducted by governmental regulatory agencies. In litigation involving employment discrimination or antitrust violations, for example, it has become common for each side to hire its own statistical expert to support its case and for the judge or jury to have to evaluate different statistical analyses presented by these experts that usually lead to contradictory conclusions.

In this book, we present a collection of chapters that describe several different cases or types of cases in which statistical analyses were important elements, together with the main statistical methods that either were found to be of value in these cases or promise to be of value in future cases of these types. Many of the chapters are written by the statisticians who actually participated in the cases.

This book has been developed primarily by statisticians for statisticians. Its basic purpose is to provide useful background and information to statisticians who may serve as expert witnesses or consultants in the future, and to give statisticians a better understanding of the legal process and philosophy to which the lawyers with whom they work must adhere. More generally, the book is an attempt to inform members of the broad statistics community who are interested in statistics and the law about some of the central issues and some of the activities of their colleagues in this area.

We believe, however, that lawyers will also find this book to be interesting and informative. Indeed, an important secondary purpose of the book is to improve communication between statisticians and lawyers by giving lawyers a better understanding of the statistical methods used by their experts.

There is a fundamental difference in outlook between lawyers and statisticians, and it is essential that both groups understand this difference if they are to have a successful working relationship. Statisticians are trained in the scientific method. It is their profession to collect and analyze data in such a way as to give them the deepest and most complete understanding possible of the processes they are studying. They know that it is good practice to carry out as many different analyses of a given data set as are feasible, to try to resolve the discrepant results that are typically obtained, and to develop a coherent and balanced view of the underlying process. They are trained to recognize the uncertainties of their conclusions and to quantify these uncertainties.

Lawyers, on the other hand, are hired by one party in a legal proceeding to present as strong a case for that party as possible, consistent with whatever facts or data are available to them and ethical legal tenets. They seek the most effective way to support their case, and when they hire a statistician as a consultant or an expert witness, they do so expecting that the statistician will supply the technical arguments as to why their contentions are correct. Lawyers are advocates for the parties they represent, and their goal is to have their side win. Of course, lawyers may be involved in a case for reasons going beyond their duty to a specific client. For example, a lawyer handling a challenge to the jury system in a criminal case may be interested in improving the jury selection process for future cases.

This partisan climate can generate a difficult ethical problem for the statistician. The statistician works with the lawyers, and often with the clients who have retained the lawyers, as they develop their case. Although the lawyers may be aware of the statistician's professional obligations, their suggestions to the statistician about what kinds of analyses might be carried out and how the results might be presented are inevitably designed to put their side in a favorable light. Statisticians working in this climate of advocacy often become friendly with their co-workers and know what is expected of them. It can be difficult for statisticians to retain their professional integrity when they are under so much implicit pressure to let the goals of "their side" guide their statistical work. If a statistician's work does not meet the lawyers' expectations or satisfactorily serve the purposes of the clients, the lawyers may dismiss the statistician and hire another one. Statisticians should be made aware of the pressures to which their integrity will be subjected, and must strive to create acceptance among lawyers of the notion that they are hired to perform the best analyses that their professional training indicates. It is then up to the lawyers to decide how to use these analyses in the courtroom. There is a danger to the lawyers that, once they have paid to have a statistical analysis performed, it may be difficult to keep it out of court; but their biggest problem in practice may be the time involved in hiring a new statistical expert who might need weeks or months to complete a new set of analyses with uncertain outcomes.

This volume covers a wide variety of applications of statistical concepts in legal settings, many of which have provoked controversy. It is not surprising that a new methodology such as "reverse regression" is controversial (see the chapter by Conway and Roberts in this volume, and the associated discussion), but even the methods that are regarded as *standard* by some statisticians are regarded as inappropriate by others. Several examples can be mentioned: (a) Nonstatisticians commonly misinterpret a p-value or a tail area obtained in a significance test as the probability that certain results occurred by chance, but more important for our purposes, many statisticians themselves believe that significance tests, when properly interpreted, simply do not respond to the relevant questions in legal cases. (b) Confidence intervals are regarded by some statisticians as the backbone of applied statistics, while others are disturbed by what Solomon calls (in his chapter in this volume) their "ad infinitum" aspects. (c) The interpretation of probabilities based on sampling distributions when the data pertain to *all* employees of some organization who have a specified characteristic raises fundamental statistical issues. (See the discussion by Michelson of the chapter by Conway and Roberts in this volume.) (d) The relevance of *unbiased* estimates, despite their name, is open to dispute. (e) Although the three of us believe that Bayesian methods provide the only approach to statistical inference that both asks and attempts to answer the right questions, we recognize that the use of Bayesian methods based on subjective prior distributions in legal proceedings is controversial. Our view is that the Bayesian approach is what the court needs and often thinks it gets from the statistical analyses that it hears, whereas what it usually gets is actually a frequentist or "classical" approach to statistical inference. In summary, contemporary statistics as applied in legal settings is a controversial field. The same methods that some statisticians believe are most practical are regarded by others as most impractical.

We now present a brief summary of the contents of this volume. As we noted above, two areas of litigation that have relied heavily on the use of statisticians as expert witnesses are employment discrimination and antitrust violations. The area of employment discrimination is represented by the three chapters in this volume written by Meier, Sacks, and Zabell; Conway and Roberts; and Coulam and Fienberg. The antitrust area is represented by the three chapters written by King; Finkelstein and Levenbach; and Geisel and Masson.

Meier, Sacks, and Zabell review the use of statistics in employment discrimination cases, with a special and critical focus on the landmark

1977 Supreme Court decision in *Hazelwood v. U.S.* and on the 80% rule promulgated at about the same time by the Equal Employment Opportunity Commission and other U.S. federal agencies. They argue that the commonplace use of the binomial test of statistical significance, illustrated in the Hazelwood decision, is unsatisfactory for use in evaluating allegations of discrimination because many of the assumptions on which such tests are based are inapplicable to employment settings. Fienberg underlines some of their concerns and reconsiders the suitability of the Fisher exact test and the 80% rule.

Conway and Roberts consider the uses of regression analysis to adjust for productivity and other factors in comparing the wages of men and women in employment discrimination cases. They include a discussion of a method known as reverse regression, which has generated considerable debate in the econometrics and statistical communities. Two discussants, Goldberger and Michelson, point out problems with the reverse regression approach, and Michelson goes considerably further in a broad-based critique of regression as a suitable method in the study of wage discrimination problems. In a hearing regarding alleged employment discrimination by the Harris Bank in Chicago, Roberts was a statistical expert for the Bank and Michelson was a statistical expert for the government.

Coulam and Fienberg also consider an employment discrimination case that involved the extensive use of regression analyses by experts for both the plaintiffs and the defendant. Their focus, however, is not on regression as a statistical tool in such meetings, but rather on the use of a court-appointed expert during the pretrial discovery process.

Finkelstein and Levenbach describe the use of regression analysis to estimate damages in antitrust price-fixing cases. They focus on time-specific cases, all of which involved the use of multiple regression methodology, and they discuss various issues raised by this use.

In another chapter in this general area, King gives a broad overview of statistical aspects of both liability and damages in antitrust litigation and of how various questions have been framed, with special focus on the uses of survey sampling. He also describes the different roles that a statistician can play over the course of litigation, and considers at some length the ethical aspects raised by the statistician's involvement in the adversarial process.

A third chapter in the area of antitrust litigation by Geisel and Masson focuses on the role of capital market analyses. They review the economic theory of capital markets and describe how regression and other techniques can be used in actual litigation.

A common ingredient to much of the statistical work in both employment discrimination and antitrust litigation is the assessment of damages.

Such assessments are also of primary interest in a very different context described in a chapter by Whitmore. He presents a case study of the role of statistical experts in litigation following a 1967 strike at a Quebec aluminum smelter, which the company claimed reduced the operating lives of the aluminum-reduction cells in the smelter's pot rooms. The techniques used by the experts who testified included probability plots, the analysis of censored data, and life table methodology.

Another example involving the assessment of damages is presented by Fairley and Glen, who describe the analysis used to establish damages in a case in which Brink's employees were stealing coins collected from parking meters in New York City. Levin, who was the Brink's rebuttal expert witness, criticizes their inferences on several grounds.

Many of the general problems associated with the use of regression analysis in employment discrimination, especially those involving model specification, also occur in other types of cases. Pincus and Rolph describe a regression study of the relationship between school funding and student outcomes, and how it was used in expert testimony in a case brought against the state of Washington by several of the state's school districts.

Kadane describes his experience as a witness in a case involving whether playing video draw poker involves skill. This case is somewhat unusual in that it was ultimately decided by the Pennsylvania Supreme Court, so there are three different court decisions on the case. Lehoczky discusses some issues related to the measurement and quantification of skill.

Berry and Geisser show how statistical inference and models are used in cases of disputed paternity, where a "probability of paternity" is developed based on data from various populations. Ylvisaker, in his comment, is even more critical than Berry and Geisser about this use, or misuse, of statistics in the law.

Gilliland and Meier present a stochastic model for disputed elections, and compare their findings to a number of cases in which new elections either were or were not ordered by the courts. In his comment, Robbins compares this work to an earlier model developed by himself and Finkelstein.

Apple, Hunter, and Bisgaard take up the broad range of the use of statistics in environmental cases, with special reference to the *Ethyl* case, in which several companies challenged the EPA's decision to move toward unleaded gasoline. They describe the interplay between the level of proof required and the nature of the data available. Splitstone comments on a wide variety of issues that remain to be resolved in the area of environmental law and regulation.

Solomon discusses a variety of cases in which confidence intervals have been important, and raises the issue of how confidence intervals might be appropriately interpreted in the courts.

Because of the broad spectrum of controversy that exists in the field of statistics and the law, it should be clear that we did not invite, edit, or accept the chapters that appear in this volume on the basis of whether or not we agreed with the methods or philosophy of the authors. Nor do we make any claim regarding the completeness of the coverage of this volume. Rather, we have simply tried to present a variety of topics of current interest and to reveal many of these controversies for further consideration.

There are several other published sources of useful information on this topic, such as the encyclopedia articles by Zeisel (1978) and Fairley (1983); the article on epidemiologic data in the courts by Hoffman (1984); the elementary article by Gray (1983), which discusses some famous cases such as *People v. Collins* and *United States v. Spock;* the books by Finkelstein (1978), Baldus and Cole (1980), Loewen (1982), Barnes (1983), and Monahan and Walker (1985); the volume edited by Saks and Baron (1980); The Royal Statistical Society Meeting on Statistics and the Law [*Journal of the Royal Statistical Society, Series A,* **145,** 395–438 (1982)]; the collection of journal articles edited by Peterson (1983); and the forthcoming report prepared by the Panel on Statistical Assessments as Evidence in the Courts of the National Research Council.

The effective presentation of statistical analyses in a courtroom is not easily accomplished. Juries are often bored by statistical presentation, and find them difficult to follow. For that reason, there is a tendency for expert witnesses to use only simple methods that can be explained relatively easily to a judge or jury and to ignore or suppress the technical assumptions which underlie these methods. This view of the courtroom is a disservice to both the legal and statistical professions. We believe that the public has the right to expect only the highest standards of statistical practice and methodology, and the highest standards of integrity from expert witnesses in legal proceedings. It is our hope that this volume will contribute to the development and maintenance of those high standards. It is also our hope that the volume will be of value to both statisticians and lawyers who want to understand more clearly the relationship between their two professions.

References

Baldus, D. C., and Cole, J. W. (1980). *Statistical Proof of Discrimination*. Colorado Springs, Colo.: Shepard's.

Barnes, D. W. (1983). *Statistics as Proof: Fundamentals of Quantitative Evidence*. Boston: Little, Brown.

Fairley, W. B. (1983). Law, statistics. In S. Kotz and N. L. Johnson (eds.), *Encyclopedia of Statistical Sciences,* New York: Wiley, Vol. 4, pp. 568–577.

Finkelstein, M. O. (1978). *Quantitative Methods in Law*. New York: Free Press.

Gray, M. W. (1983). "Statistics and the Law." *Mathematics Magazine,* **56,** 67–81.

Hoffman, R. E. (1984). "The use of Epidemiologic Data in the Courts." *American Journal of Epidemiology,* **120,** 190–202.

Loewen, J. W. (1982). *Social Science in the Courtroom: Statistical Techniques and Research Methods for Winning Class-Action Suits*. Lexington, Mass.: Lexington Books.

Monahan, J., and Walker, L. (1985). *Social Science in Law: Cases and Materials*. Mineola, N.Y.: Foundations Press.

Peterson, D. W. (ed.) (1983). *Statistical Inference in Litigation,* a special issue of *Law and Contemporary Problems,* **46** (4).

Saks, M. J., and Baron, C. H. (eds.) (1980). *The Use/Nonuse/Misuse of Applied Social Research in the Courts*. Cambridge, Mass.: Abt Associates.

Zeisel, H. (1978). Statistics as legal evidence. In W. Kruskal and J. M. Tanur (eds.), *International Encyclopedia of Statistics,* New York: Macmillan, Vol. 2, pp. 1118–1122.

Pittsburgh, Pennsylvania
July 1986

MORRIS H. DEGROOT
STEPHEN E. FIENBERG
JOSEPH B. KADANE

Contents

Statistics and the Law

What Happened in *Hazelwood*
Statistics, Employment Discrimination, and the 80% Rule

Paul Meier
Department of Statistics
University of Chicago
Chicago, Illinois

Jerome Sacks
Department of Statistics
University of Illinois
Urbana, Illinois

Sandy L. Zabell
Department of Mathematics
Northwestern University
Evanston, Illinois

ABSTRACT

Tests of statistical significance have been increasingly used in employment discrimination cases since the United States Supreme Court's decision in *Hazelwood*. In that case, the Court ruled that "in a proper case" statistical evidence can suffice for a prima facie showing of employment discrimination. The Court also discussed the use of a binomial significance test to assess whether a difference between the proportion of black teachers employed by the Hazelwood School District and the proportion of black teachers in the relevant labor market was substantial enough to indicate discrimination. The Equal Employment Opportunity Commission (EEOC) has proposed a somewhat stricter standard for evaluating how substantial a difference must be to constitute evidence of discrimination. Under the so-called 80% rule promulgated by the EEOC, the differ-

The authors thank David Dolan for invaluable research assistance, and Mary Becker, Tom Davies, Morris DeGroot, Stephen Fienberg, Bernard Meltzer, and Lincoln Moses for helpful comments. This chapter is based on a previous version of the same title, which appeared in the *American Bar Foundation Research Journal* **1984,** 139–186.

ence must not only be statistically significant, but the hire rate for the group allegedly discriminated against must be less than 80% of the rate for the most favored group. This chapter argues that a binomial statistical significance test standing alone is unsatisfactory for evaluating allegations of discrimination because many of the assumptions on which such tests are based are inapplicable to employment settings; the 80% rule is a more appropriate standard for evaluating whether a difference in hire rates should be treated as a prima facie showing of discrimination.

Nobody could have predicted just what has happened at Hazelwood, and at the moment it appears as if nobody can elucidate it either. (Innes, 1970, p. 7)

1. Introduction

The present use of statistical evidence in employment discrimination cases stems in large part from *Griggs v. Duke Power Company,* 401 U.S. 424 (1971), where the Supreme Court established "disparate impact" as a basis for challenging employment selection procedures under Title VII of the Civil Rights Act of 1964. Under *Griggs,* use of a qualifications test for which the pass rate of black applicants is "substantially" less than that of whites is illegal unless the employer can show the use of the test is a business necessity. The burden of demonstrating business necessity being onerous (Lerner, 1976) and proof of discriminatory intent being unnecessary, the prima facie showing of disparate (or "adverse") impact may often be sufficient for the plaintiff to prevail. Disparate impact cases are thus uniquely statistical, with much depending on the interpretation given to the term *substantial* and on how large a disparity must be to be judged substantial.

Statistical evidence can also play a role in cases alleging "disparate treatment," but such cases require a showing of discriminatory intent, and the burden on the plaintiff is consequently greater. Nonetheless, in disparate treatment cases statistical evidence of disparity in hiring or promotion may provide evidence supporting the allegation of intentional discrimination. Here, too, the problem arises of deciding how large a disparity must be to support such an inference. In *International Brotherhood of Teamsters v. United States,* 431 U.S. 324 (1977), the Court refers to "longstanding and gross disparity" and in *Hazelwood School District v. United States,* 433 U.S. 299 (1977), to "gross statistical disparities," but in neither case does it give a precise definition of what these are.

Given the increasingly statistical nature of proof in employment discrimination cases, it is not surprising that the lower courts have turned to the formal methods of statistical inference for assistance in quantifying concepts such as *substantial* or *gross*. In *Hazelwood* the Supreme Court pointed to a specific statistical test, "the standard deviation analysis," describing it as "a precise method of measuring the significance of such statistical disparities" (although it also cautioned that "precise calculations of statistical significance" were not "necessary in employing statistical proof"). As a result, a flood of statistical tests of significance, confidence intervals, and multiple regressions thundered forth from the lower courts. In the years since *Hazelwood,* there has been a striking increase in the number of such cases and in the sophistication of statistical methodology used.

The legal resort to statistical evidence in employment discrimination cases contrasts sharply with earlier practices. Until recently, the courts have generally been wary of such evidence. In the domain of identification evidence, for example, the courts have been reluctant, despite the encouragement of distinguished authorities, to admit probability calculations purporting to establish identity. In this instance the intuitive appraisal of the courts has served well as a buffer against overenthusiastic adoption of "sophisticated" methods that were in fact inappropriate for the application urged (Tribe, 1971).

An important exception to the rejection of statistical evidence of identity is the assessment of paternity using evidence of blood type (Ellman and Kaye, 1979; Berry and Geisser, this volume), a domain in which the nonstatistical evidence is inherently inconclusive, so that a quantitative assessment of the statistical evidence seems more urgent. In addition, the statistical model on which the calculations are based has considerable support in the well-established applications of probability to the inheritance of genetic traits (although there is considerable debate over how this model is used in actual cases; see, e.g., Berry and Geisser, this volume, and the comment on their paper by Ylvisaker).

Until recently, the courts were also reluctant to accept statistical assessment in a far less controversial domain: the use of probability theory in scientific sampling (see, e.g., Sprowls, 1957). The requirements of economic efficiency and the widely recognized success of probability sampling methods in many domains have now resulted in the general acceptance of well-supported sampling evidence (Solomon, 1971).

The need for and credibility of statistical methods to assess evidence of employment discrimination are similar to that in the paternity setting. The nonstatistical evidence is likely to be inconclusive and the need for the sharpest possible assessment of the statistical evidence therefore acute.

Here, too, the courts had approached the use of statistical evidence cautiously, initially requiring nonstatistical evidence to buttress statistical evidence of discrimination. Gradually, however, the Supreme Court came to accept purely statistical proof as sufficient for a prima facie showing of discrimination: first, in *Griggs,* for disparate impact cases; then, in *Teamsters,* for disparate treatment cases involving the extreme instance of the "inexorable zero" (i.e., where *no* members of a group allegedly discriminated against had been hired); and finally, culminating in *Hazelwood,* with the statement that "gross statistical disparities" could alone constitute prima facie proof of discrimination "in a proper case."

The use of the formal methods of statistical *inference,* in particular, in assessing evidence of employment discrimination, is borrowed from their recent use in jury discrimination cases (see generally Finkelstein, 1978, Chapter 2). In *Castaneda v. Partida,* 430 U.S. 482 (1977), a jury discrimination case involving the Texas "key man" system, the Supreme Court adopted a formal statistical criterion based on the theory of random sampling from a binomial population. The justification offered for this particular criterion of disparity was the following:

> As a general rule for such large samples, if the difference between the expected value and the observed number is greater than two or three standard deviations, then the hypothesis that the jury drawing was random would be suspect to a social scientist.

In systems where jury selection is purportedly carried out by a random process, such statistical tests are appropriate (Zeisel, 1969). The key man system, whose constitutionality was not at issue in *Castaneda,* is not a random process, but the courts might well require as a matter of public policy that the operation of such systems closely approximate a random process.

Hazelwood

The statistical rule first put forward in *Castaneda* was then immediately adapted to employment discrimination in *Hazelwood*—with little in the way of argument or justification: "a fluctuation of more than two or three standard deviations would undercut the hypothesis that decisions were being made randomly with respect to race." The powerful assumption of random sampling—justifiable in paternity cases because of biological experience, imposed by the statistician in scientific sampling, or required in jury selection as a matter of public policy—was never supported or argued. Despite the apparent innocence of this transition from jury to employment discrimination, the use of statistical methods in the latter setting is subject to question. Indeed, on its face the employment process is necessarily highly nonrandom; thus, as discussed below, the statistical analysis of employment patterns is often highly questionable and of limited utility.

Although the *Hazelwood* Court discussed the use of statistical tests of significance, it did not go so far as to identify a disparity of "two or three standard deviations" as "substantial." Rather, the Court stated that "where gross statistical disparities can be shown, they alone may in a proper case constitute prima facie proof of a pattern or practice of discrimination." On the issue of whether *Hazelwood* is a "proper case," in the sense of the above quotation, the opinion is silent. Nor does the opinion present further guidance as to the relationships between "statistically significant disparity," "substantial disparity," and "gross disparity," the last being the only measure explicitly adopted as adequate to establish prima facie proof of discrimination. *Hazelwood* is at once both the source and the high-water mark of explanation for the use of statistical testing in employment discrimination cases. Statistical testing continues to be used by the courts, but judicial opinions have shed little further light on the need for a quantitative criterion of substantiality.

In contrast to court opinions, the Equal Employment Opportunity Commission's (EEOC's) *Uniform Guidelines on Employee Selection Procedures* propose a more elaborate criterion of substantiality, described variously as the "the 80% rule" or the "four-fifths rule of thumb." (The rule is quoted at the beginning of Sec. 6.) The rule provides in part that disparate impact can ordinarily only be found when (a) the difference in selection rates is statistically significant *and* (b) the selection rate for a protected group is less than 80% of the rate for the group with the highest rate. The key feature of the 80% rule is that, in contrast to the *sole* use of statistical tests of significance, it also incorporates a precise quantitative criterion for when observed disparities are to be adjudged "substantial." The Supreme Court gave its implicit imprimatur to the 80% rule in *Connecticut v. Teal,* 457 U.S. 440 (1982), when it noted that petitioners had not contested the district court's finding of disparate impact under the rule. Although the 80% rule is typically applied in adverse impact settings involving the use of "objective" employment tests or rules, we shall argue that the criterion of substantiality it introduces has pertinence in all cases where statistical evidence must be assessed.

A substantial literature of comment on the 80% rule has emerged, mostly adverse (Shoben, 1978; Van Bowen and Riggins, 1978; Cohn, 1980; Baldus and Cole, 1980; Booth and MacKay, 1980). Critics of the 80% rule frequently disagree with one another, but their criticisms appear largely based on a misunderstanding of the rule itself (a result of ignoring the statistical significance component of the rule) or on an unargued requirement of symmetry between the treatment of pass rates and fail rates.

Many of the critiques encourage the sole use of simple tests of significance but overlook their major infirmities when applied in the context of employment statistics, and fail to recognize the need for just such modifications as are exemplified by the 80% rule.

Overview of This Chapter

The statistical methodology for executing the recommended statistical tests of significance, illustrated by the Supreme Court in *Castaneda* and *Hazelwood,* has since been described in numerous lower court decisions, law reviews, and books. Unfortunately, most such discussions either fail to clarify or actually misstate the relationships between various statistical test procedures, and almost all are inadequate in their discussion of the assumptions underlying the interpretation of such tests and the extent to which those assumptions are likely to be met in legal applications. In the first part of this article (Secs. 2 and 3), we discuss the basis for the use of these methods in employment discrimination, explore the validity of the assumptions on which they are based, and attempt to dispel some of the confusion surrounding their application.

In the second half (Secs. 4–6), we discuss why these elementary statistical procedures, although inappropriate for direct application in employment discrimination, can provide useful benchmarks when appropriately interpreted. We argue, further, that the so-called 80% rule is preferable to the sole use of simple statistical significance testing as the criterion of substantiality to be met by plaintiffs in establishing a prima facie case of adverse impact.

2. Tests of Significance

Statistical tests often compare either an *observed* distribution with a distribution that would be *expected* under some model or ideal (the one-sample model) or compare two observed distributions with each other (the two-sample model). The statistical evidence typically encountered in employment discrimination cases often falls into one of these two categories. *Hazelwood* illustrates the use of the one-sample model: the observed proportion of blacks among teachers hired was compared with the proportion of blacks in the eligible population. *Teal* illustrates the use of the two-sample model: the proportion promoted in a sample of blacks was compared with the proportion promoted in a sample of whites. *Teal* represents the broad category of cases envisioned by the EEOC guidelines in which two groups (e.g., black/white, male/female) are compared with each other on a simple binary characteristic (promoted/not promoted, passes/

fails an employment test, has/does not have a high school diploma). In both situations there are underlying random sampling models, which we now discuss.

A. The One-Sample Model: *Proportion in Sample Compared to Proportion in Population*

In *Hazelwood School District v. United States,* the United States brought suit against the Hazelwood School District, alleging that it had engaged in a "pattern or practice" of employment discrimination in violation of Title VII. Hazelwood was a largely rural area in the northern part of St. Louis County, with few blacks in its student body and a corresponding dearth (under 2%) of black teachers on its staff. The government based its statistical case on the small proportion of black teachers in the district compared with the surrounding area in which the percentage of black teachers was substantially higher (St. Louis County and the City of St. Louis, the latter being encompassed by, but not included in, the former).

The district court found the approximate parity between the percentage of black staff and black students in the district an adequate defense, and found for defendant. On appellate review, the court of appeals ruled that the appropriate comparison was instead that of staff with the labor force in the St. Louis area, and reversed. The Supreme Court reviewed the case and vacated the appellate decision, holding that the court of appeals had erred in "disregarding the statistical data in the record dealing with Hazelwood's hiring after it became subject to Title VII" and in "disregarding the possibility that the prima facie statistical proof in the record" might be rebutted by those statistics. The opinion notes that of 405 teachers hired by the defendant in the two years 1972–1973 and 1973–1974, 15, or 3.7%, were black. This proportion was judged against the proportion of black teachers in St. Louis County, namely, 5.7%. The comparison is based on the implicit (and, as discussed below, highly questionable) assumption that, in the absence of discrimination, the percentage of blacks hired would not differ substantially from the percentage of blacks in the population of teachers.

The random sampling model for nondiscriminatory hiring used by the Court rests on two implicit assumptions: (a) the chance that any particular teacher hired will be black is the same as the overall percentage of black teachers—5.7%, and (b) the racial outcomes of the 405 hires are "statistically independent." Granted these two assumptions, the sample proportion $\hat{p}$ (i.e., the proportion of blacks hired) will approximate the population proportion $p = 0.057$ and deviate from it only to an extent described by the binomial probability distribution.

Calculations directly using the binomial distribution are often tedious, but may be facilitated by use of the *normal approximation,* which of course requires calculation of the standard error of $\hat{p}$, $\sigma_{\hat{p}} = \sqrt{\hat{p}(1 - \hat{p})/n}$. The Court's statement that "a fluctuation of two or three standard deviations would undercut the hypothesis" of randomness is based on the mathematical fact that there are fewer than five chances in 100 that a deviation from 5.7% as large as two standard errors will occur and fewer than three chances in 10,000 that a deviation as large as three standard errors will occur.

In *Hazelwood,* $\sigma_{\hat{p}} = 0.0115$ or 1.15%, and the difference between sample and population proportions is 3.7% − 5.7% = −2%, or 1.74 standard errors (=2%/1.15%), and under the assumption of random selection, a discrepancy this large has more than eight chances in 100 of occurring. Because the discrepancy was less than two standard errors, the Court concluded that the disparity between the observed sample proportion, 3.7%, and the standard of comparison, 5.7% might be "sufficiently small to weaken the Government's other proof, while the disparity between 3.7% and 15.4% [the standard advanced by the plaintiff] may be sufficiently large to reinforce it." The proportion of blacks among qualified teachers in St. Louis County excluding the city of St. Louis was specified as 5.7%, and thus was the standard initially cited by the Court. The corresponding proportion in St. Louis County including the city was 15.4%, and the government, representing the plaintiffs in a class action, asserted that this was the proper standard.

Thus, the probative value of the evidence in an employment discrimination case such as *Hazelwood* can depend in a very sensitive way on the "relevant" labor market chosen as a standard for comparison. In scientific sampling, where a random sample is drawn from an unambiguous and prespecified population, this standard must, of necessity, be clearly identified prior to the sample being drawn. But in cases such as *Hazelwood,* what one begins with is the sample—the hiring rate for a protected class—and the population or relevant labor market is a hypothetical construct from which the hired employees are imagined to have been drawn. As the text of the *Hazelwood* decision makes clear, the choice of the relevant labor market to be used as a standard of comparison can be a complex question not necessarily admitting of a unique, simple, or unambiguous answer.

For the data in *Hazelwood,* the normal approximation to the binomial distribution is quite satisfactory as a means of closely estimating the relevant probabilities. In other cases, where the *expected* number of black hires is small—say, five or fewer—the normal approximation is not as satisfactory (Cochran, 1952), and recourse may be had to better approximations or to exact calculations of binomial probabilities. It is worth emphasizing, however, that the various approximations are all variants of a single exact test, based on the binomial distribution, and are valid only to the extent that they give results sufficiently close to those provided by this exact test.

B. The Two-Sample Binomial Model: *Comparison of Two Samples with Each Other*

In *Connecticut v. Teal* a test used to determine eligibility for promotion was challenged on the ground of adverse impact. Of 48 blacks who took the test, 26 passed (54%), while of 259 whites who took the test, 206 passed (80%). In this case, there is no standard pass rate for comparison. There being no external standard, the only question is whether the pass rate for black applicants is different enough from the pass rate for white applicants to indicate a substantial difference. The sampling model for the testing situation, analogous to that used in *Hazelwood,* presumes that, absent discrimination, applicants in both groups have the same chance of passing and that the outcomes, pass or fail, for the 307 candidates (48 blacks plus 259 whites) are statistically independent. In *Teal* the difference in hire rates is 25%, or 3.76 standard errors, well above the "two or three standard errors" benchmark enunciated in *Hazelwood,* and is *possibly* indicative of a discriminatory mechanism.

Here again there is a single exact test, commonly referred to as the *Fisher exact test,* which is appropriate for determining whether a difference in hire or promotion rates is statistically significant. The Fisher exact test is both the appropriate conditional procedure and the UMPU, or uniformly most powerful unbiased, test (see Fisher, 1970, pp. 96–97; Tocher, 1950; Lehmann, 1959, p. 146; Kendall and Stuart, 1973, pp. 570–576; Pratt and Gibbons, 1981, pp. 238–241). Strictly speaking, in both the one- and two-sample cases the UMPU tests have randomized versions that may become appropriate if testing at a predetermined level of significance is required. This detail is of interest in the theoretical comparison of tests but has no relevance for our present purpose. In any event, the attained level of significance is the same for both the randomized and nonrandomized versions of the tests.) As in the one-sample case, there are a number of alternative tests and variant approximations. But *all* the other tests—two-sample binomial, or chi-squared with or without continuity correction, for example—may properly be considered as more or less adequate approximations to the Fisher exact test. The failure to note this has caused confusion in the literature and has led to the erroneous assertion that there are substantially different procedures available for the comparison of two samples and that these different procedures lead to different conclusions (Spradlin and Drane, 1978).

Interpretation of Significance Level

Unless an employer were hiring by quota, we should not expect *exactly* equal hire or promotion rates for two different groups. Thus we always expect to see some disparity due merely to chance or random fluctuation, and what is needed is a tool to enable us to distinguish between minor disparities resulting from such fluctuations on the one hand, and substantial disparities due to intentional or unintentional discriminatory practices on the other. Absolute certainty being seldom attainable, the most we shall be able to arrive at will be a *conditional probability* that the practice in question was fair or nondiscriminatory, based on (or "given") the evidence of the observed disparity and all other available information; symbolically, this is

$$P\,(\text{nondiscrimination}\,|\,\text{observed disparity and other evidence}).$$

It is a serious but very frequent error to confuse this conditional probability with the level of significance or "P *value*" *that results from a test of significance*. The level of significance, it will be recalled, is the chance that, assuming one is hiring or promoting at random, a disparity at least as large as the one observed would occur; symbolically this is

$$P\,(\text{observed or larger disparity}\,|\,\text{random selection})$$

The first conditional probability is what would be most relevant for evaluating whether there is discrimination; but the second conditional probability is all that classical statistical theory is able to provide. Note that there are three major differences between the two:

1. The first conditional probability refers to whether there is nondiscriminatory or fair selection, the second is a calculation based on the hypothesis of *random* selection, where "random" is used in its technical statistical sense. It is a major point of this paper, discussed at length in Section 3, that there is no reason to expect nondiscriminatory employment practices to closely resemble the outcome of a random selection of employees from a specified pool of applicants.

2. The first conditional probability refers to the total evidence available (the observed disparity *and* all other evidence), the second only to the evidence of the observed disparity. The remaining evidence in an employment discrimination case will often be nonstatistical, perhaps of an anecdotal and individual nature, and it is hardly surprising that it does not enter into the calculation of the level of statistical significance.

Despite these reservations, one might decide on pragmatic grounds to proceed, keeping these distinctions in mind, and substitute in the first conditional probability "random" for "fair" or "nondiscriminatory," and "observed disparity" for "observed disparity and other evidence"; the result would be

$$P\text{ (random selection}|\text{observed disparity).}$$

The only essential difference between *this* conditional probability and the previous one (besides that between "observed disparity" and "observed or larger disparity") is that

3. The two terms, what is taken as given and what is in question, are transposed.

It is this third difference, that between $P(A|B)$ and $P(B|A)$, which is crucial. The two conditional probabilities are not unrelated, but the relationship between them is complex and not easily stated, requiring the introduction of subjective probabilities. A simple example, sometimes attributed to Keynes, will make the distinction clear. If the Archbishop of Canterbury were playing in a poker game, the probability that the Archbishop would deal himself a straight flush, given honest play on his part, is *not* the same as the probability of honest play on his part given that the prelate dealt himself a straight flush. (The first conditional probability may be calculated to be 40 in 2,598,960 and is quite small; the second conditional probability would be, at least for most Anglicans, much larger, indeed quite close to 1.) The first of the two probabilities is a frequency or so-called "objective" probability, stating what proportion of hands will, in the long run, give rise to straight flushes; the second probability is an epistemic or "subjective" probability, stating one's degree of belief or personal evaluation of the odds for the event in question. The first probability has a unique, mathematically calculable value; the second will vary from person to person depending on his knowledge and opinions. (The confusion of the two is very common and numbers among its distinguished victims DeMoivre (1756, p. v).)

Consider a hypothetical example in an employment discrimination context. Suppose a statistical expert witness testifies that an observed hiring disparity is statistically significant at the 25% level. If this meant that there were a 75% probability that the employer had engaged in discriminatory practice, then under the preponderance of evidence standard the court would be warranted in finding against him. But what the 25% in fact means is that a *company that hires at random will produce disparities this large 25% of the time*. That is, if we confine our attention to the stratum of random (or "fair") employers, we will, *by definition,* find one out of every four such employers exhibiting disparities that are statistically significant at the 25% level. And if one were to look at different divisions within a single large company that hired randomly, one out of every four divisions would likewise exhibit disparities significant at the 25% level.

Choice of Significance Level

Using statistical tests of significance to assess disparity requires not only strong assumptions about the sampling model (for the most part invalid or unverifiable, as discussed in Section 3) but also the choice of an appropriate probability level or level of significance at which a disparity will be considered substantial.

The Supreme Court's introduction of tests of significance in *Castaneda* and *Hazelwood* leaves a large measure of ambiguity about the significance level thought to provide convincing evidence. There is good reason for this. Courts ordinarily consider far more than statistical evidence, and in one case a difference not "significant" at the 5% level may combine with collateral evidence to make a convincing case of discrimination, while in another case a stronger statistical finding, say, a result significant at the 3% level, might be overcome by contrary evidence pointing to fair and nondiscriminatory practices. However, in light of the impact of the *Hazelwood* opinion on the lower courts, we have reason to wish for more specific guidance. Thus far that additional guidance has not been forthcoming, and therefore some review of the statistical tradition to which the Court alludes in *Castaneda* may prove helpful.

The tradition of choosing a specific "level of significance," and discounting evidence that fails to reach that level is due to R. A. Fisher. The method is presented most clearly, but almost solely by example, in his classic work *Statistical Methods for Research Workers* (1st ed. 1925, 14th ed. 1970). An examination of Fisher's examples discloses that he regards any test of significance that results in a larger than 5% significance level (i.e., less than 1.96 standard errors) as unpersuasive, a difference with significance level between 5% and 2% (i.e., between 1.96 and 2.33 standard errors) as credible, and a difference with significance level more extreme than 2% (i.e., greater than 2.33 standard errors) as clearly indicative of a real, underlying difference.

Fisher thought of attained levels of significance as providing evidence against a hypothesis (in the present context, the assumption of random selection), smaller levels providing stronger evidence. He did not regard them, nor should the courts, as decision rules on which alone to accept or reject a hypothesis. As is the case with scientific questions, employment discrimination cases involve other types of evidence, and the extent or severity of relief granted by the Court will, as a practical matter, depend on how clear or well established a case of discrimination is. The level of significance is therefore best thought of as but one item of evidence, rather than as the sole basis for a decision. Even though later approaches can also accommodate such considerations, we shall adopt Fisher's view with its attendant simplicity.

Extending the ideas described above, some have argued that no magic resides in the conventional 5% level of significance; that, say, a 6% level of significance has very nearly the same persuasiveness as 5%; and that a useful mode of statistical presentation is simply to state the attained level of significance, however weak it may be. Thus, a result significant at the 20% level would be viewed as having some persuasive power, even though it would be far from convincing. Such a variant procedure departs markedly from Fisher's usage and seems more likely to confuse than clarify in the legal setting. The key point is to recognize that variation is pervasive, and that what is required is a means of avoiding distraction by inessential differences which have little or no probative value. Although an inherent arbitrariness obviously resides in the particular choice of the 5% (or any other) level as a universal minimal standard, such arbitrariness is unavoidable if one is to adopt a minimal standard, and *some* minimal standard is clearly necessary. The 5% level is the conventional choice. If a difference does not attain the 5% level of significance, it does not deserve to be given weight as evidence of a disparity. It is a "feather."

One- and Two-sided Tests

Implicit in our discussion of the simple models and their associated statistical tests is the premise that disparity would be recognized not only when the difference is negative but also when it is positive. Thus, even if blacks are favored in the sample data, disparity would be found. If one is asking whether the selection process treats blacks and whites equally, this "two-sided" view is natural enough. But if the appropriate question is whether the proportion of blacks hired is substantially *less* than some standard, then the disparity would be worthy of note only if, as in *Hazelwood,* it reflected less favorable treatment of blacks. The probability of a given disparity in a specified or "one-sided" direction is half the significance level provided by the "two-sided" perspective indicated above, and it is this one-sided formulation that is at the heart of Justice John Paul Stevens's statistical analysis in his dissent to the *Hazelwood* opinion. Stevens had quoted a calculation by one of his clerks, without specifying its details; let us reconstruct the calculation that this clerk must have performed.

If we regard the 405 hires as a strictly random sample of size 405 from a job pool with 5.7% blacks, the probability of obtaining 15 *or fewer* black hires is

$$P\{X \leq 15\} = \sum_{k=0}^{15} \binom{405}{15} (0.057)^k (0.943)^{405-k} = .0456$$

or 4.6%. The value for $P\{X \leq 15\}$ quoted by Stevens is 5.2%, presumably the result of his clerk's having used the continuity-corrected normal approximation rather than the exact calculation (see footnote 1 below). Having noted that this tail probability is approximately 5%, Stevens concluded that the evidence is persuasive that the hiring by the Hazelwood School District was not in fact random. Inasmuch as many statisticians regard the 5% level as insufficiently stringent for a one-tailed test and prefer, at a minimum, the 2.5% level, and given that in *Hazelwood* the level of significance is appreciably greater than 2.5%, the evidence of nonrandomness in *Hazelwood* might well be judged unpersuasive.

The differing conclusions of Stevens (in dissent) and Justice Potter Stewart (writing for the majority) are unrelated to the latter's (implicit) use of the normal approximation in its standard deviation analysis. The approximation is based on the fact that the distribution of

$$Z = (X - np)/\sqrt{np(1 - p)}$$

is approximately that of the standard normal. In *Hazelwood* $X = 15$, $n = 405$, and $p = 0.057$; hence $Z = -1.73$ and

$$P\{X \leq 15\} = P\{Z \leq -1.73\} \approx 0.0418$$

or 4.2%. The approximation is, in fact, fairly good (i.e., 4.2% is quite close to the exact 4.6%) and the differing interpretations have nothing to do with the quality of the approximation. The Court had noted that "the expected value of 23 would be less than two standard deviations from the observed total of 15"; thus the differing interpretations of Stevens and Stewart arise from the one-sided or two-sided interpretation of the test results, not from any problem with the normal approximation.[1]

The question whether to adopt a "one-sided" or a "two-sided" view has an ambiguous history in the statistical literature. Fisher, in his concern for scientific research, regarded one-sided tests as largely inappropriate for statistical inference; but some later commentators admit and employ one-sided tests in a prominent way. The matter is confounded

[1] The normal approximation ordinarily produces tail probabilities that are too small, and use of the "continuity corrected" Z value Z_c often improves the closeness of the approximation. For the *Hazelwood* data, $Z_c = (15.5 - 23.1)/4.62 = -1.63$, and using the normal approximation, we obtain

$$P\{Z_c \leq -1.63\} \approx 0.052$$

or 5.2%, the value calculated by Justice Stevens's clerk. Thus in *Hazelwood* the continuity correction overcorrects, giving a slightly *less* accurate approximation to the binomial tail probability than does the uncorrected Z value.

with the choice of significance level, since the one-sided significance level is typically half the two-sided level. A simple rule, which we adopt here, is to take a one-sided 2.5% level criterion (equivalent to a two-sided 5% level) and assert that an observed disparity could not have plausibly arisen from purely random fluctuations if the magnitude of the difference is more than 1.96 standard errors. While there are examples where the one-sided, two-sided issue has surfaced in court (e.g., in *E.E.O.C. v. Federal Reserve Bank of Richmond,* 698 F.2d 633 (4th Cir., 1983)), we do not regard the matter as one of great importance. For further general discussion, and some useful cautionary notes on the use of one-sided tests, see Freedman et al. (1978, pp. 494–496).

3. Critique of the Use of Statistical Models

With this statistical background in mind, we turn to a more detailed examination of the appropriateness of the statistical models used by the courts. There are two key assumptions required to permit unambiguous interpretation of the results of statistical testing, and neither is ordinarily capable of direct verification. We discuss these two assumptions under the headings of *random selection* and *relevance*. Under *random selection* we examine the issues of statistical independence and stratification; under *relevance,* we examine the relationship of the models and statistical tests to the concept of discrimination. Because both these assumptions are likely to be violated to some degree in typical employment processes even in the absence of discrimination, a careful assessment of the applicability of the assumptions is essential to decide the extent to which statistical analysis should be used as evidence of discrimination.

A. Random Selection

Both models discussed in Section 2 assume (a) that the sample(s) under review are produced by an effectively random process of sampling from some identifiable group and (b) that each such group is essentially equivalent to the population of appropriately qualified individuals. Thus, in judging the statistical significance of the disparate hiring in *Hazelwood,* the Court states explicitly that a large deviation from the given standard of black teachers, $p = 5.7\%$, "would undercut the hypothesis that decisions were being made *randomly* with respect to race." In other words, the hypothesis being tested is that each hire has probability 5.7% of resulting in the selection of a black and that choices are statistically independent, each being unrelated to the race outcome of any other choice or set of choices. This assumption leads directly to the binomial probability distribution specified by the *Hazelwood* court. The product rule based on independence is often applied, almost as a matter of course, in circumstances in which there is no clear basis for the assumption of independence.

In fact, however, random sampling, which gives rise to the applicability of the product rule, is rarely present in nature or human affairs. Indeed, statistical sampling practitioners often find they must go to great lengths to achieve it.

The Court is correct in its assertion that the recommended statistical procedure tests "randomness" in hiring to the extent that "randomness" is read to mean random selection from a specified pool of teachers. The Court then appears to equate "randomness" with "fairness" or "nondiscrimination" in employment, however. The latter concepts are relevant to the legal issue, but only the former is tested statistically; since hiring will seldom be truly "random" even if it *is* fair or racially nondiscriminatory, the nexus between the two sets of concepts is more problematic than is usually recognized.

Indeed, there is a long history in the legal literature of misconceived notions about the applicability of the assumptions of statistical independence and random sampling. Thus, in the context of employment testing, Shoben states: "Independence in performance can be assumed if no person is allowed to take the test more than once and there are no opportunities for cheating or passing on information about the test to later test takers" (Shoben, 1978, p. 801). While these are certainly necessary conditions for independence, they are by no means sufficient, since they provide no constraints on the origins of the sample of applicants being tested. Two relevant issues here are the effects of *clustering* and *stratification*.

Clustering

As Smith and Abram aptly point out, varying recruitment procedures inevitably cause departures from the simple random sampling model:

> Actual hiring practices, however, rarely approximate a random process. For instance, an employer may vary its recruitment practice over time. The employer might recruit in the suburbs one month and the inner city next, or advertise job openings in a Spanish-language newspaper one week and a newspaper with a predominantly black readership the next. Thus, the selection of applicants resulting from these recruitment efforts may not, for any given time period, be truly random because the applicants may not be representative of the population in terms of ethnicity or socio-economic background. Because an employer's selection process cannot, by definition, be a random

> procedure, statistical inference cannot play as decisive a role in employment discrimination cases as it has in jury selection decisions. That an employer's selection process may not be random for any particular time period underscores the fact that economic concepts of labor supply and demand apply to employment patterns developed over the long term. Reliance on evidence from a narrow time period may be misleading, precluding reliable inferences about the nature of an employer's behavior. (Smith and Abram, 1981, p. 42)

It is easy to suggest other scenarios as well: a person recruited may recommend or otherwise bring along friends into the company; success with applicants from a given school or other organization may set up a short-term "pipeline" of future applicants. Any of these scenarios might be entirely compatible with nondiscriminatory hiring and even with overall long-term equality of selection rates for whites and blacks. They are not, however, compatible with the binomial model on which the previously discussed statistical tests are based. The problem is that this type of "clustered sampling" effect is the rule in real-world hiring, not the exception.

In some circumstances it might be possible to take clustering into account in the statistical analysis. Since the allowance for error (the standard error) is inversely proportional to the square root of the sample size, it is clear that clustering, by lowering the *effective* sample size, will necessarily increase the magnitude of the proper allowance for error. The resulting difference in sample sizes can have a profound effect because, as a general rule, *the smaller the sample size, the larger the disparity in rates can be without reaching statistical significance*.

To be specific, consider a hypothetical example: 100 blacks and 100 whites apply for jobs and do so in clusters; for example, groups of classmates apply together. For simplicity, suppose each cluster has 4 individuals, leading to 25 clusters of 4 blacks and 25 clusters of 4 whites, and that clusters are hired or rejected as a group. Thus, if 10 black clusters are hired, then 40 blacks are hired, and if 15 white clusters are hired, then 60 whites are hired. Supposing that hiring decisions are made completely at random, we have a random sample from the 25 clusters for each of the two groups. It follows that we should compare the cluster rates, which are 10/25 and 15/25 for blacks and whites, respectively, to determine if a significant disparity exists. Even though the individual hire rates themselves (40/100, 60/100) are equivalent to the cluster rates (10/25, 15/25), the sample size of the cluster data is only one-quarter as large as the number of individuals. Since decisions were made by clusters, not by individuals, comparison of the individual hire rates could not properly be based on the theory of random sampling of individuals. In this example, the smaller cluster sample doubles the standard error that would result if the analysis were made using individual data. Thus, what might appear to be a significant difference in the individual rates assessed using the binomial standard error can (and in this example does) turn out to be a nonsignificant difference in cluster rates when clustering effects are taken into account.

Where clustering is measurable and can be taken into account in the analysis, or where it is a result of illegal practice (e.g., if only the sons of existing union members are admitted to apprenticeship), the strain on the assumption of random sampling as a standard for judgment is not a critical matter. The existence of undetected and "natural" clustering, however, which is bound to exist to some degree and which can rarely be measured or modeled, substantially diminishes the credibility of the random selection assumption in most other situations.

Stratification

A second source of variability not accounted for in the simple random sampling model results from the aggregation of several distinct but related job categories into a single broader one. Thus the category of "craft" worker might include carpenters, plumbers, and electricians as well as other specialties, but data on the racial composition of the pool of craft workers may not provide so much detail. If there are different proportions of blacks in the several subgroups, or strata, an employer operating "fairly" in his hiring from each stratum separately might nonetheless appear to be hiring a substantially lower proportion of blacks overall than the proportion of blacks in the combined employment pool.

An example of this effect is provided by an analysis of graduate admission rates for men and women at the University of California at Berkeley (Bickel et al., 1975). Overall, the percentage of male applicants admitted in 1973 was significantly *higher* than the percentage of female applicants admitted, but a survey of the admissions to individual departments showed that for virtually all departments the admission rate for male applicants was *lower* than the admission rate for females.

This paradoxical situation, sometimes referred to as *Simpson's paradox,* actually has a very simple explanation. At Berkeley, women had applied preferentially to departments such as English and history, with large numbers of applicants and corresponding low rates of admission, while men had applied preferentially to departments such as mathematics and physics, with fewer applicants and hence much higher rates of admission. Simpson's paradox should not be thought of as a rare or freakish event: other instances can easily be found. To cite an example in a perhaps more familiar setting, it will come as no surprise that from 1976 to

1979 the overall rate of personal income paid out as federal income tax increased, although surprisingly, the rate in each bracket actually decreased (Wagner, 1982). The explanation: because of inflation, taxpayers had been pushed into higher brackets. (As a simple illustration, consider two hypothetical brackets in 1976, say, 20% and 40%, with 90% of the taxpayers in the 20% bracket, 10% in the 40% bracket. If, in 1979, the brackets decline to 19% and 39%, but now 90% of the taxpayers are in the higher bracket, the overall rate of income paid out in taxes will change from 22% in 1976 to 37% in 1979.)

Stratification can occur when different divisions of a firm make hiring decisions independently of one another. Such decisions may appear fair at the divisional level but unfair when viewed in the aggregate. Like clustering, stratification (i.e., aggregation of relevant strata), may occur in nondetectable ways or, even when detectable, in ways not amenable to appropriate measurement or calculation. It is possible, of course, that an employer may create irrelevant or unnecessary strata in an attempt to explain away the appearance of discrimination (Kruskal, 1977), but the existence of real strata beyond the reach of analysis is pervasive and, once again, undercuts the simple random sampling model.

In a typical hiring situation there may be some elements of clustering and some of stratification. Taking these considerations into account, we must reject the proposition that hiring is a random selection process. Although the facts of a given case may indicate that there are gross departures from random selection (e.g., resulting from an affirmative action plan, as in *Washington v. Davis,* 429 U.S. 299 (1976)), the absence of such an indication does *not* ensure that the departure from random selection is either trivial or negligible.

The Failure of Independence

Given this variety of ways in which the simple random sampling model for selection of employees can fail to reflect the realities of a nondiscriminatory employment process, why has it so often been taken for granted? The answer, in part, is that the associated notion of statistical independence is elusive and subject to frequent misapplication (see, e.g., Meier and Zabell, 1980). A brief typology of the reasons for confusion about statistical independence might include the following:

1. its meaning is easily confused with other, everyday notions of independence;
2. it is often overlooked or forgotten as a necessary precondition for the applicability of the "randomness" model or the use of the product rule in calculating joint probabilities;

3. use of the product rule qualitatively agrees with intuition;
4. the difficulties inherent in calculating a joint probability, absent independence, make it tempting to introduce independence as a "simplifying assumption."

The first point deserves amplification. Independence, as it is used in statistics and probability, is not simply a statement of the absence of causal connection. Tosses of a coin are deemed independent, for example, in part because they are causally unconnected, but the mere absence of direct causal connection, or apparent causal connection, does not ensure that two events will be independent (Keynes, 1921, Chapter 16). In the sinking of the *Titanic,* it is often asserted that the tragedy was the unfortunate and highly improbable concatenation of a large number of "independent" events, each not improbable in itself: nighttime conditions make it difficult to see icebergs, the speed at which the *Titanic* was sailing made it impossible to stop in time, the radio of a nearby ship was unmanned, there were insufficient lifeboats on board, etc. But the tragedy was really due to a small number of factors responsible for all the others (Machol, 1975): the accident took place at night (so that visibility and attentiveness were at lowered levels), and little credence was given to the possibility of disaster (hence the absence of enough lifeboats, cruising at dangerously high speeds, etc.). The nuclear mishap at Three Mile Island occurred, reactor safety reports to the contrary, because the necessary concatenation of system failures arose from a set of untoward events that were *not* independent.

B. Relevance

As the Supreme Court implicitly notes, the consideration of statistics bearing on a charge of discrimination must begin with the question of what should reasonably be expected in a situation where discrimination is absent. In *Teamsters,* the Court states:

> [A]bsent explanation, it is ordinarily to be expected that nondiscriminatory hiring practices will in time result in a work force more or less representative of the racial and ethnic composition of the population in the community from which employees are hired. Evidence of longlasting and gross disparity between the composition of a work force and that of the general population thus may be significant even though §703(j) makes clear that Title VII imposes no requirement that a work force mirror the general population.

One need not call on the special insights of the statistician or the scientist to recognize that the first assertion in this statement is untenable, and it can only be given a reasonable interpretation as a statement of the impartiality with which the Court must approach a case of this kind. Two major effects that preclude an expectation of absolute representativeness in the work force are the *correlation of characteristics* and *applicant self-selection*.

Correlation of Characteristics

It is a matter of everyday experience that individuals distinguished by some observable characteristic—not only sex or race or religion but also height or hair color—are likely to differ to some degree in many other characteristics, although the *direction* of the difference is not always intuitively evident. For example, we are not at all surprised to learn that longevity varies with extent of obesity, but it is an unexpected finding that individuals 10% or 20% overweight have somewhat greater longevity than do those at or lower than ideal weight (Dyer et al. 1975). Again, it is well known that females tend to be shorter than males, but assuming that they receive comparable education, it is a surprise to learn that young females are generally less proficient at geometric visualization than are young males but more proficient at language skills. In view of long-term social deprivation enforced on some racial groups, it is hardly surprising that for tests requiring high levels of education or extensive familiarity with certain aspects of the dominant culture, performance is less favorable for some groups than for others. Thus, one might well expect average performance in IQ tests to vary among groups, depending on extent of language acquisition, education, length of residence, and cultural assimilation, although few could predict in any detail the direction of such intergroup differences. (For example, one might predict that the dominant culture group in a country would always outperform any other, but, in fact, Orientals in the United States have, on average, higher IQ scores than do comparably educated Caucasians.)

The point intended under this heading is not that *substantial* differences will always exist between groups but merely that typically *some* differences do, that the relation is often one of association or correlation rather than one of direct causation, and that for that very reason neither the size nor the direction of such differences can usually be reliably predicted.

Applicant Self-Selection

A second major limitation on the interpretation of comparisons between hire or promotion rates for different groups is the self-selection of individuals in applying for a position. This question was considered by the Supreme Court in *Washington v. Davis,* 426 U.S. 229 (1976). The Court took note in this case that the D.C. Police Department was engaged in an affirmative action program, seeking to increase the number of black recruits. Its campaign urged blacks to take a qualification test, even if they doubted that they would succeed. Not surprisingly, the pass rate for blacks during this period fell, but, in view of the recruitment campaign, the Court did not find this disparity of pass rates evidence of an "intent to discriminate" against blacks. Since this case was judged by constitutional and not Title VII standards, and therefore required a finding of *intent,* the Court did not indicate how its discussion would apply in a Title VII case brought under the doctrine of *Griggs*.

The selective effect in *Davis* was obvious, but absent a carefully executed sampling procedure, analogous selective effects will always be present to some degree. The issue arose in *Hazelwood* (did the affirmative action plan of the city of St. Louis deter city black teachers from applying for jobs in the county?), but the opinion deals with it only in a superficial way.

A substantial contribution to self-selection arises from cultural diversity in the population. Thus, a group in which there is general disapproval of women who work in any but a few professions may contribute its most intelligent and able women to those few approved professions—for example, teaching. At the same time, the abilities of male applicants in those fields will be more varied, and thus female applicants might be expected to test better than males.

An additional, perhaps extensive, force for differential selection is the effect of discriminatory practices elsewhere in the labor market. Qualified women finding themselves largely excluded from such professions as engineering will compete for positions in teaching, nursing, and the like, and the quality of female applicants for positions in those professions will therefore be high. A substantial fraction of qualified men, however, will be drained off into engineering, and the average ability of the males applying in the other areas mentioned would be correspondingly reduced.

A pervasive source of selection is simple proximity. If an employer operates from a location central for group A but inconvenient for members of group B, it is plausible that the employer will find a wide spread of abilities among applicants from group A but be limited to the less able members of group B—those unable to find employment more conveniently located for them.

Admittedly, conventional expectations of the direction and magnitude of differences do not form a suitable basis for legal adjudication in a case alleging discrimination. Considerations of equity and justice may indeed require a neutral stance on the part of the law. The point we emphasize here, however, is that "neutrality" is not equivalent to an assumption of "no difference." This is distinct from our point about the failure of randomness. If there is a failure of random sampling in the form of clustered sampling, the use of simple statistical tests would incorporate an underestimate of the standard error, possibly by a large factor, but if the groups compared are inherently of equal ability, we would still expect—with a sufficiently large sample—a nondiscriminatory employment test to provide very nearly equal pass rates. The point we catalog under the term *relevance* is that small differences in characteristics of all kinds, including ability at different tasks between identifiably distinct groups, are the norm and not the exception, and that with very large samples—those so large that sampling error itself is negligible—we should not expect to find exactly equal performance. All experience points the other way.

Thus, a reasonable allowance for "nondiscriminatory disparities" (i.e., a degree of disparity ordinarily to be expected where no discrimination is at work) cannot reasonably be based on random sampling models, nor is it sufficient to multiply the standard error by some factor to allow for the effects of nonrandom sampling. We must to some degree allow for the pervasive fact that groups do indeed differ from one another, and that a small difference in average performance, albeit a real one, is not reasonably to be regarded as prima facie evidence of prohibited discrimination.

The key issue is most obvious in cases involving large employers; in a case involving thousands of test takers, it is almost certain that the pass rate for blacks and whites will be statistically disparate, simply because the large sample sizes will tend to make *any* difference *statistically* significant. In smaller samples the failure to allow for real but unimportant differences is largely masked, because the standard errors based on the random selection model are large enough to swamp them. In larger samples the issue cannot unfortunately be thus avoided. In a comparison with, say, 2500 black and 2500 white applicants, a pass rate of 61% for one group and 64% for the other would result in a finding of a statistically significant difference. If the finding is controlling and the lower pass rate goes with whites, no consequences would typically ensue, but if it goes with blacks it would result, under the methodology advanced in *Hazelwood,* in a finding of prima facie evidence of discrimination against blacks. Thus a firm employing large numbers is almost certain to find that, under the *Hazelwood* standard, its tests or hiring criteria show adverse impact against one group or another. It must then be prepared to mount studies to validate these hiring criteria as predictive of employee productivity (and face the enormous obstacles so well described by Lerner, 1976) or abandon the hiring criteria or employment tests altogether and operate by quota, hiring and promoting equal proportions—in effect, using different standards for different groups. Whether this results in good or bad policy and whether it conforms to congressional intent in passing the Civil Rights Act are questions we do not address.

4. Alternative Outlooks

What should the Court do, then, given this gap between mathematical theory and practical reality? There are two polar positions.

Abolitionists and Strict Constructionists

This position, represented by legal scholar Laurence Tribe (1971) and statistician David Freedman (Freedman et al., 1978), is that one should entirely eschew the use of *inferential* statistics in situations such as employment discrimination cases.[2] Tribe has argued that quantitative methods are easily misused: they often hide unstated assumptions, focus attention on those elements of a case most readily quantified but not necessarily most important, and imperil the values of our society by making explicit the implicit fallibility of our legal system. Apart from the last objection, which has troubling paternalistic and elitist implications, these are all legitimate concerns, but they are equally applicable to *any* type of expert testimony: experts often make unequivocal assertions later contradicted by other experts, state conclusions based on a large number of theoretical assumptions seldom articulated and often inapplicable, and carry undue weight with the jury. Just as with any form of technical or scientific evidence, the use of statistics calls for caution rather than neglect.

Freedman, coauthor of a recent and influential textbook, argues that statistical methods of inference should only be used when a random sampling model is clearly applicable. There is again much merit in such a position. The academic literature, as Freedman's text amply illustrates, is rife with examples where no model makes sense and the statistical application is dubious or inappropriate. Numbers, like words, must have a definable and relatively precise meaning to inform; statistical levels of significance refer to the frequency with which certain events occur when samples are drawn at random from a prespecified population, and, absent the framework of a genuinely random process, the onus lies on the user of statistical techniques to justify their appropriateness. *Wovon man nicht sprechen kann, darüber muss man schweigen.*

The Benthamites

The holders of the opposite position find the sirenlike attraction of mathematical precision too powerful to resist. If we can succeed in framing a

[2] We cite Tribe and Freedman as representatives of a stance but we do not describe their actual positions with precision. Neither specifically addresses the employment discrimination problem.

question in a form admitting an exact mathematical solution, we shall have at least escaped some of the uncertainties and fallibility with which the law is usually afflicted. The difficulty is, in forcing our old problem into this new form, is the resulting problem truly relevant? As John Tukey has remarked, an approximate answer to the right question is far better than an exact answer to the wrong question.

A closely related stance is that decisions have to be made, life is complex, and one should take advantage of anything that aids or assists us. Quibbles about the relevance of random sampling models are just that—we are trying to do the best we can. Although in the best of all possible worlds the model assumptions would be met, at least we are using an objective methodology, which is better than nothing. The trouble with this argument is that if the model departures are serious, one risks giving credibility or undue weight to an erroneous conclusion; when the mathematical abstraction becomes a caricature rather than a mirror of reality, it is best to simply face up to that fact and acknowledge a situation that does not admit of simple or clean answers. And as a disciple of Bentham himself observed,

> It is obvious, too, that even when the probabilities are derived from observation and experiment, a very slight improvement in the data, by better observations, or by taking into fuller consideration the special circumstances of the case, is of more use than the most elaborate application of the calculus to probabilities founded on the data in their previous state of inferiority." (John Stuart Mill, *Logic,* Book II, Chapter 18, Section 3)

Our own approach to these questions lies between the two polar positions (although it is certainly much closer to the first than the second). Statistics per se cannot ascertain whether employment discrimination exists. Nor is the question answered by the standard significance test immediately relevant: hiring does not typically proceed by random selection from a pool. But statistical testing is, we would argue, a useful *benchmark.*

Recognizing the difficulties inherent in proving discriminatory intent on an individual basis, the courts have permitted plaintiffs to proceed by impeaching the *process* by which employment takes place. Although hiring is not a process of random selection, it is not unreasonable to compare it to such a benchmark process. If the disparity between black and white hire rates is so small that it could have arisen by random sampling from an appropriate population, then it seems reasonable to conclude that the statistics, by themselves, provide no evidence of discrimination.

But what if the differences in hire rates are "statistically significant?" As we have pointed out, it would be inappropriate to view this alone as establishing a prima facie case of discrimination. The first reason for being wary of a statistically significant disparity is that the basic assumptions for applicability of the random sampling model do not apply in employment settings, and may cause the perceived significance of the data to be unduly inflated. We have shown how this arises in the discussion of random sampling in Section 3. The second reason is that, even if sampling were random from an identifiable pool, *statistical* significance does not imply *practical* significance. To prove that *some* difference exists is not to prove that the difference is the result of legally cognizable discrimination. If we roll a carefully balanced pair of dice enough times, we shall inevitably find some departure from the expected outcomes, since perfection in the manufacture is impossible. With a large enough number of rolls, such departures may exhibit statistical significance, but they are certainly not of practical significance. We have noted in the discussion of "relevance" in Section 3 why some differences between groups must be expected as the rule and why it is indispensable that some allowance should be made for departures from a model that is not entirely credible.

The issue of practical versus statistical significance has been discussed in the legal literature (see, e.g., Baldus and Cole, 1980, pp. 317–318; Smith and Abram, 1981, p. 53). The matter does not seem to carry weight with Shoben (1978, p. 806), who states that the "flaws in the four-fifths rule can be eliminated by replacing it with a test of the statistical significance of differences in pass rate proportions." The Supreme Court has not addressed the problem, but the lower courts have recognized the issue. For example, in *Moore v. Southwestern Bell Tel. Co.*, 593 F.2d 607, 608 n.1 (5th Cir. 1979), a test used to qualify clerks for promotion resulted in 248 of 277 blacks passing (a pass rate of 0.895) while 453 of 469 whites passed (a pass rate of 0.966). The difference in pass rates is 3.94 standard errors and therefore highly statistically significant. The district court did not find the difference to be substantial and was upheld on appeal.

We are led by these considerations to use statistical analysis as a benchmark and guide but at the same time to use a methodology that incorporates a measure of practical significance *as well as* formal statistical significance. Two decisions must be made in order to select an appropriate standard of practical significance. First, it is necessary to choose an appropriate measure of disparity; generally the choice is between a comparison of selection rates or the computation of a ratio of the selection rates. Second, it is then necessary to choose an appropriate cutoff point or benchmark for the magnitude of disparity that can reasonably be expected to be indicative of discrimination rather than simply normal group differences. We discuss the merits of possible measures of disparity in Section 5; in Section 6 we discuss the use of the so-called 80% rule as a reasonable benchmark for a magnitude of disparity that has practical significance.

5. The Measure of Disparity

Any methodology for statistical analysis must begin with a quantification of "disparity" either as a measure of the extent of discrimination or as an indicator of the persuasiveness of the evidence of discriminatory behavior. While consensus on a standard measure is desirable, there are many plausible ways to quantify disparity. The simplest candidates for the measurement of disparity are the difference in selection rates and the ratio of selection rates. If the selection rate for blacks is 35% and for whites is 50%, the difference in selection rates is 35% − 50% = −15%, and the ratio of selection rates is 35/50 = 0.70—a selection rate for blacks that is 70% of that for whites. The important point is that no single measure can capture *completely* the information provided by the two measures combined, and each has infirmities if used alone.

Difference in Selection Rates

Thus, referring to the difference in selection rates, the Court in *Davis v. City of Dallas,* 487 F.Supp. 389 (N.D. Tex. 1980), observed that a "7% difference between 97% and 90% ought not to be treated the same as a 7% difference between, e.g., 14% and 7%, since the latter figures evince a much larger degree of disparity." In contrast, however, Shoben (1978) urges the virtues of using the difference and emphasizes the equivalence of the absolute difference of selection and rejection rates and the non-equivalence of the ratio measure when applied to selection and rejection rates. (If $p_B = 0.70$, $p_W = 0.80$ are the black and white selection rates, then the absolute difference is 10%. The rejection rates $q_B = 1 - p_B = 0.30$, $q_W = 1 - p_W = 0.20$, also have an absolute difference of $0.30 - 0.20 = 10\%$. On the other hand, $p_B/p_W = 0.70/0.80 = 0.875$, whereas $q_W/q_B = 0.20/0.30 = 0.67$.)

The argument favoring the use of the absolute difference in selection rates on grounds of symmetry between selection and rejection rates has some superficial appeal—it acquires mathematical elegance and avoids the need to consider whether Title VII regards selection rates and rejection rates equivalently. However, neither of these attractions justifies the primacy of "difference" as the measure of disparity.

Ratio of Selection Rates

If symmetry is required, there are any number of indices based on the ratio of selection rates that can serve equally well. For example, consider the most extreme of the two ratios, selection ratio and rejection ratio. In the example at the beginning of this section, the rejection rates are 65% for blacks and 50% for whites, giving a ratio of rejection rates 50/65 =

0.77. The more extreme of 0.77 and 0.70 is then 0.70. Another measure of disparity derives from applications in epidemiological studies, namely, the *odds ratio,* which is simply the product of the selection rate and the rejection rate—here 0.70 × 0.77 = 0.53. Symmetry is not a barrier to using a measure based on the ratios of rates.

An example suggests why the ratio of selection rates may indeed be preferable to the difference. Consider employer A, who selects from a pool of applicants consisting of 500 blacks and 500 whites. A selects 75 blacks and 100 whites, yielding a black selection rate of 75/500 = 15% and a white selection rate of 100/500 = 20%. The absolute difference is 5%, and the selection rate ratio is 75%. Next consider employer B, who selects from an applicant pool of 1000 blacks and 1000 whites. B selects 75 blacks and 100 whites. Now the selection rates are 75/1000 = 7.5% and 100/1000 = 10% for blacks and whites, with an absolute difference of 2.5% and a selection rate ratio of 75%.

The point is transparent: Both employers need the same number of new employees, both are faced with applicant pools having the same racial composition (half black, half white), and both hire the same number of blacks (and the same number of whites). The ratio of hire rates is the same for both, but the difference is halved for the larger employer, B, solely because twice as many applicants appear. (If we pursue a statistical analysis here, then the impact of the different numbers of applicants will be felt no matter which measure—the ratio or the absolute difference—is used. In any case, the choice of index necessarily precedes any statistical considerations.) Both the *extent* and the *appearance* of discrimination are the same, and the statistical significance of the disparity in each case is also virtually the same. Thus, we conclude that ratios of selection rates will typically be more useful as measures of disparity than simple differences in selection rates.

Selection Rates Versus Rejection Rates

A remaining issue is whether it is more appropriate to base measures of disparity on selection rates or rejection rates. Absent a showing of business necessity, *Griggs* prohibits practices with adverse impact that thus deny groups a fair share of the employment pie. This approach suggests that a comparison of selection or pass rates is more relevant than comparison of rejection rates. Consider, for example, a situation where the selection rates are nearly equal, 95% for blacks, 98% for whites. The ratio is 0.95/0.98 = 97%. The rejection rates, 5% and 2%, are highly discrepant since their ratio is 0.02/0.05 = 40%. With respect to "share of the pie," the effect of such a disparity is negligible since, if there were equal numbers of blacks and whites in the applicant pool, blacks would receive

$0.95(0.95 + 0.98) = 49\%$ of the available positions—nearly half the positions. Such data do not indicate an underrepresentation of blacks in the work force; thus, neither an inference of adverse impact nor of discriminatory intent would be warranted.

When a discrepancy in rejection rates should raise an issue of illegal discrimination is not self-evident. If hiring is based on a practice whose primary goal is to select applicants, then pass rates would appear appropriate; however, a practice that emphasizes rejection of applicants may call for inspection of rejection rates. The context in which the 80% rule usually operates is one where selection for employment or promotion is emphasized, and such cases rely on a disparity measure based on pass rates. Such cases as *New York City Transit Authority v. Beazer,* 440 U.S. 568 (1979) and *Green v. Missouri Pacific Railroad Company,* 523 F.2d 1290 (8th Cir. 1975), on the other hand, deal with practices emphasizing rejection and lead to examination of the rejection rates. (In *Green,* for example, of the 3282 black applicants and 5206 white applicants, 174 blacks and 118 whites were rejected on the grounds of having a prior conviction. Here $q_B = 174/3282 = 0.05$, and $q_W = 118/5206 = 0.02$, i.e., the data have the same rejection (and selection) rates as in the example of the last paragraph. The court found the use of a conviction record to be arbitrary and found for the plaintiff, whose conviction resulted from refusal to be drafted into military service.) Ultimately, of course, the measure of disparity relevant to a Title VII action must be sought primarily in Title VII and the legal interpretations that followed, not in the principles of statistics.

6. The 80% Rule

Having concluded that the ratio of selection rates will usually be the best measure of disparity, let us turn to a standard to evaluate the practical significance of disparity. One standard that seeks to measure the practical significance of disparity is the so-called 80% rule. The rule states in part:

> A selection rate for any race, sex, or ethnic group which is less than four-fifths ($\frac{4}{5}$) (or 80%) of the rate for the group with the highest rate will generally be regarded by the Federal enforcement agencies as evidence of adverse impact, while a greater than four-fifths rate will generally not be regarded by Federal enforcement agencies as evidence of adverse impact. Smaller differences in selection rate may nevertheless constitute adverse impact, where they are significant in both statistical and practical terms or where a user's actions have discouraged applicants disproportionately on grounds of race, sex, or

> ethnic group. Greater differences in selection rate may not constitute adverse impact where the differences are based on small numbers and are not statistically significant, or where special recruiting or other programs cause the pool of minority or female candidates to be atypical of the normal pool of applicants from that group. Where the user's evidence concerning the impact of a selection procedure indicates adverse impact but is based upon numbers which are too small to be reliable, evidence concerning the impact of the procedure over a longer period of time and/or evidence concerning the impact which the selection procedure had when used in the same manner in similar circumstances elsewhere may be considered in determining adverse impact. (29 C.F.R. §1607.4(d) (1983))

The 80% rule has sometimes been misinterpreted by the courts, and by legal scholars as a simple and naively misguided substitute for a statistical significance test.[3] As recognized by the direct court in *Wilmore v. City of Wilmington,* 533 F.Supp. 844 (D. Del. 1982), however, the rule is, in fact, a far more sophisticated instrument, incorporating the very significance test proposed as a substitute, and consideration of its strengths and weaknesses must proceed on a higher level.[4]

Statistical evidence is now accepted by the courts as relevant and, in some cases, strongly persuasive standing alone. However, as discussed above, the simple models, on which traditional statistical analyses rely, fail to reflect reality and do not properly represent the neutrality that must guide legal considerations. What is required is a rule that can provide a reasonable balance between the right of a plaintiff to seek redress and the right of a defendant to be free of unreasonable harassment. Attractive as it might be for the courts to find a solution to this problem in the practice of statistical methods in scientific research, it would be an abdication of the

[3] And even on occasion by the EEOC itself. In contrast to its original clear and direct statement of the rule, a "Questions and Answers" supplement later published by the EEOC is confusing to the point of incoherence.

[4] Historically, the 80% rule, as it appears in the EEOC agency guidelines, evolved from a California regulation that required both the 4/5 disparity in selection rate ratios *and* statistical significance for an administrative finding of adverse impact. Subsequently, compliance officers experienced difficulties in properly performing statistical significance tests in the field and the role of statistical significance was much reduced in the application of the regulation. This experience and lengthy discussion led, at the federal level, to the adoption of the present rule, which reaffirms the role of statistical significance but couched in terms permitting generally easier administrative application. The sources of misinterpretation of the rule have been the error in focusing *only* on the 4/5 disparity and the failure to recognize that at no time was it suggested that the 4/5 disparity alone be adopted as having probative force in the courtroom. We are grateful to William Burns, David Rose, and Philip Sklover for this information.

law's responsibility to borrow thoughtlessly a criterion not suited to its purpose. Simple statistical significance tests, such as that used in *Hazelwood,* are clearly not well adapted to the legal situation and should not alone be used as a basis for a prima facie showing of discrimination.

Since an allowance for disparity broader than that provided for by a simple significance test is necessary, the question of what particular allowance for nondiscriminatory disparity should be made—whether a ratio of selection rates of 85%, or 67%, as some courts used earlier, or 90%—is a matter of practical judgment on which detailed evidence is lacking. It is at least arguable that the courts should defer to the administrative agencies, and their expertise, in making this choice.

We argue that we should settle for a reasonable rule, recognizing that it may not be the only possible choice. The unmodified two-sample binomial model has the merit of simplicity, but no responsible qualified observer believes that it comes close to representing the statistical fluctuation to be expected in a hiring situation. The 80% rule, arbitrary as some of its elements are, does make allowance for the limitations of statistical models and, by so doing, exhibits characteristics that make it preferable to the two-sample binomial test. Moreover, by incorporating a measure of practical significance, the 80% rule avoids detecting a disparity too small to be meaningful, a characteristic not shared by rules based on the usual statistical significance test operating alone.

Comparing the 80% Rule and the Two-Sample Test

In many circumstances (mainly when pass rates are low or sample sizes are small), the 80% rule and the two-sample significance test produce equivalent conclusions. For example, if the pass rate for the favored group is 5% and there are equal numbers of applicants from each group, then the two rules come to the same conclusion unless there are 3300 or more applicants in each group. An instance like this appears in one of the myriad of data sets in *Cormier v. P.P.G. Indus.*, 519 F.Supp. 211 (W.D. La. 1981), where there were 2499 white and 1050 black applicants for utility crew jobs in the year 1977, of which 86 whites and 24 blacks were hired. The hire rates are $p_B = 0.023$, $p_W = 0.034$ with a ratio $p_B/p_W = 0.664$. The two-sample tests shows that this disparity is not statistically significant; thus, both the 80% rule and the two-sample binomial test would not find adverse impact.

Generally, if both pass rates are below 10%, both rules lead to the same conclusion unless the number of applicants is very large. When pass rates are moderate or high, however, the result of applying the 80% rule is noticeably different from that achieved by applying the two-sample binomial rule. For example, if the pass rate for the favored group is 90% and there are at least 37 applicants in each group, or the pass rate for the favored group is 50% and there are more than 190 applicants in each group, the two rules can fail to agree. In *Jackson v. Nassau County Civil Serv. Comm'n,* 424 F.Supp. 1162, 1167 (E.D. N.Y. 1976), 113 identified whites and 55 blacks took an examination for the position of community service assistant. Of these, 99 whites and 40 blacks passed. The pass rates are $p_B = 0.727$, $p_W = 0.876$ with ratio $p_B/p_W = 0.830$. Thus the 80% rule would not find substantial disparity. The two-sample binomial test, however, is significant (with $Z_c = -2.178$).

A case in this domain but where the sample sizes are less and disparity is great, resulting in a borderline situation, is *Reynolds v. Sheet Metal Workers Local 102,* 498 F.Supp. 952, 960, 965 (D. D.C. 1980). In *Reynolds,* of 44 black and 80 nonblack applicants to an apprenticeship training program, 14 blacks and 41 nonblacks were selected. The selection rates were $p_B = 0.318$, $p_W = 0.513$; the ratio is 0.621 (the reported calculations in the opinion are in error). The one-sided two-sample binomial test shows significance at the 0.028 level, which is slightly greater than the 0.025 standard. Strictly speaking, neither the 80% rule nor the two-sample binomial test of significance would find disparity. The court agonized about the 0.025 standard and decided on the basis of collateral evidence and the inessentiality of such a precise standard that a prima facie case was established. Thus there are common circumstances (both hypothetical and actual) where conclusions drawn from application of the 80% rule differ from those drawn using the two-sample binomial alone.

Properly understood, the 80% rule has the potential to rationalize much of the case law to date (see appendix). As a practical matter, even when statistically significant differences have been noted, the courts have been reluctant to find adverse impact when the differences lack what is variously described as "practical," "substantive," or "constitutional" significance. And conversely, substantial disparities have been found insufficient to establish a prima facie case when the sample sizes are so small as to make statistical significance unlikely. The 80% rule appears to be a reasonable articulation of a statistical criterion to determine whether statistically significant differences are substantial enough to warrant legal liability.

It also remains open to consideration whether the particular form of the rule articulated in the EEOC guidelines sets forth an appropriate standard for suspicion of illegal discrimination. A variety of alternatives suggest themselves immediately. For example, one might require that the ratio of selection rates be significantly more discrepant than 80%. One might also introduce a factor to increase the standard error to allow for the likely effects of clustering. It is improbable, however, that any rule can be devised to completely satisfy all requirements. Such an ideal rule would have to be based on far more detailed knowledge of the underlying relationships and employment behavior than the social sciences are able to provide.

7. Conclusions

The twentieth century has seen a dramatic growth in the development of statistical methodology and its increasingly pervasive application in science and society. The recent and expanding resort to statistics by the courts can be seen as merely the reflection and logical outcome of the ubiquitous use of statistics in accounting, marketing, political science, psychology, economics, and throughout the social sciences generally.

In its proper domain of application the modern science of statistics has achieved a number of impressive successes: the calculation of celestial orbits by the method of least squares, the use of scientific sampling in public opinion polls, the design of agricultural experiments, and the randomized, double-blind clinical trial, to name but a few. Such successes encourage the exportation of the successful methodology to other areas: just as the achievements of mathematics in the physical sciences led to the increasing quantification of the social sciences, so the achievements of statistics led to a qualitative reorientation of particular disciplines such as psychology and sociology within the social sciences.

But mathematics, as the court wrote in *People v. Collins,* 68 Cal. 2d 319 (1968), "while assisting the trier of fact in the search for truth, must not cast a spell over him"; an understandable enthusiasm for a precise and "objective" methodology must be tempered by a sober assessment of that methodology's appropriateness, accuracy, relevance, and informativeness for the law. In this chapter we have attempted a case study of the legal use of some simple statistical methods in the domain of employment discrimination. This is *not* an area where the statistical applications are either immediately applicable or obviously relevant, although we have argued that, viewed as *benchmarks,* tests of statistical significance do have a useful, if limited, role to play.

During the past decade there has been a sharp increase in the resort to statistics in employment discrimination cases. Initially this use was almost entirely confined to the simple methods discussed in this chapter, but in recent years more complex methods have been employed, most notably those of multiple regression. Just as with tests of statistical significance, serious reservations must be raised here as well, both about the statistical appropriateness of the models employed and the legal relevance of the information obtained. Furthermore, the assumptions implicit in multiple regression models are substantially more difficult to state and interpret than those for the simple models we have described. As a result, the (potentially ever present) dangers of erroneous implementation, uncritical acceptance, and numerical obscurantism may reach critical mass and lead to a case that has a great deal to do with statistics but very little with the issue ostensibly at hand (did discrimination, intentional or unintentional, take place?).

In particular, the whole issue of when to shift the burden of proof in an employment discrimination case becomes especially acute when complex multiple regression models are used. A substantial and statistically significant difference in simple pass rates for blacks and whites may indeed seem reasonable ground for shifting the burden of proof to an employer to justify the reason for such a disparity. But a difference in average salary for male and female academics resulting from a complex multiple regression adjustment, based on assumptions that are false on their face, is a very different matter.

These reservations are not intended as a counsel of despair. In the case of hiring rate comparisons, the difficulties of underestimating standard errors and distinguishing "substantial" (as opposed to merely "statistically significant") disparities can both be dealt with by employing the EEOC's 80% rule. As such, the rule serves as a paradigm for how statistics *should* be employed by the courts. Some critics of the 80% rule have argued that it is a crude rule of thumb and contrast it unfavorably with the apparently more precise two-sample binomial test of significance. But in fact the whole point of the rule was that it arose not from ignorance of the obvious statistical procedure, but from an appreciation of that very procedure's shortcomings and limitations. And so, too, in any contemplated legal application of statistics, the questions that should immediately come to mind are: Is the model employed credible? Are the findings relevant? If not, then attention must shift to alternative or additional methodologies that make fewer assumptions, that allow for failures in the model, and that convey legally relevant information. The development of such methodologies for the application of multiple regression models to employment discrimination cases is a fertile but as yet largely uncultivated area for future research.

Appendix: The Treatment of Small Samples in Case Law

As noted above, the 80% rule rationalizes a good deal of the case law developed on disparate impact. With evidence from large samples, showing small but statistically significant disparities, courts have been reluctant to make a finding of disparate impact. Thus, for example, in *United States v. Test,* 550 F.2d 577 (10th Cir. 1976), the Court observed: "The mathematical conclusion that the disparity between these two figures is 'statistically significant' does not, however, require an a priori finding that these deviations are 'legally significant.' "

The courts have shown more ambivalence and less consistency in their reliance on statistical evidence when small samples are involved. The reasons advanced are sometimes confused, and despite a considerable number of cases that have dealt with the issue over the past decade, no clear consensus has emerged or logic been enunciated. (The Supreme Court has twice given its imprimatur to such concerns but in each case, characteristically, has not elaborated. See *Mayor of Philadelphia v. Educ. Equality League,* 415 U.S. 605, 621 (1974); *International Bhd. of Teamsters v. United States,* 431 U.S. 324, 340 n.20 (1977)). The reservations most often stated involved technical confusion about the meaning of statistical terms, understandable distrust of the random sampling model, and intuitive (often erroneous) assessments of statistical significance.

Statistical Confusion

It is sometimes thought that when a sample is sufficiently small, no test of significance is possible. It is also sometimes thought that statistical significance cannot occur in small samples (the latter most notably in *Williams v. City and County of San Francisco,* 483 F.Supp. 335, 341-342 (N.D. Cal. 1979)), where the court misread the Department of Justice guidelines and a large number of cases as equating small sample size with lack of statistical significance). Both are misconceptions: although tests based on the normal distribution in the one-sample or two-sample models involve approximations that break down with very small samples, in each case the approximation is to an exact test (binomial or Fisher) that *is* always valid in the presence of random sampling, no matter how small the sample size. And while statistical significance is, of course, harder to achieve in small samples, it is not impossible.

Example 1: *Lee v. City of Richmond,* 456 F.Supp. 756 (E.D. Va. 1978)

	Promoted	Not	
Black	0	4	4
White	4	2	6
	4	6	10

$\frac{1}{2}P = 0.0714.$

Here the court was under the misapprehension that "thirty is a generally accepted minimum of persons for a sample used in statistical testing." (Although there was once a general "rule of 30" presumed to be necessary for a valid test of the sample mean, Student's discovery of the t distribution in 1908 ended the need for choosing a minimal sample size to validate statistical tests, and thus the rule is no longer relevant. In any event, the rule of 30 has no application to the problem at hand.) In *Lee,* the probability of a sample disparity as large or larger than that in the table above is $\frac{1}{2}P = 0.0714$, so the disparity does not reach the conventional 2.5% one-sided level of significance. (The notation $\frac{1}{2}P$ is used to emphasize that the significance probability in question is one-sided.)

Example 2: *Chicano Police Officers Ass'n v. Stover,* 526 F.2d 431 (10th Cir. 1975), *vacated,* 426 U.S. 944 (1976)

	Pass	Fail		
Spanish surnamed	3	23	26	
Other	14	50	64	$\frac{1}{2}P = 0.204$
	17	73	90	

Here the district court thought the "2 to 1 ratio . . . statistically insignificant because of the smallness of the sample"; the court of appeals reversed partly on the ground that "the group tested was not too small to be evaluated." Both are in part correct. The significance level can indeed be evaluated, but as shown above, the difference is not statistically significant at anywhere near the conventional 2.5% level.

Distrust of the Random Sampling Model

Despite the availability of the binomial or Fisher exact tests for doing statistical inference with samples of any size, other reasons have been given by some courts for approaching small samples with special caution. These were clearly stated in *Chance v. Board of Examiners and Board of Education of City of New York,* 330 F.Supp. 203 (S.D. N.Y. 1971), *aff'd* F.2d 1167 (2d Cir. 1972):

> [S]uch a sample is less reliable than analyses based on larger samples, even after making allowances for greater margins of error in use of the small sample. Use of the small sample involves more extrapolation and theory superimposed on less fact. We prefer the greater fact content found in the larger sample.

The court's preference for large samples is understandable but not justifiable in the terms just quoted. The increased "extrapolation" resulting from a smaller sample is precisely what the "greater margins of error" allow for, and the imposition of theory on large samples is no more likely—indeed, often less likely—to lead to justified conclusions. In sum, the doubts raised are entirely justified, but they should not be limited to the appraisal of small samples.

Intuitive Assessment of Significance

The courts have also remarked that in a case with small samples selection rates are often very sensitive to small changes in the numbers accepted (see, e.g., *Mayor of Philadelphia v. Educ. Equality League,* 415 U.S. 605 (1974)). The reluctance to rely on small samples in such cases seems to be based on an intuitive appraisal of statistical significance, expressed in one instance as an aversion to concluding that "an inference of racial discrimination is raised by the discharge of one or two extra minority employees when the total number of dismissals is so small" (*Wade v. New York Tel. Co.,* 500 F.Supp. 1170, 1180 (S.D. N.Y. 1980)). Another example is *Bridgeport Guardians Inc. v. Members of Bridgeport Civil Serv. Comm'n,* 354 F.Supp. 788 (D. Conn. 1973), where the court wrote:

> While this probability may have some statistical significance, the numbers on which it is based are too small to have constitutional significance. If only one more non-White candidate had passed the exam, the comparative passing rates would be 68% and 50%, too slight a difference to establish a prima facie case of discrimination. Even if the 68% to 40% comparison is sufficient under *Chance,* these figures cannot be relied upon when a different result achieved by a single candidate could so drastically alter the comparative figures.

To the extent the pure randomness is taken to be the appropriate standard, this plausible-sounding concern is again without merit—a formal significance test is more effective in judging departures from randomness than is the court's intuition.

Whatever the actual bases for these various statements of reluctance to accept the implications of tests of randomness, it is worth noting that in almost all these cases the 80% rule would not establish a prima facie case of discrimination. Thus, in *Bridgeport Guardians, Inc. v. Members of Bridgeport Civil Service Commission* the "same statistical significance" is 0.071, which is far above the 0.025 standard; in *Jackson v. Nassau County Civil Service Commission* the ratio of pass rates exceeds 80%. Of the cases cited, only *Dendy v. Washington Hospital Center* and *United States v. City of Chicago* would be found significant by the 80% rule. In *Dendy* the court plainly rejected the possibility of statistical evidence (but was later reversed):

> To be persuasive, statistical evidence must rest on data large enough to mirror the reality of the employment situation. If, on the one

TABLE 1. Selected Cases

Case	Data	Selected	Rejected	Pass Rates p_B	Pass Rates p_W	Pass Rates p_B/p_W	Significance Probability
Bridgeport Guardians, Inc. v. Members of Bridgeport Civil Service Commission, 354 F.Supp. 778 (D. Conn. 1973)	Black	4	6	0.400	0.681	0.588	0.071
	White	130	61				
Chicano Police Officers Association v. Stover, 526 F.2d 431 (10th Cir. 1975)	Black	3	23	0.115	0.219	0.527	0.204
	White	14	50				
Connecticut v. Teal, 457 U.S. 440 (1982)	Black	26	22	0.542	0.795	0.68	0.0003
	White	206	53				
Cormier v. P.P.G. Industries, 519 F.Supp. 211 (W.D. La. 1981)	Black	24	1,026	0.023	0.034	0.66	0.041
	White	86	2,413				
Davis v. City of Dallas, 487 F.Supp. 211 (N.D. Tex. 1980)	Black	88	808	0.098	0.153	0.644	0.000
	White	437	2,427				
Dendy v. Washington Hospital Center, 431 F.Supp. 873 (D. D.C. 1977)	Black	4	5	0.444	1.000	0.444	0.0004
	White	26	0				
Jackson v. Nassau County Civil Service Commission, 424 F.Supp. 1162 (E.D. N.Y. 1976)	Black	40	15	0.727	0.876	0.83	0.016
	White	99	14				
Reynolds v. Sheet Metal Workers Local 102, 498 F.Supp. 952 (D. D.C. 1980)	Black	14	30	0.318	0.513	0.621	0.028
	White	41	39				
United States v. City of Chicago, 21 Fair Empl. Prac. Cas. (BNA) 200 (N.D. Ill. 1979)	Black	5	7	0.417	0.742	0.562	0.019
	White	417	145				
Wilmore v. City of Wilmington, 533 F.Supp. 844 (D. Del. 1982)	Black	1	10	0.091	0.131	0.693	0.584
	White	8	53				

Notes: p_B and p_W are the pass (or "selection") rates for blacks and whites, respectively; if the ratio is less than 0.8, then one part of the 80% rule is violated. Significance probabilities are one-sided.

(*Black* and *white* are used as generic terms for least-favored and most favored groups, respectively.)

> hand, the courts were to ignore broadly based statistical data, that would be manifestly unfair to Title VII complainants. But if, on the other hand, the courts were to rely heavily on statistics drawn from narrow samples, that would inevitably upset legitimate employment practices for reasons of appearance rather than substance. The courts must be astute to safe-guard both of these conflicting interests. . . .
>
> In the instant matter, the Court is convinced that the data offered by plaintiffs represents too slender a reed on which to rest a weighty remedy of preliminary relief.

In *United States v. City of Chicago* it is not clear from its opinion whether the court would have found differently were it to learn the outcome of a test of significance. In this case 562 whites and 12 blacks took the 1973 Fire Captain Examination, and 417 whites and 5 blacks passed. The ratio of pass rates is 0.562. From the opinion one gathers that no significance test was performed, but the $\frac{4}{5}$ part of the 80% rule was appealed to by the plaintiffs. At the same time the court expressed concern about the small number of black applicants "especially in the light of the absence of thorough testimony on the statistical significance." In fact, the Fisher exact test establishes statistical significance at the 0.019 level.

Table 1 summarizes a number of cases discussed in this chapter.

References

Baldus, D. C., and Cole, J. W. L. (1980). *Statistical Proof of Discrimination,* Colorado Springs, Colo.: Shepard's.

Bickel, P. J., Hammel, E. A., and O'Connell, J. W. (1975). "Sex Bias in Graduate Admissions: Data from Berkeley," *Science,* **187,** 398–404.

Booth, D., and MacKay, J. L. (1980). "Legal Constraints on Employment Testing and Evolving Trends in the Law," *Emory Law Journal,* **29,** 121–194.

Cochran, W. G. (1952). "The χ^2 Test of Goodness of Fit," *Annals of Mathematical Statistics,* **23,** 315–345.

Cohn, R. M. (1980). "On the Use of Statistics in Employment Discrimination Cases," *Indiana Law Journal,* **55,** 493–514.

DeMoivre, A. (1756). *The Doctrine of Chances,* 3rd ed. Reprinted 1967, New York: Chelsea.

Dyer, A. R., Stamler, J., Berkson, J. M., and Lindberg, H. A. (1975). "Relationship of Relative Weight and Body Mass Index to 14-Year Mortality in the Chicago Peoples Gas Company Study," *Journal of Chronic Diseases,* **28,** 109–123.

Ellman, I. M., and Kaye, D. (1979). "Probabilities and Proof: Can HLA and Blood Group Testing Prove Paternity?," *NYU Law Review,* **54,** 1131–1162.

Finkelstein, M. O. (1978). *Quantitative Methods in Law: Studies in the Applica-*

tion of Mathematical Probability and Statistics to Legal Problems, New York: Free Press.

Fisher, R. A. (1970). *Statistical Methods for Research Workers,* 14th ed., New York: Hafner Press.

Freedman, D., Pisani, R., and Purves, R. (1978). *Statistics,* New York: Norton.

Innes, Michael (1970). *What Happened at Hazelwood,* New York: Penguin Books.

Kendall, M., and Stuart, A. (1973). *The Advanced Theory of Statistics,* Vol. 2, 3rd ed., New York: Hafner Press.

Keynes, J. M. (1921). *A Treatise on Probability,* London: Macmillan.

Kruskal, W. H. (1977). Letters. In W. Fairley and F. Mosteller (eds.), *Statistics and Public Policy,* Reading, Mass.: Addison-Wesley, pp. 127–130.

Lehmann, E. L. (1959). *Testing Statistical Hypotheses,* New York: John Wiley & Sons.

Lerner, B. (1976). "Washington v. Davis: Quantity, Quality, and Equality in Employment Testing," *Supreme Court Review,* **1976,** 263–316.

Machol, R. (1975). "The Titanic Coincidence," *Interfaces,* **5,** 53–54.

Meier, P., and Zabell, S. L. (1980). "Benjamin Peirce and the Howland Will," *Journal of the American Statistical Association,* **75,** 497–506.

Pratt, J., and Gibbons, J. D. (1981). *Concepts of Nonparametric Statistics,* New York: Springer-Verlag.

Shoben, E. W. (1978). "Differential Pass-Fail Rates in Employment Testing: Statistical Proof Under Title VII," *Harvard Law Review,* **91,** 793–813.

Smith, A. B. Jr., and Abram, T. G. (1981). "Quantitative Analysis and Proof of Employment Discrimination," *University of Illinois Law Review,* **1981,** 33–74.

Solomon, H. (1971). "Statistics in Legal Settings in Federal Agencies," *Federal Statistics: Report of the President's Commission,* Vol. 2, Washington, D.C.: U.S. Government Printing Office, pp. 497–525.

Spradlin, B. C., and Drane, J. W. (1978). "Additional Comments on the Application of Statistical Analysis to Differential Pass-Fail Rates in Employment Testing," *Duquesne Law Review,* **17,** 777–783.

Sprowls, R. C. (1957). "The Admissibility of Sample Data into a Court of Law: A Case History," *UCLA Law Review,* **4,** 222–232.

Tocher, K. D. (1950). "Extension of the Neyman-Pearson Theory of Tests to Discontinuous Variates," *Biometrika,* **37,** 130–144.

Tribe, L. H. (1971). "Trial by Mathematics: Precision and Ritual in the Legal Process," *Harvard Law Review,* **84,** 1329–1393.

Van Bowen, J. Jr., and Riggins, C. A. (1978). "A Technical Look at the Eighty Per Cent Rule as Applied to Employee Selection Procedures," *University of Richmond Law Review,* **12,** 647–656.

Wagner, C. H. (1982). "Simpson's Paradox in Real Life," *American Statistician,* **36,** 46–48.

Zeisel, H. (1969). "Dr. Spock and the Case of the Vanishing Women Jurors," *University of Chicago Law Review,* **37,** 1–18.

Comment

Stephen E. Fienberg
Departments of Statistics and Social Sciences
Carnegie-Mellon University
Pittsburgh, Pennsylvania

Meier, Sacks, and Zabell (MSZ) provide a thoughtful and exhaustive review of the Supreme Court's opinion in *Hazelwood*. They also trace some of the subsequent uses and abuses of statistical methods in federal court opinions on employment discrimination cases, especially those involving simple two-group comparisons. They present arguments for the use of Fisher's exact test as an appropriate benchmark in such settings and they conclude that the courts would be better off using the so-called 80% rule as a substantiality criterion of discrimination rather than relying only on tests of statistical significance. This comment (a) reconsiders the appropriateness of Fisher's exact test as the sole approach to statistical significance in a "two-binomial" problem, (b) reiterates MSZ's concern about the oversimplification of the "two-binomial" model for employment discrimination problems, (c) raises some questions about the suitability of the 80% rule as a measure of disparity, and (d) reflects briefly on the possible consequences of MSZ's recommendations for the federal courts' response to the use of statistical methodology by expert witnesses.

In order to set the stage for my remarks, I need to be explicit about my position on what MSZ correctly take to be a fundamental issue of statistical philosophy. I had never consciously thought of myself as a Benthamite before. But given the choice between MSZ's two polar positions, I find my philosophy on the relevance of inferential statistics to settings

The preparation of this paper was supported in part by National Science Foundation grant No. BNS-80494 while the author was a Guggenheim Fellow at the Center for Advanced Study in the Behavioral Sciences.

that are not obviously stochastic in nature to be far closer to the Benthamite view of the utility of mathematical (and statistical) models than to the strict constructionist or abolitionist view attributed to Tribe and to Freedman. This will not come as a surprise to anyone familiar with my Bayesian perspective (e.g., see Fienberg and Kadane, 1983) or who has read my recent exchange with Freedman on the topic (see Freedman, 1985a, 1985b; Fienberg, 1985). Even though MSZ claim a position closer to the strict constructionist end of the philosophical spectrum, I still agree with their assessment of the general state of disarray in the federal courts with regard to the use of statistics in employment discrimination litigation. What remains at issue is what path the courts should follow in order to make more effective use of statistical evidence in such litigation.

On Fisher's Exact Test

MSZ state that "there is a single exact test, commonly referred to as the *Fisher exact test,* which is appropriate for determining whether a difference in hire or promotion rates is statistically significant. . . . there are a number of alternative tests and variant approximations. But *all* the other tests—two-sample binomial, or chi-squared with or without continuity correction, for example—may properly be considered as more or less adequate approximations to the Fisher exact test." To see that this is an overstatement of the consensus in the statistical community, one need only read the discussion following Yates (1984). This paper marked the golden anniversary of Yates's original paper introducing the exact test and the approximation to it by the continuity corrected chi-squared test statistic.

I strongly agree with MSZ that the failure of federal judges and legal and statistical commentators to take note of the virtual equivalence of the various procedures available for the comparison of two samples rates has led to great confusion and has contributed to the statistical disarray found in court opinions. In particular, except for cases where n_1 and n_2 are quite small (or, by symmetry, the numbers selected and not selected), there just is not much difference between inferences based on the exact test and those based on the usual chi-squared test for 2×2 tables. For these very small sample cases, I, too, would recommend, and do in practice use, the exact test. But otherwise I advocate the use of uncorrected chi-squared statistic and its likelihood ratio counterpart. For me, this use has both a technical and a practical rationale.

The technical rationale is straightforward. Referring the uncorrected chi-squared statistic to the chi-squared distribution on 1 d.f. gives a significance test that, for modest and large samples, achieves nominal levels under either a multinominal or a two-binomial sampling model. In this sense, the chi-squared test does not involve conditioning on both sets of marginal totals and it does not derive its validity from the extent to which it gives results sufficiently close to the exact test. From this unconditional perspective correcting the chi-squared statistic for continuity makes the resulting test conservative, since the correction is designed to improve the approximation to the tail probabilities of the hypergeometric distribution, and not to the chi-squared distribution on 1 d.f.

Moreover, the chi-squared statistic for 2×2 tables is a special case of the test statistic for a much broader class of statistical problems, in larger two-way tables involving more than two groups or two outcomes and in multiway tables, representable in terms of log-linear and logit models. Such problems often arise in employment discrimination cases, and the Neyman–Pearsonian or Fisherian arguments for conditioning do not really carry over in a simple and direct fashion to these more general settings; however, the chi-squared and likelihood ratio tests do. In addition, for tests involving multiple d.f. carried out in connection with log-linear or logistic models, we often wish to partition the chi-squared statistic or the likelihood ratio statistic into 1 d.f. components. The use of the standard statistics without correction and the corresponding chi-squared reference distributions seems preferable in such circumstances. As a practical matter, I think that it is not necessary for a statistician to clutter up courtroom testimony with exact tests for 2×2 tables, and in almost the next breath use chi-squared statistics for a related problems. Besides, if we are to argue that the tests of significance should only serve as benchmarks, as MSZ do, then we need not concern ourselves with the minor differences in achieved levels of significance.

I cannot help noting that there is a pernicious aspect to the use of tests of significance in such legal settings. While courts almost always are presented with such tests and related p-values, what they want and what they often think they get are posterior probabilities (what MSZ refer to as conditional probabilities). Thus, if we as statisticians are to avoid the confusion associated with this common misinterpretation of p-values, we would do well to give the courts those quantities that, in MSZ's own words, "would be most relevant for evaluating whether there is discrimination." That tests of significance are all that classical statistical theory is able to provide is, for me, a good reason to abandon the classical theory in favor of the Bayesian approach.

On Statistical Models of Employment Processes

MSZ do well to remind us that the assumption of random selection or sampling is rarely appropriate in nature or human affairs. This assumption is inherent in the use of the exact and chi-squared tests, and courts and statistical witnesses often confuse matters even further by equating "fairness" or "nondiscrimination" with "randomness." If the process used by an employer for hiring departs from random selection or sampling, then the relevance of the usual tests can and should be called into question, even if they are used only as benchmarks. As MSZ note, clustering and stratification are important aspects of many recruiting processes and they can be incorporated into statistical models of "fair" or "nondiscriminatory" selection, before tests of significance are carried out. For example, if selection is carried out at the department level, then an appropriate statistical model should take this stratification into account by conditioning on department. For such a situation, it is still quite easy to carry out an overall test of significance for departures from random selection within a department.

Expert witnesses and judges often fail to take cognizance of the extent to which statistical models used to analyze employment data fail to capture the relevant aspects of the employment process. When they do so, the standard tests of significance that ignore these aspects may be of little or no use, even as benchmarks. The burden of demonstrating the relevance of a statistical model of the employment process should rest on the shoulders of those who would use the model for statistical purposes.

On the 80% Rule

Statistical significance does not imply practical significance. What then should the measure and level of discrimination be in order for the courts to take legal cognizance of it? MSZ suggest that the 80% rule, promulgated by a collection of federal agencies in the late 1970s, should be used to measure substantiality associated with a difference in employment selection rates. In doing so they fully recognize that any such choice has infirmities if used alone and that the measure must be sought primarily in the context of the law and its interpretation, not in the principles of statistics. In making their argument MSZ examine alternative measures, such as differences and ratios, and they consider the symmetry of measures with respect to the focus on selection or rejection rates. In the end they conclude that the 80% rule is a fairly sophisticated instrument and is as reasonable a choice as any, combining as it does a measure of practical significance with a consideration of statistical significance.

I think that in coming down in favor of the 80% rule they give too much credit to its creators and downplay its shortcomings. The 80% rule was originally proposed in a testing context in which the pass rate was of primary interest. The translation of the rule to hiring situations required an orientation—selection or rejection—that, according to my reading of the law, really is arbitrary. Thus the lack of symmetry of the conclusions that result from the application of the rule in the two alternative modes is disturbing, at least to me. Moreover, the same rule is used to examine promotion and termination rates where Title VII, the law at issue, clearly gives us no guidance on what the relevant orientation should be. Of course, there is a measure that is symmetric in this regard and retains the relative feature of the ratio of rates, the odds or log-odds ratio. Although MSZ note this, they seem to imply that we should accept the ratio aspect of the 80% rule because it already exists and because judges and administrators not trained in statistics will find it easier to understand than a more complex rule based on odds or log-odds ratios.

Moreover, there are other features of the 80% rule that have given courts and administrators difficulty. Why 80% and not 75%? Or 85%? What does the rule mean by the "most favored group"? Although this may be clear in a lawsuit involving only issues of sex, what happens if race is superimposed and we have six or seven groupings for it? Shouldn't we worry about issues of multiplicity of comparisons, especially when we come to examining statistical significance? Finally, although the 80% rule allows us to combine considerations of both statistical and practical significance, it fails to tell us how to do so. Who is to provide the needed clarifications—the courts, government agencies, or statisticians? All these problems have led, not surprisingly, to confused behavior on the part of judges and administrators when they attempt to lean on the 80% rule to resolve employment discrimination disputes.

Basically I agree with MSZ on the need for an approach that combines a statistical assessment of the data with a practical measure of substantiality. I am just not ready to accept the 80% rule as the approach of choice.

On Possible Consequences of Accepting the 80% Rule

There are few examples of litigation in which the statistical issues are actually as simple as those encapsulated in the *Hazelwood* decision and described in MSZ's chapter. There are usually a variety of variables available to measure qualifications and productivity, albeit imperfectly, and outcome measures other than selection or rejection come into consideration. When asked as a statistical consultant or as an expert witness to suggest statistical approaches for modeling employment processes in Title VII and other discrimination litigation, I have rarely found that random selection models and methods for the analysis of 2×2 tables suffice. Statistical techniques involving such approaches as multiple regression analysis, log-linear and logistic models for multiway contingency table analysis, and even Cox's regression model for survival analysis are often of some help—sometimes justifiable in their own right and other times justifiable simply as benchmarks.

I share MSZ's concerns about the assumptions implicit in the use of such methods and in their interpretations, and I wonder how courts are to deal with substantiality issues in such contexts. Perhaps, even more than in simple two-group comparison situations, they need measures of disparity and some notion of substantiality. In addition, I would make a strong argument in favor of measures that fit together. For example, whatever statistical method is used to "adjust for qualifications," the measure of substantiality chosen for that setting needs to be designed so as to reduce to the simple measure for MSZ's problem when the "qualifications have no effect."

Who should decide what statistical methods and what measures of disparity are legally relevant? I am reluctant to leave the choice of measures and the resolution of issues such as consistency solely for the courts or administrative agencies to legislate. The next step would likely have courts ruling, as some have suggested they should, that only certain statistical techniques can be used in discrimination cases or that the proportion of variability explained by a *logistic* regression analysis must exceed one-half (whatever that means) in order for the analysis to be admitted into evidence (see the related suggestion of substantiality for multiple regression in Campbell, 1985). If such rulings and prescriptions come to pass, then MSZ would truly have grounds for presenting a counsel of despair.

Additional References

Campbell, T. J. (1985). "Regression Analysis in Title VII Cases: Minimum Standards, Comparable Worth, and Other Issues Where Law and Statistics Meet," *Stanford Law Review,* in press.

Fienberg, S. E. (1985). Comments and reactions to Freedman, statistics and the scientific method. In W. M. Mason and S. E. Fienberg (eds.), *Cohort Analysis in Social Research,* New York: Springer-Verlag, pp. 371–384.

Fienberg, S. E., and Kadane, J. B. (1983). "The Presentation of Bayesian Statistical Analyses in Legal Proceedings," *The Statistician,* **32,** 88–98.

Freedman, D. A. (1985a). Statistics and the scientific method. In W. M. Mason and S. E. Fienberg (eds.), *Cohort Analysis in Social Research,* New York: Springer-Verlag, pp. 343–366.

Freedman, D. A. (1985b). A rejoinder to Fienberg's comments. In W. M. Mason and S. E. Fienberg (eds.), *Cohort Analysis in Social Research,* New York: Springer-Verlag, pp. 385–390.

Yates, F. (1984). "Tests of Significance for 2 × 2 Contingency Tables (with Discussion)," *Journal of the Royal Statistical Society* (*A*), **147,** 426–463.

Rejoinder

Paul Meier / Jerome Sacks / Sandy L. Zabell

We are grateful to Fienberg for his thoughtful commentary, and we confine our response to those few points on which our agreement is not complete.

In our discussion of the Fisher exact test for the 2 × 2 table (which we affirm as the standard to which other procedures approximate), we believe we reflect a consensus in the domain of the statistical theory. Of the three major schools of inference—Neyman–Pearsonian, Fisherian, and Bayesian—the two that accept tests of significance both single out the exact test as being the appropriate procedure, in one case because it is uniformly most powerful and unbiased, in the other because it is the conditional test. (The controversy over the use of the Yates correction arises because the conditional and unconditional attained levels of significance differ; although the Yates correction typically improves the accuracy of the chi-squared approximation to the former, it frequently does not do so for the latter.) The conditionality of the test we view as a virtue rather than a defect: in a typical employment discrimination case the 2 × 2 contingency table at issue does not arise from random sampling, and an unconditional analysis that distinguished between the multinomial, two-sample binomial, and hypergeometric cases would be Proscrustean. Indeed, one can adopt an entirely nonstochastic point of view and regard the conditional attained level of significance as merely stating, given the margins, what fraction of tables exhibit discrepancies as large or larger than that observed.

As a *practical* matter we agree that for purposes of courtroom presentation, it may often be simpler to summarize an analysis in terms of the familiar chi-squared test. When the difference in hiring proportions is clearly significant, this will make little difference. But in borderline cases where different methods can result in differing conclusions, it is a merit for the courts to be able to point to one procedure as the standard to which all others approximate. And, indeed, Fienberg himself recommends the exact test when dealing with small samples.

For the rest, we agree that there are limitations and difficulties in the application of the 80% rule. The rule is not the solution to the world's problems, but, attempting as it does to grapple with the issue of substantiality, it clearly represents an important, albeit first, step forward from the sole, naive use of tests of significance. Eschewing dogmatism, we urge those who agree with us on the need for an approach that combines statistical and practical significance but who are "not ready to accept the 80% rule as the approach of choice" to formulate and defend an alternative. Our choice of the admittedly arbitrary 80% in preference to alternatives is based on its established status in regulation and its conformance with judgments made by courts in other circumstances.

The asymmetry of the rule when applied to selection and rejection rates has been noted before (Shoben, 1978), and we argue that this is not a defect—the protection provided in Title VII shares that asymmetry. That the rule requires the finder of fact to make judgments about the choice of index and about the "practical significance" of small differences reflects, in our view, an appropriate reluctance to lock the courts into purely numerical decision making. (The simplicity of the 80% rule compared to one based on odds ratios is an important secondary benefit, but it is not the primary basis for the choice.)

We agree that cases requiring complex statistical analyses abound in discrimination law, but we note that the simple type of case discussed in our chapter arises frequently. We chose to focus on the use of simple dichotomous two-sample comparisons because it provides a case study that illustrates many of the substantive legal and statistical issues that arise in the analysis of employment discrimination, while it avoids clouding the issue with the technical complexities that can occur when dealing with more intricate cases.

Statistics in Antitrust Litigation

Benjamin F. King
Educational Testing Service
Princeton, New Jersey

ABSTRACT

The antitrust laws were made to discourage business firms from improper use of economic power. They were intended to keep markets competitive and to ensure that the behavior of businesses toward other firms, as well as toward consumers, is fair. Cases brought under the antitrust laws involve price-fixing, improper allocation, control of production, resale price maintenance, tying arrangements, and price discrimination, among other specific practices that may have adverse effects on competition in the marketplace. These cases often fall in the category of complex litigation. Because much of the complexity results from issues of measurement, estimation, and inference under uncertainty, attorneys may hire statisticians as consultants or expert witnesses. This chapter deals with the role of the statistician as a consultant or expert witness in cases of antitrust litigation.

1. Introduction

The antitrust laws were made to discourage business firms from improper use of economic power. They were intended to keep markets competitive, and to ensure that the behavior of businesses toward other firms, as well as toward consumers, is fair. These laws are embodied in the Sherman Act (1890), the Clayton Act (1914)—part of which was amended by the Robinson–Patman Act (1936)—and Section 5 of the Federal Trade Commission Act of 1914 (as amended in 1938). They speak of monopoly, conspiracy in restraint of trade, and unfair or deceptive acts affecting commerce. Cases brought under the antitrust laws involve price-fixing, improper allocation, control of production, resale price maintenance, tying arrangements, and price discrimination among other specific practices that may have adverse effects on competition in the marketplace. Because the alleged improper activities may cover years of business operations with very large numbers of transactions, and because the arguments may involve difficult issues of economic theory as well as matters of empirical fact, these cases often fall in the category of complex litigation (Federal Judicial Center, Board of Editors, 1978). Since many of the complexities result from issues of measurement, estimation, and inference under uncertainty, attorneys may hire statisticians as consultants or expert witnesses. Among attorneys, however, as with the population in general, understanding of what statisticians "do" is not as widespread as one might hope. Thus, in a recent article on hiring and dealing with experts, the author recommends economists to assist in matters of antitrust liability and professors of accounting or finance for estimation of damages, but statisticians are conspicuously absent from his list (Jeffers, 1980). Nevertheless, the use of statisticians as experts in antitrust cases is not uncommon, and is probably increasing along with general awareness of the importance of statistics in the law.

The procedures of complex litigation are not totally rigid—the plan of conduct of a case is determined by the trial judge after discussion with the attorneys and after initial consideration of the facts and the general situation. There are certain common elements, however, that are usually present in complex litigation that may seem foreign to the statistician accustomed to the relatively easy give and take of academic discussion and research. For example, the adversary process itself may be particularly troublesome because it prevents all informal communication between the experts for the defendants and those retained by the plaintiffs—thus one cannot merely pick up the telephone and ask the statistician on the other side why a certain point in a scatter plot does not agree with the values given in a table. It is possible that the matter could be discussed in an official conference call between the opposing experts (with attorneys on every available extension), but it is more likely that the question would be raised during the adversary expert's deposition, or else the attorneys may want to save the issue for its possibly devastating effect during the trial. In short, the curious analyst may have to wait for many weeks or even months before the anomaly in the data is explained, if ever. Similarly, errors in calculation or graphical display that would ordinarily be forgiven (if ultimately corrected) in the process of academic research can be blown up by opposing attorneys into an indication of failed expertise. In an adversary proceeding the statisticians for defendants and plaintiffs can hardly be considered a band of brothers striving together for the same truth. It is to be hoped, however, that each side will try to attain that truth independently.

The process of *discovery,* the legal mechanism by which the two sides request information from each other concerning documents, witnesses, data files—anything or anyone that may ultimately provide evidence—is also highly proceduralized in complex litigation. The statistician who has been designated an expert witness may find that he or she must turn over all analytical notes, memoranda, and charts to the other side if they bear on the opinions to be given at trial, and furthermore, that he or she must state those opinions, perhaps long in advance of trial and without all of the facts yet available. Although the adversary experts have the same obligation to provide an indication of their position, the process becomes somewhat of a cat-and-mouse game with neither side wishing to be very specific for fear that later reversals of judgment may again be taken as weakness of expertise.

The consulting statistician should be prepared for depositions, an ordeal that some have compared to the defense of one's dissertation or to an audit by the Internal Revenue Service. In antitrust litigation the statistician may be assisting an expert economist and may not necessarily be deposed personally during the pretrial period. It is common, however, for attorneys to request such pretrial testimony from all experts, especially if the arguments for and against liability and damages involve issues that are primarily statistical. In that case, the statistician will be subjected to formal questioning by the attorneys for the other side in a manner similar to the elicitation of trial testimony. The experience may be grueling—it may even seem abusive—but it does force the deponent to make the theoretical and empirical arguments understandable and convincing. Thus depositions serve as valuable practice for actual testimony at trial.

There are fairly rigid procedures governing discovery that make it desirable that the expert statistician be retained early in the pretrial period. The expert can thus advise the attorneys in the framing of their formal written questions and requests for the production of documents aimed at uncovering all relevant data. The formal questions, addressed by each side to the other, and to which the respective recipients are required by law to respond, are called *interrogatories*. If a particular set of data is overlooked—it is unlikely that the other side will volunteer to produce it—there may be no further opportunity to obtain it before the deadline for completion of all pretrial analysis. In other words, a number of weeks before the scheduled trial date, it is possible that the court will prohibit any introduction of new theoretical arguments in the case, which is usually interpreted to mean no new data gathering or analysis. It follows that the earlier the statistician is involved in the planning of discovery, the better founded it will be. The same considerations apply to the limitation of the time period in suit, that is, at some time during the pretrial proceedings the judge will probably circumscribe the years for which data will be admitted as evidence in the trial. If the statistical consultant believes strongly that information from certain years should be used—say, for example, it is desirable to include a "preconspiracy" period in order to demonstrate by contrast with later years the presence of the alleged violations—then early advice will give the lawyers more time to prepare their arguments for admitting data for those years into evidence.

In antitrust litigation, as in any consulting situation, it is difficult to "put out fires" caused by the ill-planned research efforts of others. There are situations in which a statistician is retained only as a passive consultant to help attorneys understand statistical arguments and to respond to questions of a hypothetical nature without getting involved in actual analysis or testimony. If one is approached, however, to help actively in the development of either attack or defense and to serve as an expert witness, then it is wise to establish first that there is time remaining in the pretrial process to make a significant impact on the shape of the statistical analysis that will take place. Under ideal conditions there should be no difference in the general approaches of the investigations undertaken whether the statistician is working for the plaintiff or for the defense. The plaintiff's expert will, of course, be compelled to explore all avenues that may help to prove the alleged violations and ensuing damages. On the other hand, the expert for the defendant should be free to consider all possible analytical approaches that might be used by the opponent side (within restrictions of economic feasibility) so that adequate defenses may be prepared if such are possible. It is precisely the chance that something indefensible may be uncovered that causes some defendant's lawyers to rein in their experts and to restrict the field of investigation, but there are also those who believe that the expert should be aware of all of the "facts" on the theory that it is better to know in advance what is coming than to have adverse evidence read at trial with the question of whether it changes the expert's opinion (Jentes, 1979). Failure to examine easily available data for fear of "learning too much" can hardly be considered ethical for a scientific researcher.

In the following we shall examine some of the basic ingredients and matters for consideration that are common to many applications of statistics in antitrust litigation, but it will be impossible to treat all situations satisfactorily. Because it is, perhaps, the most common alleged violation, and because many of the other activities in restraint of trade are closely related, our discussion will often center on cases of alleged price fixing, but there will be references to other violations.

2. Proving or Disproving Liability

The two broad issues present in all antitrust cases are those of *liability* and *damages*—the former raising the question of whether the defendants in fact engaged in anticompetitive practices to restrain trade, and the latter, the determination of the monetary losses suffered by the plaintiffs as a result of these practices. With respect to liability, if there were strong documentary evidence of an agreement among defendants to fix prices or to commit whatever violation is alleged, then it is unlikely that it would be necessary to retain a statistical consultant to analyze market data to further confirm the liability. Thus it is primarily in cases of uncertainty and debate about the existence of a conspiracy or other violation that the plaintiffs must convince the jury that it should infer collusion, predatory acts, or monopolistic intent from the behavior of market phenomena such as prices, and the defendants must demonstrate that whatever restrictive effects appear to be present in the market are explained by nonconspiratorial causes.

The first task facing the statistician is to construct a picture of the market for the period in suit, and possibly for periods preceding and following the time of the alleged conspiracy. By *market* we mean the milieu in which transactions between buyers and sellers take place. Decisions must be made concerning the geographic boundaries of the market—whether it is regional or national, or whether international transactions should be included.

This task of statistical description can generally be executed at two levels, the *aggregate* level and the level of *specific* firms and individual transactions. Especially when the commodity is fungible, there often will be aggregate statistical series available from government sources, and in the case of less homogeneous products, a trade association may provide similar information. These data will aid in describing general conditions of price, volume, destination of shipments, and similar measures of the surrounding business environment. The value of this first-stage descriptive activity at the aggregate level should not be underestimated. Often, the business managers who are involved do not themselves have a clear picture of the relationship between their firms and other entities in the same market, and thus cannot be relied on to provide the background information that is essential to attorneys and consulting economists and statisticians for the planning of research. For example, in a recent case, a firm in a regional market consisting of only a few like firms had experienced a decrease in market share of roughly 50%, and in a complaint it accused its competitors of causing the decline by illegal dealing with retailers. Analysis of industry expenditures on promotion and advertising, obtained from published documents, showed that prior to and during the decline the plaintiff had drastically cut promotional outlays while overall industry outlays, including those of the plaintiff's competitors, were increasing. Subsequent evidence concerning reduction of the plaintiff's sales force helped to strengthen the defense's argument that the decline in share was due to poor management and ineffective marketing strategy, rather than restrictive activities by the defendants.

This rather mundane and unexciting preliminary step of market description in the aggregate is analogous to the routine gathering of intelligence on disposition of forces, troop strength, numbers of tanks and aircraft, and the like that is essential to preparation for a military operation. It serves the additional function of familiarizing expert witnesses with the general market environment so that greater evidence of their knowledge of the industry may be provided during depositions and trial testimony. There are opportunities for considerable creativity in graphic display of the factors that describe the market, and the attorneys may decide that it will be useful to present such materials to the judge and jury during the first part of the experts' testimony at trial. This step serves to set the scene for later descriptions of analytical results of a more detailed and technical nature. If the court permits it, the usual overhead projection or wall charts in large scale may be supplemented by individual graphics packets for each member of the jury so that the panel may better follow and visualize the experts' description of the market milieu in which the plaintiff and defendant firms operate.

Figure 1 (Exhibit 4 in Preston and King, 1979) is an example of a multiple time series plot of monthly average wholesale prices for a certain commodity reported by the Bureau of Labor Statistics. The graph, prepared by the defense, is intended to show that average price for the region in suit tended to move in concert with average prices in other geographically distant regional markets, suggesting that if aggregate prices were affected by a regional conspiracy, it would have to be part of a larger nationwide conspiracy (which had not been alleged). Furthermore, any evidence from price behavior of a conspiracy applying only to the region of interest would have to be demonstrated by firm-specific data. On the other hand, the multiple plot shown in Figure 2 (Exhibit 1 in Preston and King, 1979), prepared by plaintiffs in another case, demonstrates that regional price is remarkably higher than the national average during a period of essentially constant cost, and suggests noncompetitive behavior.

Although statistics appearing in official government documents are generally considered to be authoritative and may even be judicially noticed (in effect, declared by the court to be valid), there are pitfalls in these data—shortcomings that are usually admitted by the agencies involved. First, there is the problem of using aggregate measures to support or disprove allegations that involve microeconomic phenomena. Time series of monthly, weekly, or even daily average prices may mask anomalies that would be revealed on a transaction-by-transaction basis; averaging across firms may obscure individual firm behavior; aggregate data may yield ecological correlations that overstate associations at the level of individual entities.

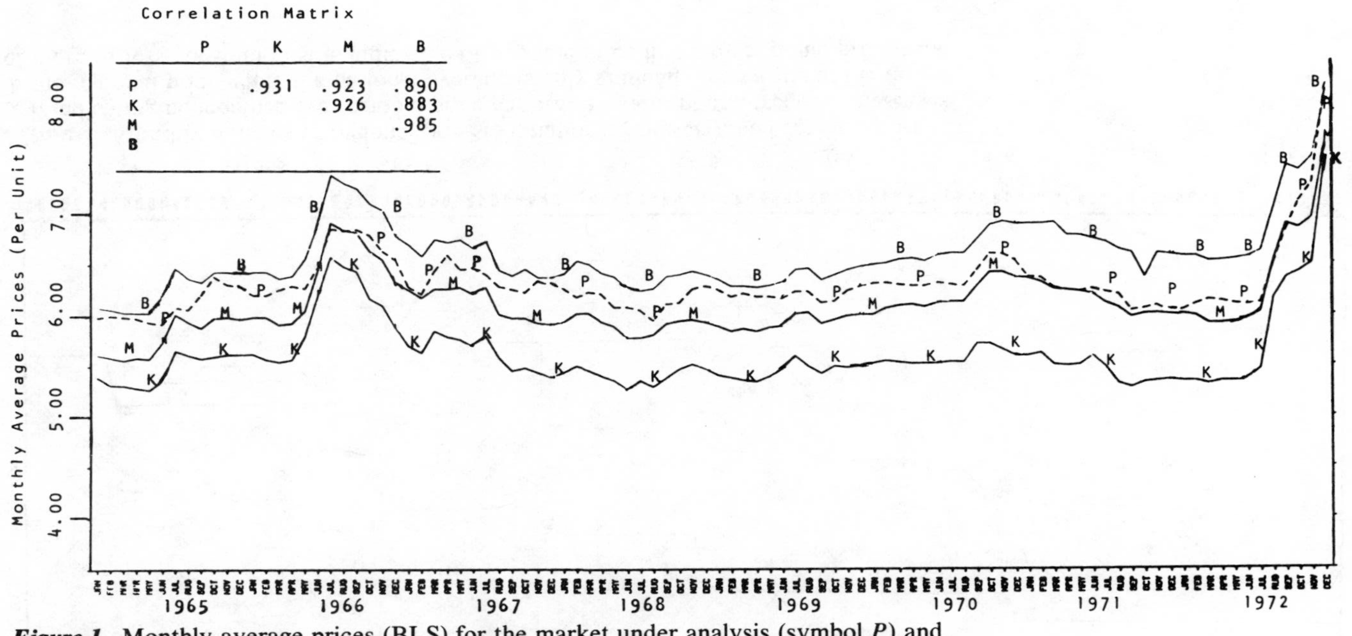

Figure 1. Monthly average prices (BLS) for the market under analysis (symbol *P*) and three other regional markets (B, M, and K). The correlations show the high degree of cross sectional association among the series.

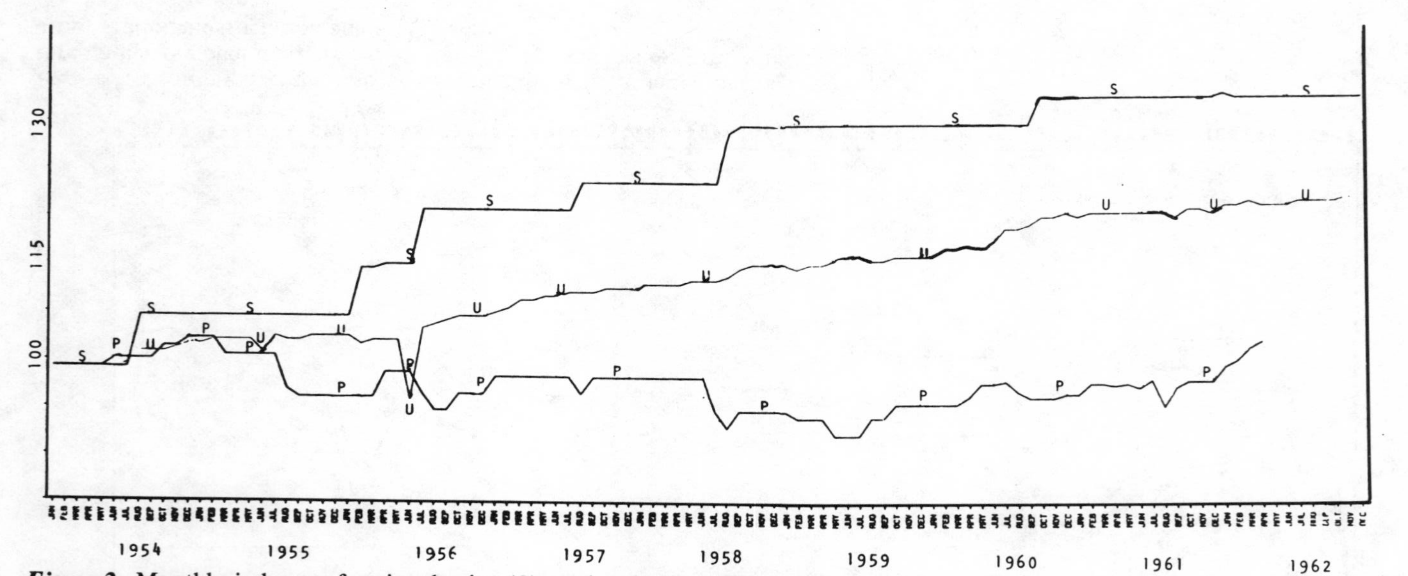

Figure 2. Monthly indexes of regional price (S), national price (U), and regional cost (P), indicating noncompetitive regional price pattern. The abrupt and frequent increases in the regional price while the national average is only gradually increasing in the face of more or less constant cost is indicative of possible price fixing by regional producers.

In addition to the obvious aggregation dangers, however, the procedures used to obtain the measures that go into the aggregate may be faulty. For example, before the recent methodological revisions leading to the Producer Price Indices, some wholesale price series were based on prices reported without verification by nonrandom samples of producers of the commodities—in effect, the answer to the question, What is the price of ______?, without the necessary follow-up question, Have you actually sold any at that price? One can easily imagine various incentives for the reporting sellers to state prices for a standard unit and quality of product that were either above or below the price actually obtained in the market. Similarly, sellers of a farm product, when queried by the U.S. Department of Agriculture, may overstate daily F.O.B. prices in order to shore up a fast-dropping market. Although the methods used in preparing government statistics are usually described in broad terms in the appendixes of the publications, because of the lack of technical detail it may be necessary to interview the agency personnel who are responsible for collecting the information to determine exactly how the series are constructed. Sometimes these investigations reveal methods that can be described as ad hoc at best, which is not to say that better measures can be obtained at any reasonable price. The situation may be as bad or worse with trade association data—for example, in a case involving antibiotic drugs, a widely used and cited measure of the relative frequency of prescription of medications for various illnesses was found to be based on a highly self-selected and biased sample of physicians. The usual argument, with some merit, is that partial information is better than no information. The important point is to recognize possible sources of bias and the restrictions that are placed on inference when biased figures are the only ones available.

The proof or disproof of liability may be accomplished with aggregate data alone, but when entering a case, one should consider as soon as possible the feasibility of gathering and analyzing data at the level of individual buyer–seller transactions. If it becomes apparent that the question of whether there were violations of the antitrust laws can only be answered by analysis of microdata, then as much lead time as possible is required to set up the necessary clerical and data management system. The range of circumstances is large, as illustrated by the following examples:

1. At one extreme is the case of a consumer class action where the plaintiff purchasers of a product are a very large and fluid group, perhaps numbering in the millions. (Consider, for example, purchasers of retail gasoline in several states of the United States.) The idea of studying individual transactions is probably out of the question. It would be possible, in principle, to analyze credit-card purchases in the case of gasoline, but even there the data management problems are formidable. Thus one may be forced to work with average prices and quantities whether or not it is ideal to do so.

2. At the other extreme is a relatively complex, expensive, and infrequently purchased item (e.g., a piece of heavy equipment used in construction projects) where the number of transactions (or bids with stated prices) is manageable.

3. An example of a situation that falls between these extremes is that where a small number of manufacturers of a product (say, fewer than 12) purchase an essential ingredient from a small number of suppliers. The purchases are made in large quantity for delivery every few weeks in various plants. The numbers of transactions per year are in the hundreds, and there are detailed invoices covering every sale. The whole set of records for a period of years can be stored on a single computer tape.

According to the principle that it is better to be somewhat extravagant than to regret later that the information is not available, one should contemplate, at least, the creation of a data base of records of individual transactions. If this luxury can be afforded, it gives the economists and statisticians a view of the dynamic functioning of the market mechanism—a view that may never have been available to any of the individual parties in the lawsuit when the violations are alleged to have occurred. If there is no other evidence of intent to fix prices, the pattern of individual prices may aid the plaintiffs in convincing the jury of collusion, or, alternatively, it may help the defendants to show the existence of competition. Figure 3 (Exhibit 6 in Preston and King, 1979) is an example of a display based on individual sales. As described by the authors, "the coded data reflect contracts involving six suppliers (five defendants and one other, . . .) and four major buyers, coded A, B, C, D. Symbols for the trading pair involved in each contract are plotted at the level of the indexed contract price; contract quantities are shown along the left-hand margin. Sixteen different contracts at six different prices are shown, although specific prices accounted for five and six contracts each." The authors discuss the way that the patterns of price and quantities bear on the issue of competition in that market. It is further observed that when gross profit margins are computed for each of the transactions, their wide variability is consistent with independent selling behavior by all defendant parties.

Figure 4 is from a case in which plaintiff growers of an agricultural commodity charged that defendant packer-shippers of the product who act as sales agents for the growers conspired with the retailers to lower the prices received by the growers, and thus increase retailer profits. In other words, we have the relatively unusual situation of an alleged buyers' conspiracy to fix prices *downward*.

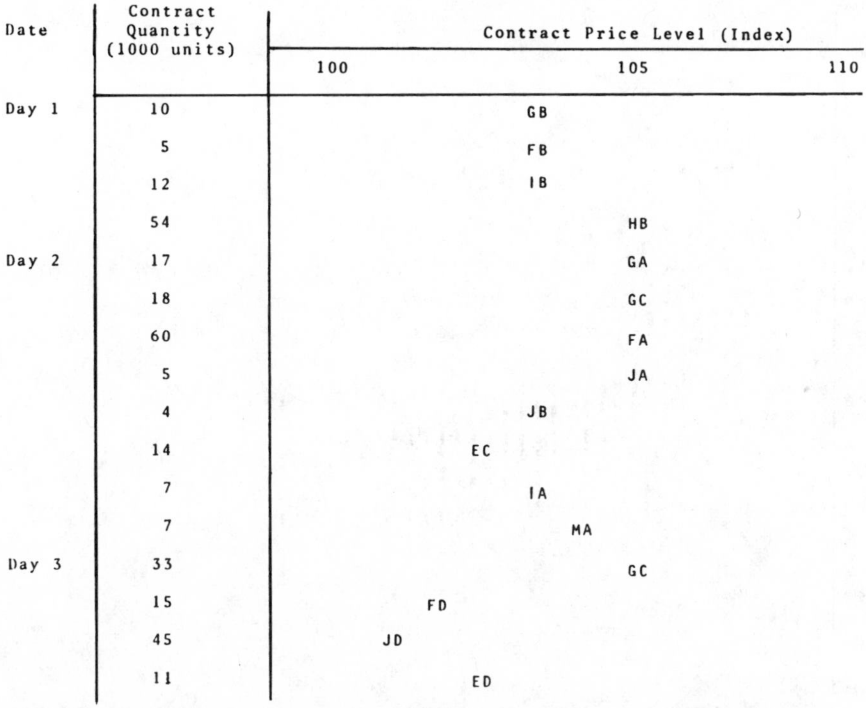

Date	Contract Quantity (1000 units)	Contract Price Level (Index): 100 — 105 — 110
Day 1	10	GB
	5	FB
	12	IB
	54	HB
Day 2	17	GA
	18	GC
	60	FA
	5	JA
	4	JB
	14	EC
	7	IA
	7	MA
Day 3	33	GC
	15	FD
	45	JD
	11	ED

Figure 3. Detailed market contract data, showing seller, buyer, price and quantity. Example of how, when the analyst can combine data from various buying and selling sources, a complete picture of market activity can be constructed. First letter of symbol pair indicates supplier and second letter indicates purchaser of quantity shown at left at price shown on scale across top of chart.

The price received by a grower for a product is the net return per pound—that is, the sales price less charges for packing and selling. The industry practice is for a sales agent to give the same net return to all growers whose product is packed on a given day. Thus a symbol in Figure 4 represents the return paid by a single agent on the day indicated on the horizontal axis. The plot shows that returns to growers followed the characteristic pattern shown by average F.O.B. selling price over the course of the season—that is, high returns when the product is first available, much lower returns after the volume in the market has reached a peak, and then higher at the end of the season as volume available diminishes. The advantage of the plot of multiple prices over a single aggregate selling price series is the ability to show the dispersion of returns paid by different sales agents on a given date. This variability demonstrates that, although F.O.B. prices may have been the same from agent to agent, different sales agents paid different net returns to growers. This fact, coupled

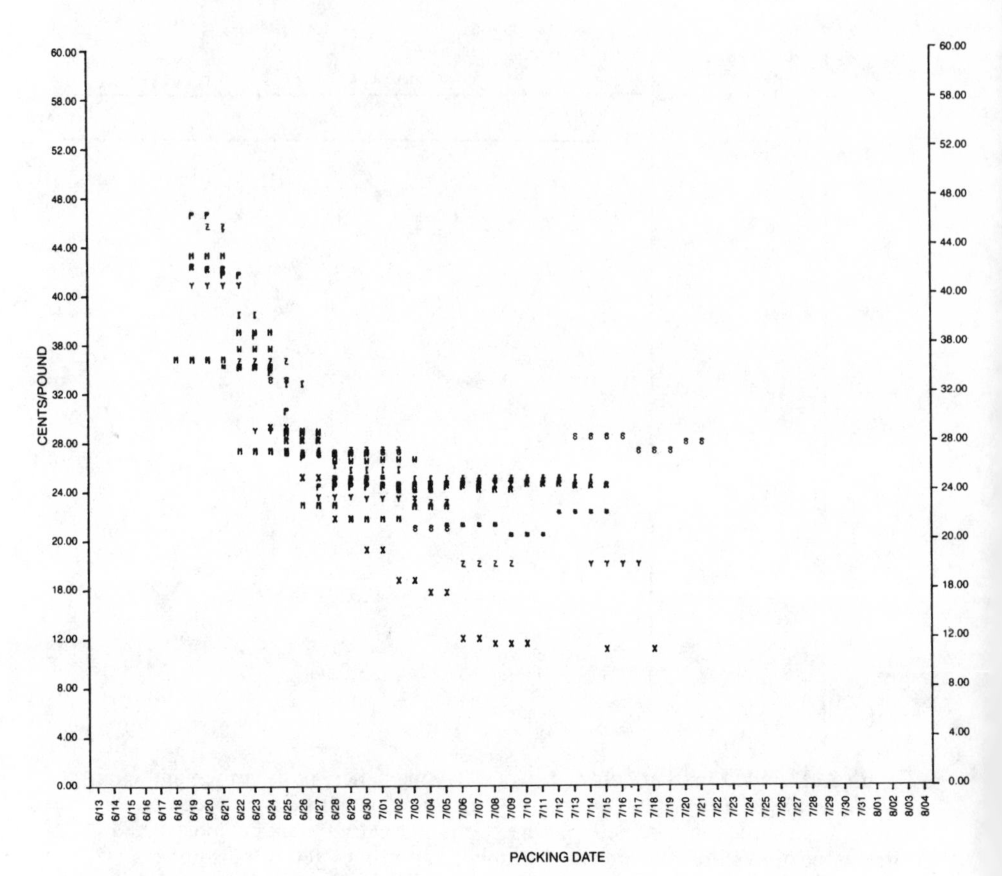

Figure 4. Returns to growers for various sales agents. On a given date a sales agent pays the same return to all growers. Different symbols indicate the returns paid by various agents on each date. The cross sectional dispersion of returns indicates presence of competition.

with the freedom of growers to switch to a new sales agent (the following season) if dissatisfied, is evidence of competitive behavior. A final caution:—in examining "prices," one must take into consideration all the variables that the seller may use to compete for a purchase. Examples are credit, terms of sale, advertising allowances, and advisory services. Thus in the present example *net return* means "net of the costs of *all* such competitive factors."

An adversary could claim that diversity of net returns is not necessarily proof of competition, but to argue this point successfully would require a "ducks flying in formation" type of conspiracy theory. The switching over time of the positions of individual sales agents in the rank structure of returns paid to growers would be evidence against that theory.

On the other hand, constant transaction prices are not necessarily evidence of a conspiracy on the part of sellers to fix prices. A price is the culmination of a bargaining process between buyer and seller, and in a competitive market where information flows freely among buyers it is usually necessary for a seller to meet the prevailing price or sell no product. This is especially true when the prospective purchasers are powerful in terms of the quantities that they buy and the sellers are numerous so that purchasers' threats to go to another seller are effective. For this reason, if a statistician must show competitive behavior among a group of defendant sellers, it is desirable to have records of *prices asked* as well as the actual transaction prices. Unfortunately, records of the give and take of the bargaining process in terms of "bids" and "asks" are seldom retained. If this writer has one piece of advice for honest sellers who want to defend in advance against future charges of price fixing, it is to keep good records of telephone conversations and correspondence with prospective buyers showing the give and take of negotiations that takes place before a final price is achieved.

Other examples, far from exhaustive, of statistical results that bear on the question of liability, and that are less open to conflicting interpretations than the behavior of prices, are as follows:

1. Easy entry by sellers into the market and frequent switching by buyers from one seller to another in response to price changes is evidence of competition. A simple table or chart showing who bought from whom over time may suffice.

2. A content analysis of communications within defendant firms that shows frequent mention of the need to maintain the level of price may lend support to allegations of pricing fixing. [For an example of the use of content analysis in another area of the law see Zeisel and Stamler (1976).]

3. It may be possible to analyze records of telephone calls between defendant firms. Evidence of frequent contact by the person responsible for pricing in one firm with his or her counterpart in another, especially during times of intense bargaining activity in the industry, could be incriminating.

The point that bears repeating with respect to all these examples is that the analysis of one party's data by the other side is unlikely to take place if the records are not obtained during pretrial discovery. The reason is not only procedural; thorough analysis requires adequate lead time.

3. Sampling and Computers

The discussion of files of transactions consisting of extremely large numbers of records invites consideration of sampling methods and the use of computers in the management of the data. Considerable space is devoted to these topics in *Manual for Complex Litigation* (Federal Judicial Center, Board of Editors, 1978) with the discussion aimed at alerting attorneys and judges to the technical and logistic implications for the conduct of a trial. The following direct quotation of six of the seven recommendations made in Sections 2.712–2.717 of the manual conveys the thinking of some members of the legal profession with regard to these issues:

> . . . Scientifically designed samples and polls, meeting the tests of necessity and trustworthiness are useful adjuncts to conventional methods of proof and may contribute materially to shortening the trial of the complex case.
>
> . . . The underlying data, method of interpretation employed and conclusions reached in polls and samples should be made available to the opposing party far in advance of trial.
>
> . . . When computer maintained records and computer analyses of raw data are valuable sources of evidence, their use and admissibility should be promoted and facilitated.
>
> . . . Discovery requests relating to the computer, its programs, inputs and outputs should be processed under methods consistent with the approach taken to discovery of other types of information.
>
> . . . Computer maintained records kept in the regular course of business should be admitted when it has been shown that the criteria required for the admission of non-computer maintained business records have been met, the court finds that reliable computer equipment and techniques have been used, and the material is of probative value.
>
> . . . Summaries and analyses of masses of data made by a computer should be admitted on the same basis as other summaries or analyses. Computer inputs and outputs, the underlying data and the program method employed should be made available to the opposing party in advance of trial as a condition of admissibility.

With the widespread use of computers in accounting by business firms it is possible that much of the data for sales by the defendants will be stored on tape, but when there are multiple defendants one should expect a hodgepodge ranging from faintly legible copies of invoices for some of the parties to machine readable records for others. To create a data base of the transactions for a total market, it is likely that many records must be encoded according to a uniform format onto worksheets and subsequently keypunched (or directly entered into the computer via a terminal), whereas the computer records that are obtained through discovery will have to be converted to the general format. In short, a time-consuming and costly data processing operation may be required.

When the purpose of examining individual transactions is to establish or disprove liability in a price-fixing case, simple random sampling of sales may not be adequate for analysis even though the cost savings may be great. Since one is generally dealing with multiple time series of prices, the sampling of events could result in failure to detect important time patterns of price behavior as well as relevant cross-sectional relations among firms. It is possible, however, that sampling of *segments* of time may be efficient. If, for example, certain months are more active than others or certain years are more critical in the allegations, it may be possible to statistically demonstrate that which requires demonstration without total coverage of the time period in suit. If the number of time segments of minimal size is not large, judgmental selection is probably preferable to probability sampling so that critical segments are not omitted by chance.

There are only a few cases of survey sampling in antitrust litigation mentioned in *American Law Reports, Annotated* (*76 ALR 2nd*), a well-known legal case digest, and not all of those examples apply to the question of liability. In *United States v. J. I. Case Co.* (D.C. Minn. 1951) a survey in Iowa was used to show that farmers had ready access to different makes of farm machinery throughout the agricultural areas of the state (presumably bearing on the issue of whether competition had been restrained). The Court declared the survey to be admissible but of doubtful relevancy or materiality. In *United States v. E. I. du Pont de Nemours & Co.* (D.C. Del. 1953) a survey was part of the evidence used in developing the "market setting" for the sale of cellophane. There are doubtless other more recent examples that bear on liability, but in this writer's opinion the potential for survey sampling is greater in damage estimation, to be discussed in Section 2.

Because surveys involve statements made by third persons to interviewers, who in turn pass the information on to an analyst, there have been objections on hearsay grounds to admitting the results as evidence. The hearsay objection and other aspects of the use of surveys are discussed by Zeisel in his admirable paper (Zeisel, 1960). One answer to the hearsay problem has been to allow the adversary to call survey respondents as witnesses. Zeisel's criticism of that suggestion (p. 337) is that "cross-examination of selected survey interviewees is likely to be misleading. It may lend the aura of reliability to an incompetent survey, or it may destroy confidence in a survey which deserves better." King and Spector (1963) regard the divulging of respondents' names (so that they may be contacted by the adversary) as a violation of the ethical standards of survey research, as would most other survey researchers.

An exception to the hearsay rule applies when the results of the survey are considered to show the "state of mind" of the respondents and not necessarily the truth of the matter asserted. If required, the personnel responsible for the sampling and the interviewing can be cross-examined on the procedures used in order to establish that the respondents' collective state of mind was correctly measured. It is likely, however, that most surveys in antitrust litigation are designed to obtain facts: What quantity of product was purchased? When? For what price? Thus the "state of mind" exception would not apply. Zeisel (1960, p. 324) views surveys as

> . . . [gaining] admittance . . . under the . . . approach . . . which weighs the necessity of utilizing the evidence against the dangers of hearsay. Unless, therefore, a survey exception to the hearsay rule emerges at some future time, the admissibility of survey evidence, as a rule, will hinge upon the assessment of its worth.

More than 20 years have passed since Zeisel's statement, and it is probably fair to say that the hearsay objection is no longer so serious a problem in the admission of evidence as it was then. As Fienberg and Straf (1982) observe,

> Arguments regarding admissibility are now more likely to revolve around the relevance of the data collected to the points of law at issue and the appropriateness of the methods used to analyse the data.

An important concern for the expert statistician is establishing to the satisfaction of the court and of the jury the validity of the methods and the precision and accuracy of the resulting estimates. In his article on statistics as legal evidence Zeisel (1978) comments:

> The discovery of but one serious flaw may endanger the entire piece of evidence; the doctrine of *falsus in uno, falsus in omnibus* is sometimes the ground for disbelieving a witness's entire testimony if it is found to be untrue in a single instance, and such a flaw may also hurt the expert witness who presents the survey evidence. His task will be facilitated if every step in the survey procedure is meticulously documented: the definition of the universe to be sampled, the details of the sample design, the process of sample selection, the communications to the interviewers, their control in the field, and finally, the analysis of their reports—all should be documented by the respective research instruments.

Note the statement by Zeisel immediately following the above: "To avoid potential bias, the nature of the issue—and if possible, its very existence—should not be divulged to them. If they must learn the purpose of the survey, they should not learn which side the sponsor is on." Although the concern for response bias implicit in this statement is understandable, the ethical acceptability of withholding information in this manner is questionable. See, for example, Hobbs (1978), especially the discussion of "informed consent" (p. 277).

A detailed list of points dealing with reliability that should be covered in the statistician's testimony and report is provided by Deming (1965). The potential payoff from the exercise of extreme care in the application of survey procedures and the ability to show that the results are reliable is shown in the following excerpt from *United States v. E. I. du Pont de Nemours & Co.* (D.C. I11, 1959). In overruling a government motion to exclude expert testimony based on a survey, the court said:

The evident care and objectivity with which the survey was conducted, and which were not criticized by the Government's own statistical expert, assure a high degree of trustworthiness. The record discloses that various agencies of the Government itself have used surveys of this type, and indeed have employed this same research organization.

Although written with sample surveys in mind, the prescriptions of Zeisel and Deming apply equally to materials produced by a computer. For the same reasons mentioned by Zeisel, it is imperative that users of computer results be able to document fully all procedures from the encoding of raw data (especially when arbitrary classification decisions are involved) through the writing of analytical programs. The technique of sampling and verifying records for purposes of auditing and quality control, discussed by Deming for his well-known textbook (Deming, 1960), is particularly useful for the statistician who must respond to an adversary's questioning of data accuracy.

4. Estimation of Damages

The second major task for the expert consultants in an antitrust case is the calculation of the dollar amount of damages to the plaintiff caused by the alleged violation. It is possible that a judge will rule that the case be bifurcated in the sense that the question of liability will be heard first, with the estimation of damages contingent on a finding that the defendants are liable. It is not unusual, however, for the plaintiff's "damage theory" and matters aimed at proving liability to run side by side in pretrial testimony as well as in the trial itself. Indeed, it is sometimes difficult to classify a statistical argument as bearing primarily on one issue (i.e., liability or damages) to the exclusion of the other. For example, in a time series regression analysis of prices the existence of significantly larger-than-expected residuals for a post-conspiracy period might be put forward as evidence of liability, and the size of the same residuals might form the basis for damage estimation.

The statistician involved for the first time in a price-fixing case may be surprised to discover that the lofty term *theory of damages* is applied by the attorneys on both sides to what may be only a rather ad hoc method of calculation of damage amounts—a method that the plaintiff's economic expert puts forth as a reasonable approach in the absence of agreement about the theoretically "just price" under competition. The basic problem is to determine the competitive price that should have obtained during the time period affected by the antitrust violation and to take the difference between the actual price and the ideal price as a measure of damages per unit sold. In the well-known Electrical Conspiracy Cases of the 1960s, for example, there were two basic methods of determining the "just price" (Shuchter, 1968), (a) the list-price method, and (b) the cost-of-production method. The former approach is based on the industry practice of selling electrical equipment at a percentage discount from the list price. The "theory" states that during the conspiracy period the discounts from list price were smaller than in a competitive period. By identifying certain time periods as "conspiracy" and "nonconspiracy" and comparing average discounts, the plaintiffs were able to estimate a reduction of percentage points in discount to call the "damage factor."

In the cost-of-production method, average prices during the period of competition are observed to exceed an index of production cost by a certain percentage. Presumably, during a conspiracy to fix prices and extract greater profits, the percentage by which prices exceed costs would be greater than when the conspiracy was not in effect. Thus an estimated excess in gross margin for the conspiracy period can be taken as the "overcharge" or damage factor.

Without providing details of the construction of his production cost indices, Shuchter (1968) claims to have applied the cost-of-production method of damage calculation to various industries, including electrical equipment. The "overcharges" that are produced by the cost-of-production method appear to be inversely correlated across industries with the concentration ratio (a measure of oligopoly), a result that is counter to both theory and intuition. This negative association leads Schuchter to suggest that in *Washington, State of, v. Federal Pacific Electric Co.* (W.D. Wash. 1966) the cost-of-production method was not chosen for its theoretical superiority, but rather because it produced remarkably higher damages than the list-price method.

When considering competing methods of damage estimation, it is often tempting for plaintiff attorneys to be attracted to the "theory" that produces the highest dollar amounts. Economic and statistical consultants should not be surprised to encounter the question, Isn't there some way for the data to show that this method is best? The need to stay objective when faced with such questions (even though they are raised at least half in jest) is one of many ethical problems that the expert witness must deal with.

To illustrate the wide differences of opinion that can exist between plaintiff and defendant experts in matters of damage estimation, the following scenario is presented. (The situation is hypothetical and fictitious although it is loosely based on events in a real case that was settled before trial. The reader may infer a bias in this illustration toward the defendant's position, but the situation could well be reversed in other circumstances.)

1. The setting is one in which sellers of a farm product are alleged to have fixed the price during recent years. The plaintiffs' expert introduces Figure 5 and explains that the straight line is based on a linear regression analysis of annual average price (deflated by an index of change in nominal prices) versus quantity (volume of the product shipped in each year) with data for the years *before* the conspiracy was in effect. The points

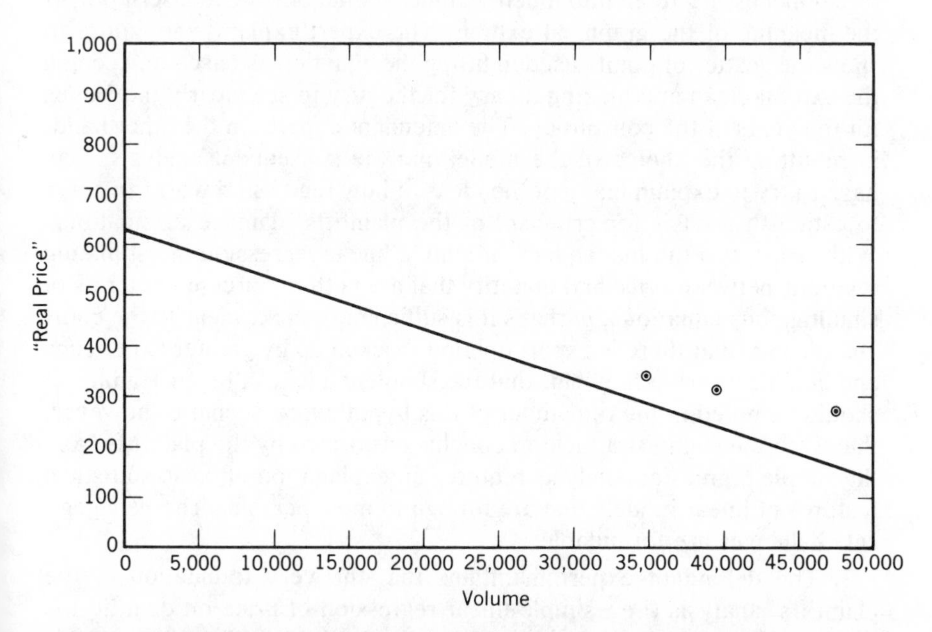

Figure 5. Exhibit introduced by plaintiffs in hypothetical scenario. Regression line was fitted with data from alleged "preconspiracy" period. Three points shown (not used in fitting) are from "postconspiracy" period (1977–1979).

shown are for the three years of the conspiracy (not used in fitting the regression line). That all three points lie above the extended line is, according to the plaintiffs' expert, (a) an indication that the conspiracy was effective in raising actual prices above the "just" price under competition, and (b) a basis for assessing damages (by taking the vertical deviation from the line as the damage factor).

2. The first response of the defendant expert is to criticize the plaintiffs' use of regression analysis on purely theoretical grounds. Citing Fisher (1980), it is explained that the proper model is not a single equation with price on the left-hand side and quantity on the right, but rather a pair of simultaneous equations, one for demand and one for supply. Thus the approach used by the plaintiffs' expert is likely to be biased in its estimation of the expected value of price under competition for the years during which the conspiracy is supposed to have been in effect.

Descending from the question of truth and ethics to the level of tactics, we observe here that the plaintiffs' expert has a distinct advantage in dealing with a technically unsophisticated jury. The regression analysis has been introduced as a time-honored method of damage estimation in cases such as these, one that involves a statistical technique (i.e., least squares) that dates back to the end of the eighteenth century. It is therefore unnecessary to go into much technical detail beyond a description of the meaning of the graphical exhibit. The expert explains the failure to show the scatter of points used in fitting the equation as based on keeping the exhibit clean and making it easy for the jury to see clearly the values for the years of the conspiracy. The defendant expert, on the other hand, in rebutting the choice of the model and the subsequent analysis, may have to try to explain many of the ideas of how regression works in order to establish a basis for criticism of the plaintiffs' damage calculations. With respect to the inadequacy of simple linear regression in estimating relations between price and quantity that are better expressed in terms of simultaneous equations, perhaps it is sufficient to make clear to the court and the jury that there is expert opinion, backed up by greater experience and qualifications in the field, that the simple model can be misleading. It should be noted in the remainder of this hypothetical scenario, however, that even the slightest attack on conclusions drawn by the plaintiffs from the simple regression analysis requires an explanation of basic statistical features of linear models that are foreign to most persons. The pedagogical challenges are formidable.

3. The defendant expert maintains that the very foundation of the plaintiffs' analysis (i.e., simple linear regression of price on quantity) is inappropriate, thus implying that damage estimates based on that analysis are likely to be erroneous and misleading. Nevertheless, to demonstrate further the plaintiff expert's lack of understanding of the statistical theory underlying linear regression and a failure to properly match the statistical model with economic theory, the defendant expert accepts the single equation model purely "for the sake of argument."

The next step is to introduce Figure 6, in which the full set of points used in fitting line of Figure 5 is displayed, as well as the three points for the alleged conspiracy period. Especially if the presentation is before a jury, the introduction of Figure 6 requires a brief and simple explanation of the method of least squares, discussing in nontechnical terms the ideas of sampling variability, uncertainty concerning the true orientation of the line, and the standard error of the residuals as it bears on the prediction of values not used in the estimation of the parameters. Whether intentionally or not, by suppressing the scatter in Figure 5, the plaintiff expert has conferred a degree of sanctity and finality on the fitted line that it does not deserve. Figure 6 shows that there are residuals for other than the three "conspiracy" years that are larger and below the line as well as above. (The latter point is obvious to statisticians, but not necessarily to jurors.) The defendant expert continues in explaining that only one of the three points on which overcharges are based is as great as one standard error from its expected value under the hypothesis that the statistical conditions have not changed from the "preconspiracy" year. A further brief explanation of statistical significance is provided.

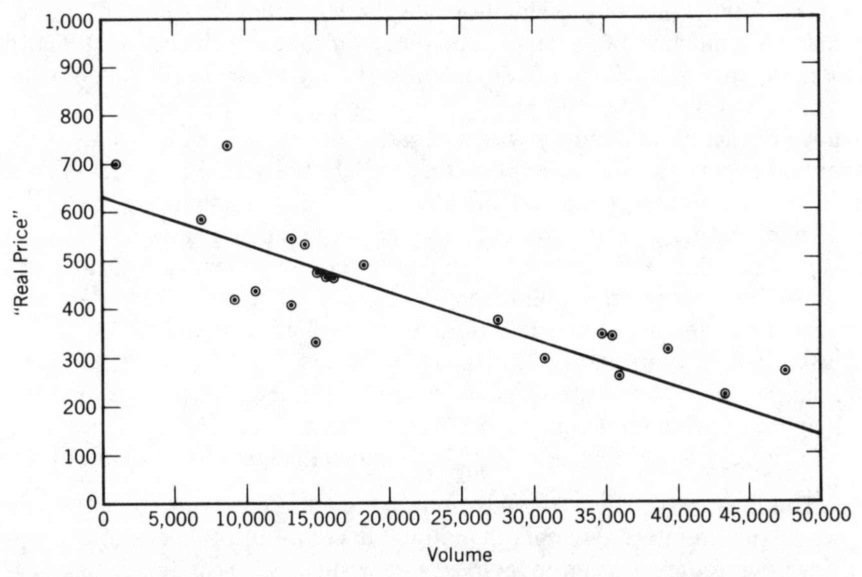

Figure 6. Plot of full data set (including three points shown in Figure 5) produced by defendants. Scatter plot accompanied by estimated standard errors shows that the three points are not so unusual as claimed by plaintiffs.

4. A second important point of criticism, still under the assumption that the single equation analysis is appropriate, is that the model is unlikely to be fully specified with only quantity on the right-hand side. A temporal shift in the level of prices may be due to shifts in underlying demand or cost conditions that are not accounted for in the simple linear model. One example of such an error in specification in the present case is a preseason arrangement with a group of foreign importers to sell them the commodity at a price that is considerable higher than the prevailing U.S. market price for the season. Since the price for those sales was not determined by the ordinary forces of supply and demand in the U.S. market, the effect of those transactions on the annual average should be removed by adjustment. It happens that the arrangement with the foreign buyers coincides with the years that are labeled "conspiracy" by the plaintiffs; hence the effect of adjustment is to reduce the ostensible overcharge amounts. Alternatively, if the plaintiff expert had introduced a zero–one indicator variable for the "conspiracy" years and had found a significant effect, the two sides could argue endlessly over whether the significance is due to a conspiracy or to other unspecified variables such as the export arrangement with the foreign buyers.

5. Finally, the defendants introduce Figure 7 to show that the linearity feature of the model is not sacrosanct. Again, ignoring the fundamental inappropriateness of a single equation relating price and quantity, an economic implication of the simple linear model is that the elasticity of demand is a function of the level of quantity. In contrast, the curve in Figure 7 results from the fitting of a double logarithmic model to the data for the "preconspiracy" period. Thus price is assumed to be approximately a power function of quantity with a negative exponent. The defendant economic expert steps in to explain that the elasticity in this model is constant, an assumption that has been widely used in econometric studies of demand for agricultural products. In fact, the estimated value of elasticity is close to that observed for other similar commodities. It is further observed that of the three points that were not used in curve fitting, one lies almost exactly on the new function, whereas one is actually below. [For a discussion of regression estimation of damages, and what in the authors' opinion would constitute sufficient evidence for defendants to counter the plaintiffs' contention that a shift in prices indicates the presence of a conspiracy, see Finkelstein and Levenbach (1983 and this volume).]

The illustrative scenario ends here, although the plaintiffs could come back with counterarguments that would decrease the chances of the two sides ever coming to an agreement concerning the basis for damage assessment. In cases such as these where the judge and jury are faced with conflicting interpretations of statistical and economic theory, the court

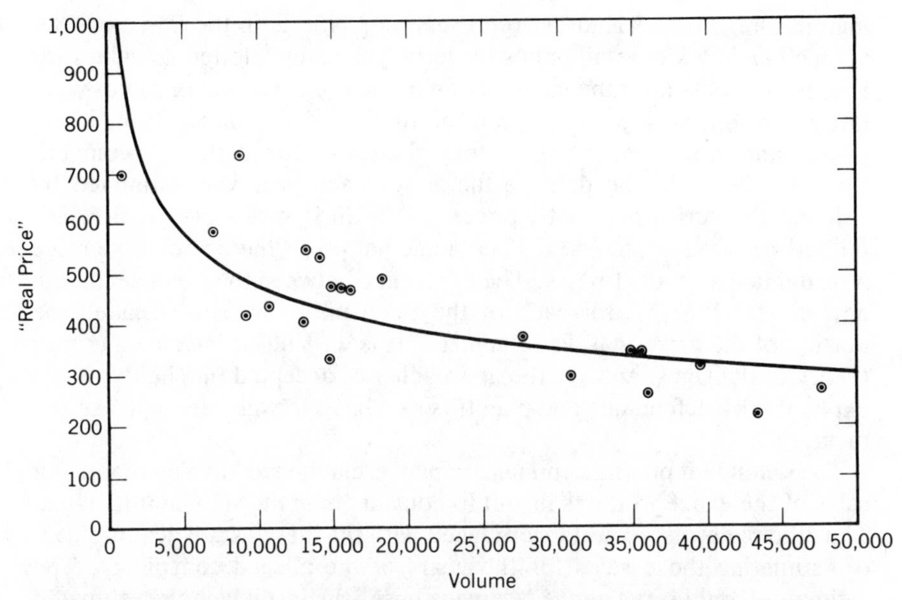

Figure 7. Plot of alternative function relating real price to quantity. Produced by defendants to demonstrate that plaintiffs' model is neither sacred nor theoretically sound. Observe that "postconspiracy" points are no longer all above the curve.

may appoint its own experts to help it weigh the merits of the respective technical arguments (Federal Judicial Center, Board of Editors, 1968, pp. 92–94). An example of the appointment of court experts during the pretrial process is found in the *Antibiotic Drug Case* (Bartsh et al., 1978, pp. 13–14), where an expert economist and an expert statistician acted as mediators among the experts for the two sides and served as chairmen of pretrial conferences designed to clarify the basic economic and statistical issues and to narrow areas of disagreement. It is difficult to assess the effectiveness of the court-appointed experts without a controlled experiment, but it is likely that their presence contributed to the settling of the case before trial—the ultimate goal in these matters. [Rubinfeld and Steiner (1983) also arrive at recommendations concerning the use of external experts, as well as suggesting that "serious consideration ought to be given to allowing appellate review of technical evidence."]

5. The Damage Base

In the consumer class action referred to as the *Antibiotic Drug Case* (Bartsh et al., 1978) a team of experts for the states of California, Washington, Oregon, and Kansas, representing plaintiffs in the litigation, developed an index of retail prices for tetracycline and plotted its values for successive years after the introduction of the drug. The index values were fairly constant during the 13 years of an alleged conspiracy to maintain prices, and in subsequent years they sharply declined to the "competitive" retail level. The damage factor for each year was estimated by defining the series of "just" prices as the time series of actual prices pushed backward. That is, if Y_t is the actual price, the competitive price X_t is defined as equal to Y_{t+7}. The difference between the original and the new series, $Y_t - X_t$, for each of the "conspiracy" years is called the portion of the price that was "unjust." It is difficult to imagine a cruder theory of damages, and yet the approach was accepted (implicitly, if not explicitly) by defendants and plaintiffs as a basis for the negotiated settlement.

The defendant pharmaceutical companies claimed to have no records of sales of the exact products in suit to consumers in the six plaintiff states. Thus the plaintiffs' experts were faced with the messy statistical problem of estimating those sales for the years of the alleged conspiracy. The estimates would serve as the "damage base" against which the estimated damage factors would be multiplied to produce a total dollar figure for damages. In many antitrust situations the damage base may be readily available from precise records of total purchases from the defendants, but in cases such as a class action with a large number of members that cannot be easily identified the value of total purchases during a given time period may be quite nebulous. Although not the most desirable solution, it is often satisfactory for the purposes of settlement negotiations to have an "order-of-magnitude" estimate (Mosteller, 1977).

In the *Antibiotic Drug Case* the plaintiffs' problem was mitigated somewhat by the existence of acceptable figures from market research sources on total annual sales of tetracycline in the United States. The principal question was how to allocate a part of total U.S. sales to each of the individual states in the lawsuit. Using data from the *Census of Business* for 1963 and for 1967 it was possible to establish a relationship between *state population* and *sales of all prescription drugs*. That is, simple regression was used to show a linear relationship between state sales of all prescription drugs (expressed as a percentage of the total U.S. sales) and state population (as a percentage of total U.S. population). The regression was run separately for 1963 with 41 reporting states and for 1967 with 47 states (Bartsh et al., 1978, pp. 49–50. Note typo in Table 2-8, p. 50; r for 1963 is 0.972.). Since the percentage of sales of all drugs appeared to be proportional to state percentage of total U.S. population, it was judged reasonable to allocate sales of tetracycline in the same manner. The experts for both sides discussed this method of estimation of the damage base in pretrial hearings under the chairmanship of the Court's experts (referred to above), and the defendants claimed that the method would be biased against them because of the "well-known" lower incidence of upper respiratory ailments in California and other western states. The plaintiffs introduced morbidity statistics that showed that the lower incidence on the West Coast was not so well known and that, in fact, upper respiratory ailments were *more* prevalent in California, Washington, and Oregon. Thus, if there was a bias in the population allocation approach, it worked in favor of the defendants. Before the final settlement figures were agreed on, it was necessary to make further adjustments to the estimated sales figures for purchases made by the respective state departments of welfare (covered by separate calculations) and for death and out-migration of residents since the period in suit. (Thus there was a brief prepared by demographers, an indication of the wide scope of expertise that is called for in these large-scale lawsuits.) Note that an analysis of the total purchases covered by persons who actually made claims for damages in the subsequent consumer refund process demonstrates that the assumptions made in the settlement concerning the demographic shifts were essentially borne out (Bartsh et al., 1978, pp. 51–53).

6. Missing Data Problems

In the process of obtaining records by discovery to use in the calculation of the damage base, one may find that whole segments of the data are not available for one reason or another. For example, in the *Antibiotic Drug Case* there was a separate class of plaintiffs designated "City, County, and State Institutions," and although in most cases complete records of purchases of tetracycline were available, there were instances of missing figures for certain years. The missing data were estimated through methods of regression that related purchases of antibiotic drugs to other variables (in the case of hospitals, to patient-days or to purchases of *all* drugs) and although sometimes the fits were rather poor, with large residual variance, the methods were deemed reasonable.

One of the consumer claimants in the *Antibiotic Drug Case* was the Department of Labor and Industries, of the State of Washington, a purchaser of large quantities of tetracycline during the period in suit on behalf of workers in the state whose medical bills were covered under unemployment compensation. To arrive at a figure for total purchases for the 13-year damage period, it was necessary to perform a sample survey of files for individual workers. By means of multistage, stratified, cluster sampling, storage racks, boxes, and ultimately individual folders were probabilistically selected and photocopied receipts were examined for evidence of purchases of antibiotic drugs. The resulting estimate of total purchases by the department for the period in suit had an acceptably small standard error.

7. Further Use of Survey Sampling

The example cited above suggests that greater use of survey sampling methods could be made in cases where total purchases, profits, or costs are an essential element in the calculation of estimated damages. Clearly there will be situations in which there are accurate records available but where the files are so voluminous that survey methods are justifiable. Obviously, all the cautions concerning precision of estimation and thorough documentation of procedures mentioned in our earlier discussion of sampling apply here.

In the *Antibiotic Drug Case* it would have been desirable to have an estimate of total purchases that was obtained directly from the consumers in the six plaintiff states. One could therefore contemplate a sample survey of the approximately 11 million households in the relevant frame, but it is likely that a survey of adequate precision would have been very expensive. When the increased costs of large-scale litigation resulting from the absence of good information are weighed against the costs of a high-quality survey, the survey is not necessarily ruled out on that basis alone, but there is also the planning and execution time to consider. It is doubtful that the two sides in the *Antibiotic Drug Case* could have arrived at sufficient agreement on the ground rules for the survey for it ever to have been carried out in time to be useful.

This does not mean, however, that preliminary surveys should not be promoted as a desirable statistical feature of antitrust litigation in the future. A relevant example, although in an area different from antitrust, is found in *Byrnes v. IDS Realty Trust* (D.C. Minn. 1975), described in King (1980). In that case, with the approval and underwriting of costs by both of the parties, an independent research group surveyed the members of the class of plaintiff shareholders who held stock in the IDS Realty Trust during a period of alleged wrongdoing. The sampled shareholders were asked to report details on their holdings that made possible the calculation of losses under various assumptions. The resulting estimates, made with considerable precision and based on a high rate of response, became the basis for subsequent settlement negotiations. With this court-approved measurement by an unbiased third party, much of the haggling over figures by the adversaries that is usually present in these situations was eliminated and the settlement process was greatly facilitated.

As a final note on the use of survey sampling in litigation, we mention two applications that occurred in the *Antibiotic Drug Case* after the settlement was agreed upon and approved by the Court:

1. The first was with respect to the processing of claims for a refund in the consumer class action. In order to distribute the approximate $20 million in settlement funds to the persons who purchased tetracycline during the period of alleged damages, the plaintiff states set up Operation Money Back, in which claim forms were mailed to potential claimants and the returned forms processed for payment. It was necessary to audit the forms to screen out erroneous claims as well as those that were clearly fraudulent, but because there were nearly one million such claims, sampling methods with selection rates based on expected losses due to missed errors were employed (King, 1977).

2. The second application involves a purely academic survey to study the claims process in the *Antibiotic Drug Case* as a sociological phenomenon (Bartsh et al., 1978). With funds that were left over in Operation Money Back from uncashed refund checks, a sample survey of the nearly 900,000 consumers who received refunds was undertaken. The purpose of the survey was to reveal characteristics, attitudes, and motivations of the participants that would facilitate the management of large-scale class actions in the future. In addition, information was obtained concerning the effectiveness of certain features, such as the initial notification to potential class members, that are required in all class sections. Rule 23 of the Federal Rules of Civil Procedure requires that potential members of a class in a class action lawsuit be notified of the action and of their rights (e.g., the right to "opt-out" of the suit). In the *Antibiotic Drug Case,* the original notice in 1971 was mailed to 9,391,000 households at a cost of $375,673 (Bartsh et al., 1978, p. 101); at today's first-class mail rates the cost would have been much greater. An indication of the effectiveness of the notice, a rather forbidding document phrased in the language of attorneys, is that less than half of the respondents in the survey of claimants (43.8%) could recall having received the notice, *even after* examining the document in a packet sent in advance of the interview.

It is ironic that while the survey results were being analyzed the U.S. Supreme Court handed down a decision, *Illinois Brick Co. v. State of Illinois* (1977 U.S.), that in effect rules out future consumer class actions of the type that occurred in the *Antibiotic Drug Case* (i.e., those involving indirect purchases from the defendant through wholesalers or retailers). The decision is based in part on grounds of excessive complexity, infeasibility in management, and the unwillingness of consumers to step forward and stand up for their rights. The findings of the survey of consumer claimants stand in direct contradiction to this belief.

8. Problems of Ethics

We began this chapter by promising to discuss some of the basic ingredients and matters for consideration that are common to statistical applications in antitrust litigation. The reader has doubtless observed in our account, even when we have not explicitly mentioned them, certain instances where questions concerning the ethics of statistical practice may arise.

For example, in the hypothetical scenario involving the use of regression to estimate the damage amounts, the defendant expert begins by attacking the very choice of the single-equation model. Is there not something troublesome in first saying that the results of the single-equation approach are useless and then continuing to discuss the misuse of the information from the simple regression as if the estimates (e.g., the standard error) were unbiased? To be more precise, suppose that the court had ruled that the single-equation model must be adopted for damage estimation (because of its simplicity and ease of understanding). Should the defendant expert refuse to participate further in the trial?

We think that the expert should continue, remembering that one has solemn obligations to the client that should be weighed against considerations of statistical practice. Perhaps the following analogy based on medical practice will not be considered too precious: The situation is similar to that in which a member of a team of surgeons has been overruled in a decision to operate using a procedure that he believes is not optimal for the patient. He decides to do the best job possible in that wrong kind of surgery, even to the point of requesting that he be allowed to perform the intricate work. He could refuse to participate, but in his absence a greater wrong might be perpetrated. In short, a single setback should not prevent the expert statistician from continuing to keep his adversaries on the up and up with respect to the proper interpretation of statistical results. His continual hammering and criticism of weaknesses on the other side may just win the case.

There are doubtless readers who believe that any engagement in the adversarial process by statisticians is bound to sully the profession. Indeed, it sometimes appears that other professionals have damaged themselves in this way, although we do not yet see statisticians conducting late-night talk shows nor do they appear in full-page magazine advertisements endorsing vodka. (Perhaps some are only waiting to be asked.) The necessity for and the difficulty in maintaining objectivity when working for one of the sides in a lawsuit is, and should be, a topic of major concern for statisticians. [See, e.g., Meier (1986) and Roberts (1983).] To stay aloof, however, from all legal proceedings would be to fail in a part of our mission—that is, the provision of good statistical practice to those who need it.

References

American Law Reports, Annotated, 2nd Series (1961). Rochester: Lawyers' Co-operative Publishing Co.

Bartsh, T. C., Boddy, F. M., King, B. F., and Thompson, P. N. (1978). *A Class-Action Suit That Worked: The Consumer Refund in the Antibiotic Litigation,* Lexington, Mass.: Lexington Books.

Deming, W. E. (1960). *Sample Design in Business Research,* New York: John Wiley & Sons, pp. 71–74.

Deming, W. E. (1965). "Principles of Professional Practice," *The Annals of Mathematical Statistics,* **36,** 1883–1900.

Federal Judicial Center, Board of Editors. (1978). *Manual for Complex Litigation,* Chicago: Commerce Clearing House.

Fienberg, S. E., and Straf, M. L. (1982). "Statistical Assessments as Evidence," *Journal of the Royal Statistical Society, A,* **145,** Part 4, 410–421.

Finkelstein, M. O., and Levenbach, H. (1983). "Regression Estimates of Damages in Price-Fixing Cases," *Law and Contemporary Problems,* **46**(4), 145–169.

Fisher, F. M. (1980). "Multiple Regression in Legal Proceedings," *Columbia Law Review,* **80,** 702–736.

Hobbs, N. (1978). Ethical issues in the social sciences. In W. H. Kruskal and J. M. Tanur (eds.), *The International Encyclopedia of Statistics,* New York: Free Press, Vol. 1, pp. 267–278.

Jeffers, J. L. (1980). "How to Present Complex Economic Evidence to a Jury," *Litigation,* (Fall) 30–32.

Jentes, W. R. (1979). "Defining the Role of the Economist—A Lawyer's Perspective," *Antitrust Law Journal,* **48**(4), 1838–1843.

King, B. F. (1977). "Auditing Claims in a Large-Scale Class Action Refund—the Antibiotic Drug Case," *The Antitrust Bulletin,* **22,** 67–93.

King, B. F. (1980). "A Survey for Loss Estimation in a Class Action Lawsuit," *Proceedings of the Section in Survey Research Methods of the American Statistical Association,* pp. 727–731.

King, A., and Spector, A. (1963). "Ethical and Legal Aspects of Survey Research," *The American Psychologist,* **18,** 204–208.

Meier, P. (1986). "Damned Liars and Expert Witnesses," *Journal of the American Statistical Association,* **81,** 269–276.

Mosteller, F. (1977). Assessing unknown numbers: Order of magnitude estimation. In W. B. Fairley and F. Mosteller (eds.), *Statistics and Public Policy,* Reading, Mass.: Addison-Wesley, pp. 163–184.

Preston, L. E., and King, B. F. (1979). "Proving Competition," *The Antitrust Bulletin,* (Winter) 787–806.

Roberts, H. V. (1983). "Comment on 'Ethical Guidelines for Statistical Practice: Report of the Ad Hoc Committee on Professional Ethics,'" *The American Statistician,* **37,** 18.

Rubinfeld, D. L., and Steiner, P. O. (1983). "Quantitative Methods in Antitrust Litigation," *Law and Contemporary Problems,* **46**(4), 69–141.

Shuchter, J. P. (1968). "The 'Just' Price," *Antitrust Law and Economics Review,* (Summer) 103–122.

Zeisel, H. (1960). "The Uniqueness of Survey Evidence," *Cornell Law Quarterly,* **45,** 322–346.

Zeisel, H. (1978). Statistics as legal evidence. In W. H. Kruskal and J. M. Tanur (eds.), *The International Encyclopedia of Statistics,* New York: Free Press, Vol. 2, pp. 1118–1122.

Zeisel, H., and Stamler, R. (1976). "The Case Against HUAC: The Evidence: A Content Analysis of the HUAC Record," *Harvard Civil Rights-Civil Liberties Law Review,* **11**(2), 263–298.

Regression Estimates of Damages in Price-Fixing Cases

Michael O. Finkelstein
Barrett, Smith, Schapiro, Simon & Armstrong
New York, New York

Hans Levenbach
Core Analytic, Inc.
Short Hills, New Jersey

ABSTRACT

Several authors have explored the question of whether multiple regression can appropriately be used to estimate damages in antitrust price-fixing cases. While theoretical economics suggests a positive response to this question, an examination of cases in which expert witnesses have attempted to use the technique in actual litigation indicates that considerable caution is required. This chapter describes four cases in which multiple regression was used to estimate damages and discusses several of the statistical issues that either were raised by the opposing parties or should have been.

In an antitrust price-fixing case, damages are measured by the difference between the prices paid by the plaintiff purchasers and the prices they would have paid in the absence of defendants' conspiracy. Even small variations in the determination of these differences may become significant because plaintiffs are entitled by statute to receive three times their actual damages (15 U.S.C. § 15 [Supp. V 1981]). Moreover, in a class action there may be thousands of purchasers whose recoveries are based on an industry-wide overcharge formula.

The difficulty of course lies in estimating the "but for" prices. An obvious idea is to assume that competitive prices during the conspiracy period would have been the same as they were before or after the conspiracy, or in interludes of competition within the conspiracy period. Thus, the difference between the conspiratorial price and the actual price from other periods measures the damage. This estimate, however, meets the immediate objection that it is likely to be incorrect because changes in

This chapter is based on a previous version of the same title, which appeared in *Law and Contemporary Problems* (1983) **46** (No. 4), 145–169.

factors affecting price other than the conspiracy would have produced changes in competitive prices if there had been competition during the conspiracy period.

These arguments may be illustrated by *Ohio Valley Electric Corp. v. General Electric*, 244 F.Supp. 944 (S.D. N.Y. 1965), one of the well-known antitrust cases involving electrical equipment manufacturers. Plaintiffs contended that the purpose and effect of the manufacturers' conspiracy was to keep transaction prices for large steam turbines close to book prices; they proposed to measure damages by the difference between the average 11% discount during the conspiracy period and the average 25.33% discount that prevailed after the conspiracy had been terminated. Defendants replied that economic conditions for the sellers had worsened in the postconspiracy period, and this accounted for the increase in the discount. They pointed to the presence for the first time of effective foreign competition, an increase in the manufacturers' capacity to produce steam turbine generators (which caused an oversupply), a lessening of growth in demand, and a drop in manufacturing costs (which permitted defendants to offer their products at a lower price).

After a bench trial, District Judge Feinberg initially found that these factors did account for some of the increase in the discount in the post-conspiracy period. He then confronted the question of how much. Treating the matter as one of "highly subjective" and "inexact" judgment, he concluded that 4.33 percentage points of the 25.33% discount in the post-conspiracy period were due to changes in economic circumstances. Since the increase in the average discount was approximately 14 percentage points (from approximately 11% to 25.33%), he thus attributed approximately one-third of the increase to changes in economic conditions, and two-thirds of the increase to the ending of the conspiracy.

The court justified this type of subjective assessment of the discount by citing "similar" judicial problems of fixing a figure for pain and suffering in a negligence case or of selecting the percentage of negligence for which the plaintiff was responsible and hence could not recover in a Jones Act case (involving injury to seamen). However, these examples are not analogous. The judgments required in those types of cases are inevitably highly subjective, and there is no generally accepted way of translating them into dollars. By contrast, determining the influence on price of various asserted economic forces may be made by the more systematic approach of econometrics and statistics. Since these methods are currently better known to economists and statisticians—and perhaps even to judges and litigants—than they were in 1965 when *Ohio Valley* was decided, it is probable that courts will require a party arguing the effect of collateral circumstances to prove its point with this type of evidence. Thus, econometric and statistical methods may be used by plaintiffs who contend that the full effect of a conspiracy is concealed by changes in other correlated factors that also affect price, or by defendants who claim that part of a price change should be attributable to other factors.

Several writers have explored the question of whether the statistical technique of multiple regression could appropriately be used for this disentangling purpose (e.g., Fisher, 1980; Harrison, 1980; Leitzinger, 1984). Based on theoretical considerations, the academic verdict is generally favorable. However, practical experience with data sets used in several litigated cases suggests considerable caution in requiring full-fledged models as part of a prima facie case. This chapter describes the cases in which models have been presented and discusses the issues they raise. Parenthetically, we note that multiple regression analysis has been used in several other antitrust contexts that we do not discuss here. For example, regression models have been used by discontinued dealers to prove lost profits (*Shreve Equipment, Inc. v. Clay Equipment Corp.*, No. C75-242A (N.D. Ohio 1978); *Rea v. Ford Motor Co.*, 355 F.Supp. 848 (W.D. Pa. 1973); *Coleman Motor Co. v. Chrysler Corporation,* 376 F.Supp. 546 (1973), *rev'd,* 525 F.2d 1338 (3rd Cir. 1975)). *See also Trans World Airlines v. Hughes,* 409 U.S. 363 (1973) (Report of Herbert Brownell, Special Master, at 94, 95).

1. The Cases

Corrugated Containers

In 1977, following the Supreme Court's decision in *U.S. v. Container Corporation of America,* 393 U.S. 333 (1969), that an exchange of price information constituted an antitrust violation, purchasers of cardboard containers and corrugated sheets brought treble-damage class actions against the manufacturers of these products (*Corrugated Container Antitrust Litigation,* J.P.M.D.L. 310 (S.D. Tex. 1977), hereafter cited as *Corrugated Container*). The charge was price fixing between 1960 and 1976. A group of defendants settled by putting up a fund of some $300 million. However, three defendants—Alton, Mead, and Westvaco—refused to settle and stood trial before a jury. During the trial, Alton and Westvaco settled for some $6.4 million, leaving Mead as the sole defendant.

On the damage question, plaintiffs' expert, John Beyer, presented two similar multiple regression models. In one study, the monthly average price per square foot of corrugated containers and sheets shipped by members of the Fiber Box Association (FBA) was regressed on (a) price the previous month; (b) change in cost of production from the previous month; (c) level of manufacturing output in industries using corrugated containers (a demand factor); (d) the wholesale price index (WPI) for all commodities (justified as another demand factor, although the reason is unclear); (e) the productive capacity of paperboard plants (a supply factor); and (f) a dummy variable to reflect the period of price controls (a factor that was given the value 1 in the period of price controls March 1971–March 1974 and 0 otherwise). The data were all expressed in the form of index numbers. The period covered by the regression was January 1963 through December 1975 (156 monthly observations); it is during this period that price fixing was deemed to have occurred. The estimated regression equation was thus computed solely from data during the conspiracy period. The fit appears to be very good: the square of the multiple correlation coefficient, R^2, being greater than 0.99. A second model was of the same form and used the same variables, except that the price of containers was drawn from a Bureau of Labor Statistics (BLS) price series (which uses the price of a representative container rather than an average price as in the FBA price series). Variables (e) and (f) were subsequently dropped from the BLS model because they were not statistically significant in that model.

The estimated regression equation for the conspiracy period ending December 1975 was then used to predict prices in the 1978–1979 postconspiracy period. These predicted prices were higher than the actual prices, the average of the difference between the projected and actual prices as a percentage of the projected prices being 7.8% for the BLS price study and 19.1% for the FBA price study. From this, plaintiffs argued that the overcharge in the conspiracy period was at least 7.8%. Figure 1 is plaintiffs' exhibit of the BLS price data in the conspiracy and postconspiracy periods and the conspiracy prices as projected by the model.

Defendants challenged these estimates on various grounds, but did not introduce a competing model. The jury found that the overcharge was 5%. The case was then sent to a special master to compute damages by applying a 5% overcharge figure to the purchases by the plaintiffs in the conspiracy period. While the case was pending before the special master, the parties settled with Mead agreeing to pay $45 million over a 10-year period.

A second trial followed for certain plaintiffs who had opted out of the first class. Most defendants settled by putting up a fund of some $60 million. Five defendants stood trial before a jury in late 1982 and early 1983, however. Beyer's model was introduced again, this time with certain technical revisions in the explanatory variables. Among other changes, the variables for the level of manufacturing output and for the WPI were refined; a variable for inventory of linerboard was added; and costs were used instead of changes in costs. Perhaps more important, the period of collusion was ended a year earlier, this time in December 1974.

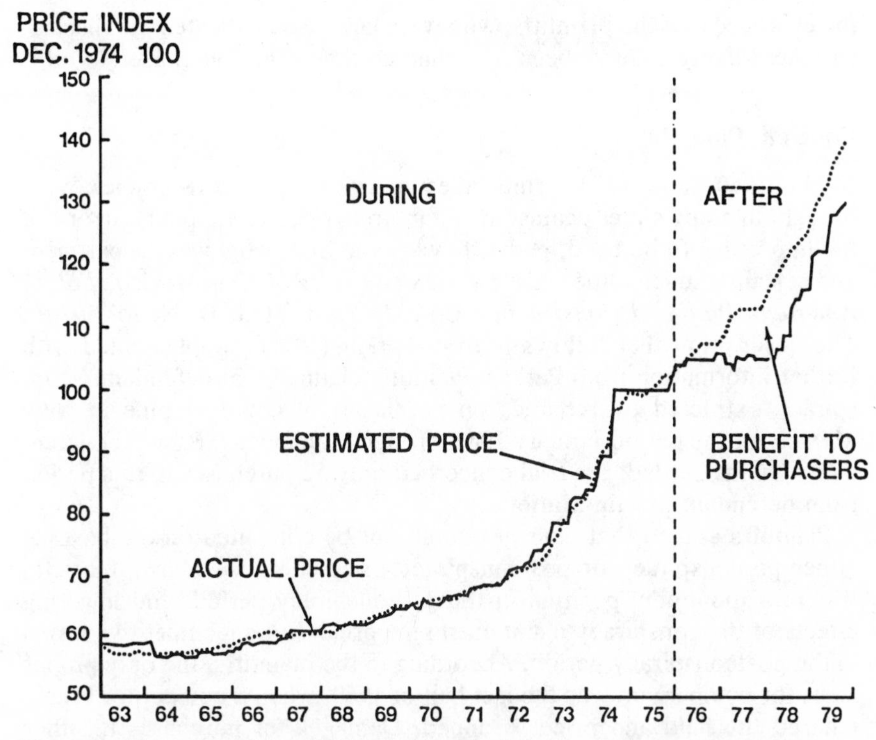

Figure 1. Time plot of the BLS price data in the conspiracy of the postconspiracy periods, along with predicted conspiracy prices for the *Corrugated Container* model.

Thus the period of transition to a fully competitive market was 1975–1978 (for which plaintiff also claimed damages) and the period of full competition was 1979 through August 1981 (the latest period for which data were available).

Using the same technique and relying solely on the FBA price series, Beyer found a remarkable 26% industry-wide overcharge. The damages estimated from the BLS price series were about half those estimated from the FBA price series. He translated this by a complex method into an overcharge for each plaintiff. This time, however, Beyer ran into a withering attack from Franklin Fisher, an economist from M.I.T., who leveled a barrage of criticism at the model and pronounced it "worthless." We discuss certain of Fisher's criticisms below. Evidently they were effective in that the jury found that there had been a conspiracy but that it had not affected the prices paid by the opt-out plaintiffs. Hence there was no recovery in the case. The jury's decision may also have been affected by the evidence that the plaintiffs, who were large, sophisticated purchasers, were less likely to have been overcharged than other purchasers.

Concrete Pipe

In the 1960s and 1970s, suits alleging price-fixing conspiracies were brought in many states against manufacturers of concrete pipes. In one of the first cases to be tried, plaintiffs were the State Highway Department and certain municipalities of New Mexico (*State of New Mexico et al. v. American Pipe and Construction Co.*, Civ. No. 7183 (D. N. M. 1970)). The description that follows is from Parker (1977), supplemented with further information from Parker. Plaintiffs claimed that defendants' conspiracy extracted overcharges on purchases of concrete pipe in New Mexico for the period January 1960 through December 1962. Other defendants having settled, the trial concerned only 52 purchases in this period from defendant Martin-Marietta.

Plaintiffs argued that damages could not be computed on the basis of either preconspiracy or postconspiracy prices because Martin-Marietta was in a monopoly position in the preconspiracy period, and lingering effects of the conspiracy maintained pipe prices at noncompetitive levels in the postconspiracy period. According to the plaintiffs, the only proper basis for comparison was the last half of 1960, when a competitor briefly entered the field and prices dropped. Damages for purchases in other periods were computed by comparing actual prices with prices in the competitive period adjusted for changes in BLS price index for concrete products. This comparison indicated a 20% overcharge.

Defendant argued that plaintiffs' simple comparison of conspiratorial and competitive periods was misleading because it failed to account for such factors as transportation costs, quantity, and type of pipe. To reinforce its argument, defendant presented the results of a regression of price per lineal foot on diameter of pipe and number of lineal feet purchased. The equation (in linear logarithmic form) was estimated using data from the nonconspiracy periods (1956–1959 and January 1963 through March 1964). Defendant then argued that any price in the conspiratory period within a 95% confidence interval of the regression estimate must be accepted as unaffected by the conspiracy. Using the difference between the actual prices charged and the upper limit of the 95% confidence interval as the measure of the overcharge, defendant obtained a negligible damage figure of $10,811.

Plaintiffs' expert, Dr. Alfred Parker of the University of New Mexico, presented a regression model to rebut the defendant's regression. Using the conspiratorial period 1960–1962, he regressed price per lineal foot of pipe on (a) diameter of pipe, (b) lineal feet of pipe purchased, (c) haul distance in miles, and (d) a dummy variable in which observations in the competitive interlude (the last half of 1960) were assigned a 1 and all other observations were assigned a 0. There were 259 observations; R^2 was 0.94. The coefficient of the dummy variable was negative and statistically significant at the 1% level. It indicated an average surcharge of $0.725 per lineal foot, or approximately 15.5% over the average price of $4.68 per lineal foot. The jury awarded $150,000; the court would have trebled that amount and then added an award for attorney fees. It is not apparent how the jury reached this figure. The case was subsequently settled for $475,000.

In his analysis, Parker used a "white-sale" period of the last half of 1960 when the conspiracy broke up because of the entry of a competitor into the market. This was more favorable to the plaintiffs as a comparison period than the pre-1960 period would have been; at that time there was no conspiracy but also no competitor and prices were almost as high as those of the conspiracy period. The white-sale period ended when the competitor was bought out. Parker's method thus questionably attributed the spread between the preconspiracy and competitive price levels to the conspiracy, whereas at least part of the spread may have been attributable to an absence of competition in the market that was disadvantageous to the plaintiffs, but not illegal:

> If a firm has taken no action to destroy competition it may be unfair to deprive it of the ordinary opportunity to set prices at a profit-maximizing level. Thus, no court has required a lawful monopolist to forfeit to a purchaser three times the increment of its price over that which would prevail in a competitive market. (*Berkey Photo, Inc. v. Eastman Kodak Co.*, 603 F.2d 263, 294 (2d Cir. 1979))

Broiler Chickens

In *Chicken Antitrust Litigation*, CCH 1980-2 Trade Cases, ¶63,485 (N.D. Ga. 1980) (opinion on fee applications), plaintiffs attacked the "recommended" price and production program of an association of chicken processors selling ready-to-cook chickens to retailers, wholesalers, and large-volume consumers. The organization, known as the National Broiler Marketing Association (NBMA), was formed in 1970 to counteract economic depression in the chicken-processing industry; its members included about half the industry. In an earlier Department of Justice lawsuit against the same defendants, the Department had elected to seek only injunctive and not monetary relief because its experts had concluded that the NBMA program had not appreciably affected broiler prices.

Edward W. Erickson of North Carolina State University, with the assistance of William Henry of Georgia State University, prepared an econometric analysis for the plaintiffs to show the effect of the NBMA program. The description that follows used the economic report of Erickson and Henry (1979), which was submitted in the proceeding on fee applications after the case was settled.

Using USDA data, the Erickson–Henry study analyzed the average monthly price of broilers in 9 cities for the 192 months between 1960 and 1975, which included the 27-month period between 1 January 1971 and 31 March 1973 when the broiler chicken producers conducted their joint economic program. These prices were regressed on (a) the monthly average prices of beef, pork, and turkey; (b) dummy variables for each quarter of the year (because the demand for broilers is seasonal); (c) the Consumer Price Index; (d) consumer disposable income (BLS data); and (e) the per-capita broiler production for each period, in pounds per person, as reported by the USDA. The study included a dummy variable for "all other influences," which took on value one in the conspiracy period (months 133 through 159) and zero otherwise.

Unfortunately for the plaintiffs, the study showed a *negative* coefficient of 1.36¢/lb for the dummy variable, which was statistically significant at the 5% level. This result suggested that prices for broilers in the price-fixing period were about 5% *lower* than would have been expected on the basis of the supply and demand factors included in the regression equation model. The fitted values from the estimated regression equation and actual prices are plotted in Figure 2. The plaintiffs' experts suggested that this perverse result could be explained by the fact that association members accounted for only about half the broiler producers and by the possibility that the recommended prices and production levels were not being adhered to by the membership of the association. These reasons would explain a statistically insignificant coefficient, but not a statistically significant negative one.

Another explanation tendered by the experts was that in July 1971, shortly after publication of a book by Ralph Nader and Harrison Wellford charging laxness in USDA meat inspection, there was a much-publicized polychlorinated biphenyl (PCB) contamination of broiler feed and broilers; the experts speculated that these two events may have reduced demand for broilers during part of the price-fixing period. Excluding the "PCB period" (August 1971–April 1972) from the time in which the dummy variable was switched on led to a new estimated coefficient of −0.43¢/lb, which was not statistically significant at the 5% level. But since the coefficient was still negative, it provided no basis for damages. To come up with a damage figure, the experts rather tentatively argued that they were justified in considering only months in the price-fixing period with positive residuals, that is, when actual prices exceeded the predicted

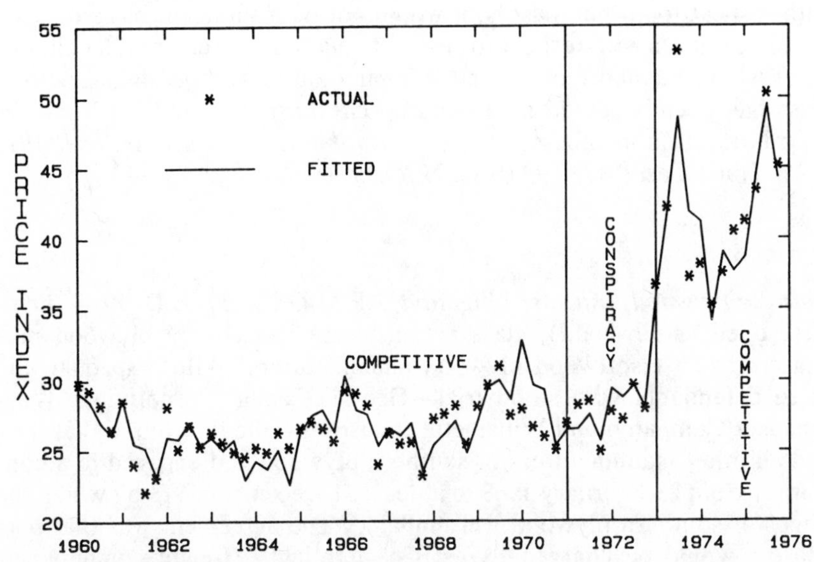

Figure 2. The fitted values from the estimated regression equation and actual prices for the *Broiler Chickens* case.

prices. No deduction was allowed for negative residual months on the theory that the price-fixing worked only intermittently or was outweighed at times by other factors such as the PCB scandal. To counter the argument that such months reflected simply chance variation, a separate calculation was made in which only those months were counted in which actual prices exceeded predicted prices by more than one standard error. This calculation yielded aggregate damages of $26 million. Defendants settled for $35 million.

The experts' ad hoc method of estimation was not well founded. By looking only at the positive residuals, a "damage" figure can be generated in any period that is long enough for random variation to generate residuals in excess of one standard error on the positive side. The record contains no indication that the experts were challenged. In its opinion on fee allowances, which followed the settlement, the court referred to their estimate as "reasonable" but it sensibly added that the damage figure was "still a matter of some speculation, despite the extensive testing done by these two experts." The court noted that recent decisions in the circuit had indicated that statistical testing would not sustain the plaintiffs' burden of proof on the issue of damages. It is unclear what was meant by this, and the cited cases do not clarify the point. The court may have meant that since the regression estimates provide only an average overcharge with respect to all purchasers, it would not be a sufficient basis for estimating damages with respect to any particular purchaser. If intended, this point would be incorrect, since a computation of average damages for a class has been accepted as an appropriate basis for individual damage awards to class members. See, e.g., *Antibiotic Antitrust Actions,* 1971 CCH Trade Cases ¶73,699 (S.D. N.Y. 1971).

Plywood

In *In Re Plywood Antitrust Litigation,* J.P.M.D.L. 159 (E.D. 1974) (hereafter cited as *Plywood*), class actions were brought by plywood purchasers against softwood plywood manufacturers. With respect to the three defendants who stood trial—Georgia Pacific Corporation, Weyerhauser Company, and Willamette Industries—the jury found that they and all other manufacturers of southern plywood had engaged in a conspiracy from 23 February 1968 to at least 31 December 1973 by which the prices of southern plywood were inflated. The agreement was that purchasers would be charged "West Coast freight" (freight computed as though the product had been shipped from the West Coast), which exceeded actual freight charges from southern shipping points (the excess being referred to as "phantom freight"), and would be charged on the basis of "standard weights" to calculate freight even though these standard weights exceeded the actual weights.

Defendants' expert, Peter Steiner from the University of Michigan, argued that in the absence of an agreement to charge phantom freight, southern suppliers would not have reduced their prices because there was sufficient demand for plywood at the higher prices. To sustain that view, he introduced a regression study of prices of plywood by himself and a colleague, Daniel Rubinfeld from the University of Michigan, which was designed to show that the market was competitive because prices responded to supply and demand factors. Details of the study are given in Rubinfeld and Steiner (1983). In the study they regressed the price of plywood (expressed in index form) on (a) housing starts by region (a demand factor), (b) income (demand), (c) stumpage price (the price for timber sold from private lands in Louisiana—a supply factor), (d) forest sales (the price for timber sold from publicly owned land in Oregon and Washington), (e) productive capacity of plywood plants (supply), (f) log drying costs (supply), and (g) gluing costs (supply).

The parameters in the model were estimated from data for the period May 1964 through February 1977, which covered the time when the defendants used the West Coast freight add-on. For that period, it appeared to track actual prices quite closely, which Steiner cited as evidence that the market was in fact competitive because prices responded to normal supply and demand factors.

Plaintiffs did not introduce a competing statistical regression model, but rather relied on judgmental expert testimony and cross-examination of the defendants' expert. The jury found that the plaintiffs had been financially injured by the amount of phantom freight and the excess of standard weights over actual weights. The court of appeals affirmed (655 F.2d 627 (5th Cir. 1981)). The Supreme Court subsequently granted certiorari (456 U.S. 971 (1982)), but before the case could be heard, the parties settled with defendants' payment of $165 million (*Wall Street Journal*, Jan. 19, 1983, §1, p. 5, col. 1).

The fitted model proposed by Steiner does not persuasively support his conclusion. The fact that prices responded to supply and demand factors did not preclude the possibility that prices included a component—like phantom freight—that drove the entire level up. Indeed, most price equations include both supply and demand factors and a dummy variable for the conspiracy on just the assumption that both may affect price. Moreover, the model did not appear to account very well for turning points even during the fit period and, more significantly, it failed to account for a sharp upward spike in prices that occurred in the post-phantom-freight period 1977–1978. For much of that period, it perversely projected phantom-freight prices that were sharply *below* actual prices. The projected and actual prices for the phantom-freight and post-phantom-freight periods are shown in Figure 3.

This brief summary of cases suggests that regression models frequently will not predict price variations with sufficient accuracy to allow an economist to estimate reliably the relative small effects of a conspiracy. We have touched on a few of the problems in some of the cases, but in most cases there was no very detailed exploration of the issues raised by the models. Perhaps the most detailed dissection of a model took place in the second *Corrugated Container* trial when the economists confronted each other. The following discussion looks in greater detail at some of the main issues that are likely to arise when models are introduced, drawing principally on the *Corrugated Container* case, and to a lesser extent on other price-fixing cases. We shall also refer to models presented in antidiscrimination class action litigation, because some of the problems are similar and there has been more experience with multiple regression studies in that context than there has been in the price-fixing cases. For a discussion of some of the earlier antidiscrimination class actions, see Finkelstein (1980).

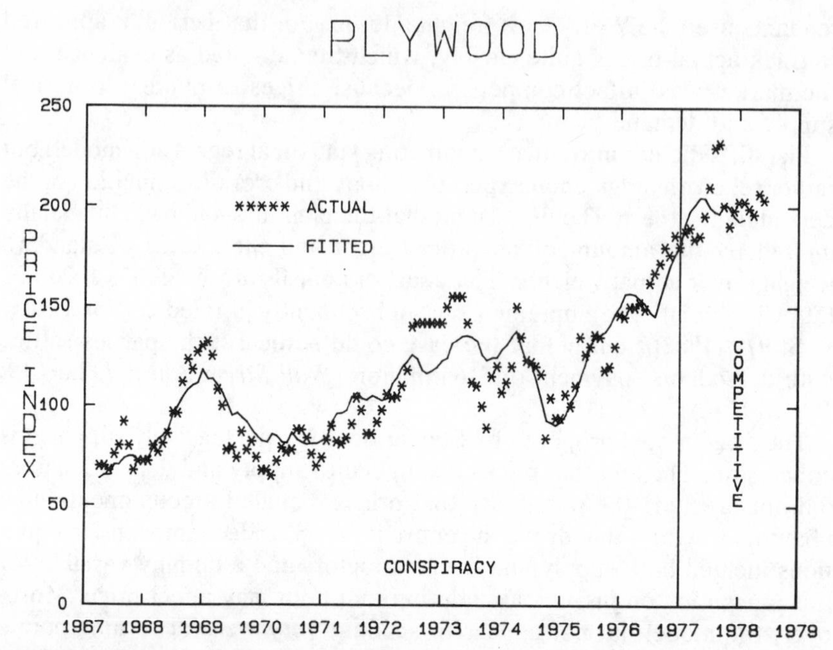

Figure 3. The projected and actual prices for the phantom-freight and post-phantom-freight periods in the *Plywood* case.

2. Issues in Model Building

Types of Models

The models used in most of the cases consist of regression equations in which the movement of price is in large part accounted for by certain explanatory factors included in the equations. Typically there are both supply and demand factors. It is not assumed that the explanatory factors account perfectly for all movement of price, but "only" that they represent the principal influences so that the regression equation gives the average price. Variations about this average are assumed to be the result of many small accidental forces so that the discrepancies between the actual and regression estimates of average price have a random character. These discrepancies are called errors when measured with respect to the "true" regression equation and are called residuals with respect to the estimated equation.

The models used in the cases previously discussed may be divided into two types. In what we refer to as a "dummy variable" model, the equation is estimated from data covering both conspiratorial and competitive periods and the explanatory factors include a dummy variable to account for the effect of the conspiracy. The dummy variable takes on a value of 1 for observations in the collusion period and 0 otherwise (or vice versa). In a properly specified linear equation, the coefficient of the conspiracy dummy variable is often interpreted as the average overcharge due to the conspiracy in the same units in which price is expressed. Plaintiffs' experts in the *Chicken Antitrust* case presented such a model. If price is expressed in natural logarithms, as it frequently is, and if a is the coefficient of the conspiracy variable, the overcharge as a proportion of the competitive price is $e^a - 1$. When a is small, the coefficient a multiplied by 100 is approximately equal to the percentage overcharge. The coefficient form of equation is the one used in most employment discrimination litigation, where one or more dummy variables may be included for sex and race. The size of the coefficients measures the extent to which the minority group or women have been underpaid relative to caucasian men.

A dummy variable model assumes that the conspiracy added a certain dollar or percentage amount to price during the conspiracy period and nothing in the competitive period. In fact, the conspiracy might have raised prices in a more complex and varying fashion. The investigator may wish to make no assumption about the effect of the conspiracy, but rather treat it as an unspecified influence that is not explained by the model. In such a case the determinants of price are the supply and demand factors dictated by economic theory and the effect of the conspiracy is appraised by looking at the differences between the actual and predicted prices, that is, the residuals. For this reason, an equation in this form is referred to here as a residuals model.

Residuals models can be estimated in several ways. One approach is to use data from both conspiracy and competitive periods and to compare the average residuals in the two periods. This method can be used to test the null hypothesis that the conspiracy had no effect, but is not useful to estimate the extent of any overcharge. To use a residuals model for that latter purpose, it should be estimated from data for the conspiracy or competitive periods and the resulting equation should then be used to project what the price would have been in the other period. It is perhaps preferable to estimate the model from competitive period data and use it to project competitive prices in the conspiracy period, because this approach leads more directly to an estimate of competitive prices in the conspiracy period and the usual supply and demand factors included in the equation are more convincingly relevant in a competitive market. Both these methods were used in the *Corrugated Container* litigation. The defendants' expert also computed a dummy variable model that showed a very small but statistically significant overcharge, but these results were not presented to the jury. In the *Concrete Pipe* litigation, plaintiff used a dummy variable model, while defendant used a residuals model.

Which type of model should be preferred? In answering this question it should be noted that the dummy variable and residuals model usually give different measures of the effect of a conspiracy. The reason is that when the conspiracy dummy variable is correlated with other explanatory factors (as it usually will be) the least-squares method will allocate some explanatory weight to each factor. By contrast, in the residuals model, price movement is first attributed to the supply and demand explanatory factors, and it is only the unexplained residual variation that is assigned to the conspiracy. In effect, the residuals model asks whether the data are consistent with an assumption that the conspiracy had no effect. Suppose that a residuals model is estimated from all the data (from the conspiratorial and competitive periods combined), and the statistical significance of the conspiracy appraised by comparing the average residuals in the conspiracy and competitive periods. It can be shown that the difference in average residuals equals the conspiracy coefficient times one minus the squared multiple correlation coefficient of the conspiracy variable with the other explanatory factors in the price equation. Thus as conspiracy becomes more and more confounded (correlated) with other factors, the difference in average residuals becomes smaller relative to the conspiracy coefficient. See Levin and Robbins (1983) at 258.

From this difference one might fairly argue that the effect of the conspiracy should be measured by the dummy variable model, since it is not tilted against a finding of a conspiracy effect, but rather is neutral as between the conspiracy and other factors. The argument for a neutral method would be strengthened in cases in which there was independent evidence of the conspiracy's effectiveness. Of course, defendants may not complain if plaintiffs use a residuals model—as they did in the *Corrugated Container* litigation—because it is less favorable to their case. The court in at least one employment discrimination case accepted a residuals model introduced by defendants, however, despite plaintiffs' argument that it was biased against a finding of discrimination (*Sobel v. Yeshiva University*, 32 FEP Cases 154 at 169-170 (S.D. N.Y. 1983), hereinafter referred to as *Sobel*).

The Choice of Explanatory Variables

The models already examined displayed high correlation between projected and actual prices over the period used to fit the equation. This goodness of fit was generally cited by the experts as evidence that the models were correct. It has been demonstrated repeatedly, however, that a high correlation in the fit period does not prove that the explanatory factors are in fact good predictors. (Nevertheless, when a lagged dependent variable is used as an explanatory variable, the equation is fitted by using the initial actual value of the lagged dependent variable and its subsequent estimated values; if the explanatory factors are weak predictors, the equation may "walk away" from the data, even during the fit period.) Nor does the fact that the projection is good when prices are smoothly trending justify a conclusion that the model will predict accurately a break in the trend. In time series models it is not uncommon to find multiple correlation coefficients exceeding 0.9 in equations that perform badly at turning points in a price trend. For an example from administrative law, see Finkelstein (1973) at 1455–1462. Moreover, the expert may have adopted the practice of "trying out" different sets of explanatory factors drawn from a large number of candidates to find the best fit. When this is done, the multiple correlation coefficient usually declines markedly in a new data set.

Sometimes defects in the explanatory variables are fairly obvious. One such case occurs when coefficients have the wrong signs. Such misspecification is not uncommon. A costs variable with a negative sign indicates a misspecified equation because it implies that prices would go down as costs go up. For example, in the *Plywood Antitrust* model, the coefficients of the variables for cost of logs and cost of gluing were negative and statistically significant and thus perverse. Cost of drying, however, had a much larger positive coefficient, and all three cost elements were highly correlated. When the regression was recomputed with the three combined, the coefficient for the combined variable was positive. That the regression was computed with them separately would not generally affect the validity of the estimates of the dependent variable.

But even when all coefficients have the correct sign, the equation cannot be assumed to be correct; its reliability must be tested by using it to make projections on another sample of data. In the *Corrugated Container* case, Fisher performed such a test by reversing Beyer's procedure. He estimated Beyer's model from data for the competitive period and then used the model to project what the competitive price would have been in the collusive period. Fisher's "backcast" was made by projecting competitive prices forward from January 1963. For the lagged price variable, the December 1962 price was used, and each subsequent lagged price was the previously projected price, not the actual price. One might object that the initial price was not securely competitive, but the effect of the starting point wears off quickly so that it becomes essentially irrelevant within a year. The projected competitive prices were far above the actual collusive prices—a result indicating that the explanatory factors were not in fact good predictors of price movement.

In the *Chicken Antitrust* litigation the model itself showed perverse results when it projected competitive prices for the collusive period that were higher than the actual collusive prices. The same indication of inadequacy appears in the *Plywood Antitrust* model, which underpredicted the rise in prices that occurred after the ending of the conspiracy.

A similar but more subtle problem arises when the model is not obviously misspecified. Most explanatory variables are proxies—and sometimes rather crude proxies—for the true quantities of interest. Data for industry-wide costs may only roughly approximate the relevant costs for a group of conspiring producers; a general price index may not reflect the true inflation for a particular industry. In the best of all econometric worlds the proxies are consistent estimators of the true factors and are highly correlated with them. But even when that is true, the fact that the proxies are imperfect will distort the model's estimates of the conspiracy's effect. This has been called underadjustment bias because the proxies do not fully adjust for the factors they are supposed to represent.

The direction of bias depends on the situation. If the proxies indicate that prices in the postconspiracy period would have been higher than in the conspiracy period (because of an increase in real costs, for example), the projected conspiratorial prices in the postconspiracy period would be lower than they would have been if the proxies had been perfect, and the effect of the conspiracy would be understated. On the other hand, if the proxies indicate that prices in the postconspiracy period would have been lower than in the conspiracy period (because of weakening in demand) the regression would understate the drop in prices and the effect of the conspiracy would be overstated.

In the simple case in which there is a single proxy (or multiple proxies may be condensed into a single proxy), the underadjustment effect may be quantified more precisely as follows: If the conspiracy had no effect, the coefficient of a dummy variable for conspiracy in the regression equation (the expected value of which should be zero) would be equal to $(\overline{P}_a - \overline{P}_b)(1 - \rho^2)$, where $\overline{P}_a$ is the average price in the conspiracy period, $\overline{P}_b$ is the average price in the competitive period, and ρ^2 is the squared correlation coefficient between the proxy and the true explanatory factor. If the proxy were perfect, ρ would be 1 and the conspiracy coefficient would be 0. Conversely, if the proxy were wholly ineffective, ρ would be 0 and the conspiracy coefficient would simply be equal to the difference in average prices. See Robbins and Levin (1983). The problem of underadjustment bias was recently recognized by Judge Goettel in the *Sobel* case as a reason for denying relief to a class of women, even though plaintiffs' multiple regression model included a statistically significant sex coefficient that apart from bias would have indicated discrimination against women (*Sobel,* 32 FEP Cases at 166).

The potential for bias from underadjustment is a problem for which there is at present no good solution. Whether the problem is serious in the context of price time series is unknown.

Lagged Dependent Variables

One type of explanatory variable deserves separate discussion because its use raises special problems. In an econometric model it is not uncommon to find that the dependent variable itself, lagged by a time period, has been included as an explanatory factor. The justification is that (a) the lagged variable picks up other influences on price not specified by the regression equation, or (b) the lagged variable reflects price "stickiness," that is, the slowness with which price responds to changes in explanatory factors. In the *Corrugated Container* case, Beyer justified the use of price the previous month as an explanatory variable by the fact that some prices were determined by annual contracts and thus could not respond immediately to changes in costs and other explanatory factors. He did not explain how this would justify the one-month lag that he used.

One difficulty with a lagged dependent variable is that, if the lag is short, that variable will tend to dominate the equation if its coefficient is estimated from smoothly trending data so that the equation becomes a poor predictor of changes in trend. These problems were evident in the *Corrugated Container* models used by Beyer in both trials. In those models the dominant explanatory factor was lagged price; no other factor approached it in importance or statistical significance. The result was that the model tended simply to project smooth trends from wherever the fit period was ended. Fisher demonstrated this by showing that the model would have projected collusive prices *below* the competitive prices in the postconspiracy period if the fit period had simply stopped at the end of 1971 instead of the end of 1974.

One technique for reducing the overprediction of trend is to enlarge the interval between successive price observations. If quarterly instead of monthly prices are used in the *Corrugated Container* data from the first trial, the dominance of the lagged price is reduced and the coefficients of the other explanatory factors are increased in relative importance. However, the estimates of conspiracy effect do not change very much.

Another technique for avoiding an overprediction of trend is a first-difference equation. In such an equation, the dependent variable is the difference in price between two successive periods and this change in price is regressed on changes in the explanatory factors. In effect, this transformation allows changes in the explanatory factors to account for departures from pure linear trend in the data. When such an equation is estimated from the *Corrugated Container* data in the first trial, the fit of the equation declines from a squared multiple correlation coefficient of 0.99 (which primarily reflects trend) to a more plausible 0.33, and the estimated average overcharge declines from 7.8% to 4.3%. In these illustrative computations, the period 1963–1975 was used to fit the equation and the forecast was made for the period 1977–1979.

Besides overprediction of trend, use of a lagged dependent variable creates the risk of another potential distortion. The most common problem in a time series regression model is serial correlation in the error term; this means that the errors in successive observations are not independent as specified by the theoretical model, but are correlated to some degree. In an equation that does not have a lagged dependent variable, the only consequence of such correlation is to make the usual calculations of standard errors unreliable and the true precision of the equation lower than that indicated by theory. If tests of statistical significance, which use standard errors, are not important in a case, then this defect in and of itself may not be crucial. But if there is serial correlation of the errors in an equation that includes a lagged dependent variable as an explanatory factor, the resulting estimates of the coefficients of the explanatory variables will become biased and inconsistent in a way that cannot be predicted. This defect goes to the heart of the acceptability of the equation estimates.

A solution to this problem is difficult to achieve. It requires either new explanatory variables that more completely capture the systematic relationships so that what remains for the error term is truely random noise, or various rather complex techniques for estimating the correlation structure of the error term and the estimation of a new equation that takes explicit account of such correlation. The latter approach is discussed under the *Time Series Analysis* heading, *infra*.

Tainted Variables

Some explanatory factors may themselves have been influenced by collusion. In the *Ohio Valley* case the court found that during the period of collusion defendants did not have to face a competitive market and consequently made less of an effort to control costs. To the extent costs are inflated as a by-product of collusive behavior, inclusion of a cost variable to explain price would conceal the full effect of the collusion. While such concealment is an issue to watch, it should not be assumed that all cost data are tainted. A similar objection was made to the cost data in the *Plywood Antitrust* case, but it seems unlikely that plywood gluing or drying costs were inflated as a result of phantom freight or standardized weights. The problem of tainted explanatory factors has arisen in much sharper form in antidiscrimination litigation. For example, the courts have divided over whether it is proper to include academic rank as an explanatory factor in regression models of faculty salary at academic institutions (in *Sobel,* FEP Cases at 166–167, the use of rank was approved, whereas in *Melani v. Board of Higher Education of The City of New York,* 561 F.Supp. 769 (S.D. N.Y. 1983) the use of rank was disapproved).

Collusive and Competitive Periods

The results of a multiple regression study will generally be significantly influenced by the choice of the collusive and competitive periods. The principal issues have centered on the length of the collusive period and the existence of a transition to full competition.

The regression coefficients of the explanatory variables represent averages over the period used to fit the equation. When this period is long, the estimate of the conspiracy coefficient is open to the objection that, being an average, it does not fairly represent the period immediately adjacent to the competitive period used for comparison. The regressions in both *Corrugated Container* cases were fitted over a long period, which raised a question whether the result would have been different if the regression had been calculated using a shorter period adjacent to the competitive period. Fisher computed regressions for subperiods within the collusive period and concluded (not surprisingly) that, on the basis of the results of a Chow test (e.g., see Johnston, 1972), the regressions had significantly different coefficients, thus indicating that the relationships did change over time.

A period of collusive activity can be compared with (a) a competitive period prior to the beginning of such activity; (b) a white-sale period within the collusive period; or (c) a period after the termination of the conspiracy. Each has its difficulties. The preconspiracy period requires a determination of when the conspiracy began, which is likely to be in doubt. The white-sale period may involve special conditions that did not persist throughout the period of collusive activities. (This difficulty was present in the *Concrete Pipe* litigation where the white sale occurred when a competitive firm entered the market and ended when it was bought out.) More commonly used is the postconspiracy period, since the ending of the conspiracy is usually a fairly dramatic event. In most cases, however, it is argued that there is a transition period in which prices are still affected by the residue of the conspiratorial activities. Whether there is such a transition is a sensitive issue because the allowance of a period of one or two years between the ending of collusion and the beginning of competition will probably change the estimated effects of the collusion considerably, and increase greatly the uncertainty of the econometric projections—which characteristically have a very short reliability life.

One technique for determining whether there is a transition period is to compute separate regressions for successive time periods (e.g., successive years) after the termination of the conspiracy. A shift in the coefficients of the regressions as the period of collusion recedes is evidence of a transition, which comes to an end when the coefficients stabilize in the postconspiracy period. Fisher tested the alleged transition period 1975–1978 in the second *Corrugated Container* case and found no statistically significant difference between the regression coefficients in this period and in the later period; he relied on this finding to reject Beyer's assertion that there was a transition.

Extrapolation

Projection by multiple regression is most reliable when the values of the explanatory factors used for the projection are the averages for those variables in the data used to fit the equation. As these values depart from the average values, the standard errors of the predicted values increase (although the standard errors of the estimated coefficients of the explanatory variables do not). More significantly, when regression estimates are based on values of explanatory factors that lie near or beyond the range of the data on which the equation is estimated, it becomes problematical whether the estimated regression relationships will continue to hold in the new range. In the *Corrugated Container* cases, the price of containers in the competitive period was much higher than in the collusive period and defendants argued that the demand for corrugated containers declined as users sought other packaging; the weakening of demand, not the ending of the conspiracy, explained the flattening of price. Cognate problems of extrapolation are likely to arise in most antitrust contexts because the price levels in periods separated by years may differ considerably, even after adjustment for inflation.

Structural Equations and Reverse Relationships

A single equation in which price is the dependent variable and supply and demand elements are the explanatory factors is what is known as the "reduced form" of a supply equation and a demand equation. In the separate supply and demand equations, quantity is the dependent variable. Price appears in both as an explanatory factor but otherwise the explanatory factors are different. A technically correct method of estimating the coefficients for price is to solve the two equations for quantity simultaneously using special methods of estimation such as two-stage least squares.

As we have seen, in most cases the parties have not estimated the separate structural equations but contented themselves with the reduced form that combines both supply and demand factors in a single equation. Although this practice is not objectionable per se, it may be desirable also to estimate the supply and demand equations separately. The structural equations disclose the relationships that are concealed by the reduced form, which is essentially a black box; showing them separately may reveal defects in the model. In the data used in the first *Corrugated*

Container case, the demand equation when estimated separately had an upward slope, a perverse result indicating a misspecification of the model. This defect was not disclosed because the separate structural equations had not been estimated by either party. The same defect did not appear in the equations used in the second trial.

Estimation of multiple equations becomes necessary if explanatory variables, in addition to influencing the dependent variable, are also influenced by it. When such simultaneous reverse relationships exist, the estimates of the coefficients generated by ordinary least-squares regression become inconsistent.

Various techniques have been developed to deal with this problem. Perhaps the most widely used is so-called two-stage least squares. In the first stage, the explanatory variable carrying a reverse relationship is estimated in a separate equation in which it is the dependent variable and other explanatory factors in the original supply and demand equations are the regressors. The effect of this procedure is to purge the explanatory variable of its component that is correlated with the error term in the final equation. In the second stage, the final equation is estimated using the estimates of the explanatory variable from the first stage rather than its original values. The result is consistent (although not unbiased) estimates of the parameters in the final equation (Intrilligator, 1978). Where there is more than one such variable, each must be estimated in a separate equation.

A discussion of a possible reverse relationship is given in the second *Corrugated Container* case. Defendants argued that the 26% overcharge was patently ridiculous because a price drop of 26% would have put defendants out of business. Beyer responded that in that event linerboard prices would also have dropped, thus reducing costs. This was a concession of a reverse influence of price on cost. The same problem arose with respect to the inventory variable, since price was likely to influence inventory, as well as inventory influencing price. While these are potent theoretical points, it cannot be assumed that a simultaneous equation approach would produce very different results from ordinary least squares applied to a reduced form equation. The effect of any such difference should be demonstrated.

Standard Error Computations

In the *Concrete Pipe* and the *Chicken Antitrust* cases, as previously discussed, parties based arguments on the computed values of the standard errors of the regression coefficients. In theory, these standard errors reflect the variation in the regression estimates that would occur if the same regression model were computed from other samples of data generated by the same process. If the average price in the postconspiracy period, for example, lies within two standard deviations of predicted values based on the regression equation, many statisticians would decline to say that there was a significant difference between the two. When a conspiracy coefficient model is estimated, a coefficient of the conspiracy dummy variable that is not significantly different from zero indicates that the hypothesis of no effect cannot be rejected. It is not, however, justifiable to argue, as defendants did in the *Concrete Pipe* litigation, that damages are properly measured by the difference between actual prices and the upper (or lower) limit of a 95% confidence interval around the regression estimate. If the model is correct the regression estimate is best and there is no justification for picking another figure with a much lower probability of being correct.

Arguments based on standard errors depend on the theoretical correctness of the calculations, which in turn depends on whether the regression model meets certain specifications. Most models based on time series do not meet those specifications and consequently standard error calculations are, at best, almost always suspect. In general, the true variability of the equation is greater than that indicated by theoretical calculation. If so, a lack of a statistically significant difference should usually be accepted, but a finding of borderline statistical significance should be suspect.

The basic requirement of the multiple regression model is that the explanatory factors reflect all major forces bearing on price so that departures from the regression estimate (i.e., the errors) are the result of many relatively small influences. This requirement is assumed to give the errors a random character. Since the errors are unobservable (because the "true" regression relationship is unknown), the regression assumptions cannot be directly observed. Instead, tests must be made on the residuals (Belsley et al., 1980).

In a regression model of time series data, the most important question is whether there is serial correlation in the errors; the existence of such correlation frequently is a warning that the model is of the wrong form, or that some important explanatory factor has been omitted. The Durbin–Watson test is commonly used to test for first-order serial correlation (e.g., see Intrilligator, 1978). If a lagged dependent variable is included as an explanatory factor, the Durbin–Watson test will tend to show no serial correlation and a different method must be substituted for a correct picture. In the *Broiler Chicken* regression, the experts did not mention that the Durbin–Watson test suggested serial correlation of the errors, indicating that the model was misspecified.

An even more important diagnostic tool, but one that is frequently ignored, is a plot of the residuals. Standard computer packages facilitate this by generating plots in which the residuals are plotted against the fitted values of the dependent variable or against values of the explanatory factors. From such plots it is relatively easy to spot patterns that suggest the linear model does not fit the data, or outlying observations that should be separately investigated. A full justification of a model should include an account of an inspection of such residual plots (see, e.g., Chatterjee and Price, 1977; Belsley et al., 1980).

A second important question in the modeling of time series is whether the variance of the error is constant. This condition will be violated if the entire scale of the equation increases with time, in which case the variance of the error will also increase. The use of undeflated dollars in a period of high inflation can create this defect. To detect such departures from a constant variance, the investigator should inspect plots of the residuals to see if they "fan out" along the time line.

Where the errors may not conform to the model, there are alternative ways of proceeding. Probably the best way is to make a direct assessment of variability by computing regressions for subperiods and observing the variation in the estimates. When this is done for the *Corrugated Container* data from the first trial, the results show much greater variation in the regression estimates than that indicated by the standard errors. Subperiod tests of the regression model during the conspiracy period showed that the difference between the predicted and actual values as a percentage of predicted values ranged up to 6.06% in the fourth year after the close of the period used to estimate the model.

Time Series Analysis

The difficulty of constructing econometric models that satisfy theoretical requirements suggests that one might use more general methods that do not assume that the explanatory variables included in the equation have accounted for everything except random error. This point of view leads to various methods of time series analysis that have found widespread applications in recent years; the so-called Box–Jenkins methods are among the most popular and successful of these techniques (Box and Jenkins, 1976; Cleary and Levenbach, 1982).

In time series modeling, the lack of independence among observations can be made explicit through the use of autoregressive–integrated–moving-average (ARIMA) models. These models, popularized by Box and Jenkins (1970), are a specialized class of linear random processes in which the random error is "filtered" so that the output represents the observed or transformed time series. ARIMA models have proved to be excellent short-term forecasting models for a wide variety of time series. The drawback of ARIMA models is that, because they are univariate, they have very limited explanatory capability. The models are essentially sophisticated extrapolative devices that are of greater value when it is expected that the underlying factors affecting conspiracy prices will behave in the future much in the same way as in the past.

Time series methods have not, to the authors' knowledge, appeared in a litigated case, although they have been used in at least one academic study of overcharges due to price fixing. A time series study of prices in the titanium industry indicated an overcharge of about 11.5%, or $8 million, during a period of collusion from 1970–1976, for which the United States brought suit in *United States v. RMI Co., et al.*, Civ. A. No. 78-1108E (W.D. Pa. 1978) and Civ. A. No. 81-4177 (E.D. N.Y. 1981). See Duggan and Narasimham (1981) for a fuller discussion.

A modest use of time series methods is to regard them as complementary to regression methods. If the errors in a regression model are serially correlated, a time series projection of the residuals, when added to the regression projection, is likely to produce better forecasts than those generated by the regression model alone, which assumes that future residuals will have an expected value of zero. When this hybrid technique is applied to the data of the first *Corrugated Container* case, the projected overcharge drops from 7.8% to about 4%. The hybrid and regression projections are shown in Figure 4. The closer fit of this hybrid model to

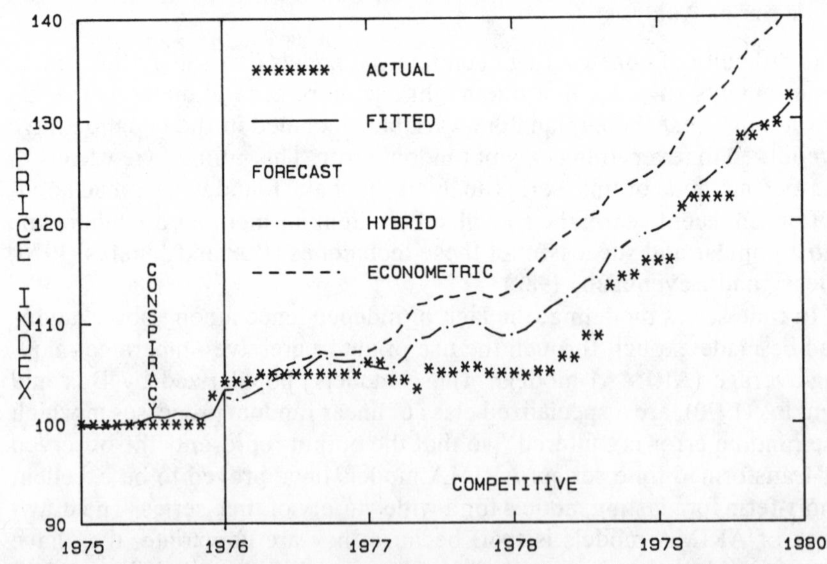

Figure 4. Projections from the hybrid regression–time series model and the econometric model for the *Corrugated Container* case.

actual prices in the postconspiracy period indicates that part of the overcharge projected by the econometric model was due to its failure to account for the correlation structure of the residuals.

A more aggressive use of time series models would involve their substitution for econometric techniques. As applied to the *Corrugated Container* data, the pure time series approach produces a straight-line projection that is closer to the actual prices than the econometric projection until 1978, but it then becomes increasingly irrelevant as it does not reflect the sharp subsequent upward turn in prices. Which type of estimate is generally superior in the short run has been much debated, and as one would expect there is no clear verdict (Armstrong, 1978). In the absence of a consensus, advocates of time series analysis argue that their methods should be preferred because of simplicity and directness. Some econometricians reply that pure time series methods are unuseable because they "explain" nothing.

3. Conclusions

The nascent excursion of the courts into econometrics and statistics suggests two perhaps conflicting observations. The first is the remarkable fact that the judicial system is presiding over the payment (mostly by settlement) of large sums as damages when the regression evidence of injury is equivocal at best. In *Chicken Antitrust* there was no credible evidence of an overcharge and the settlement exclusive of lawyers' fees was $30 million. In *Plywood Antitrust* the case was settled for $160 million and there was no econometric proof of injury. In the *Corrugated Container* cases the settlements were in excess of $300 million but a jury ultimately found the proof of overcharge unconvincing as to certain large purchases. Confronted with this record, one is tempted to say that the class action device has loaded things too heavily against the defendants, by making it too risky for them to stand trial.

On the other hand, the introduction of regression methods may have tilted things too far against the plaintiffs when there is a trial. In the *Ohio Valley* case, plaintiff proved the overcharge by showing the difference in discount in the conspiracy and competitive periods; that showing sufficed for a prima facie case. It was up to the defendants to prove—if they could—that the difference was due to factors other than the conspiracy. If plaintiffs must now produce a regression model that accounts for the other factors and isolates the effect of a conspiracy, they will have shouldered an entirely new burden. Given the great difficulties of constructing plausible models, such a burden seems unfair, particularly if there is independent evidence that the conspiracy has had some effect.

This does not mean that a simple-minded comparison of preconspiracy and postconspiracy prices or profits should be sufficient for a prima facie case. Some adjusting factors should be required, but not necessarily a full-blown model. A similar type of intermediate approach seems to be emerging in the employment discrimination cases, where the courts expressly have accepted the pragmatic notion that a multiple regression study may be sufficient for a prima facie case even if it reflects only some of the factors that plausibly may account for variations in employee productivity. The parallel in the antitrust context discussed here would be to require only the most basic of adjusting factors in a regression equation, or even to adopt a pure time series approach. If such a measure shows there was damage, the amount may be treated as determined by "a just and reasonable inference, although the result be only approximate." (*Story Parchment Company v. Paterson Parchment Company,* 282 U.S. 555, 563 (1931)). A full-fledged model would be required only to rebut an inference that would otherwise be drawn from a simple change in price level associated with the beginning or ending of a conspiracy.

How this would work in practice should depend on the context. In the situation that is most persuasive for the plaintiffs, a smoothly trending price series suffers an interruption of trend at the juncture between conspiratorial and competitive periods. Such an interruption was evident in *Corrugated Container*. Assuming that the interruption of trend is consistent with a drop in prices following the shift from conspiracy to competition, plaintiffs' prima facie case should be made by a reasonable projection of prices with the dollars deflated by an appropriate index to account for inflation. The degree of sophistication required at that point might depend on the nonstatistical evidence of the strength of the conspiracy. If defendants contend that factors other than the conspiracy caused the interruption of trend, they should produce a model that persuasively accounts for the differences by making a superior projection; explanations that are unquantified or models designed solely to cast doubt on plaintiffs' projection without being tendered as sufficiently reliable for a finding should not generally be sufficient. Of course, nothing should prevent either party from demonstrating that the class-wide figure determined by regression should not apply to any particular subclass or group of purchasers based on particular evidence with respect to those purchasers. Nevertheless, if defendants do produce a model purporting to demonstrate that the price change was due to factors other than the conspiracy, plaintiffs should be entitled to rebut by showing that the model is defective without producing one of their own.

The same format should be followed, with the parties reversed, when there is no break in the price trend associated with the beginning or ending of the conspiracy, but plaintiffs contend that prices would in fact have changed at that point if other factors had not concealed the effect of the conspiracy. This was the situation in *Chicken Antitrust*. In that event, plaintiffs would need a regression model to make their prima facie case and defendants would be entitled to rebut simply by demonstrating that plaintiffs' model was not reliable.

References

Armstrong, S. (1978). "Forecasting With Econometric Methods: Folklore vs. Fact," *Journal of Business*, **51,** 549–564.

Belsley, D. A., Kuh, E., and Welsch, R. E. (1980). *Regression Diagnostics: Identifying Influential Data and Sources of Collinearity*, New York: John Wiley & Sons.

Box, G. E. P., and Jenkins, G. M. (1976). *Time Series Analysis—Forecasting and Control*, Revised ed., San Francisco, Calif.: Holden-Day.

Chatterjee, S., and Price, B. (1977). *Regression Analysis by Example*, New York: John Wiley & Sons.

Cleary, J. P., and Levenbach, H. (1982). *The Professional Forecaster*, Belmont, Calif.: Lifetime Learning Publications.

Duggan, J. E., and Narasimham, G. V. L. (1981). "Price Fixing in the Titanium Industry: A Time Series Analysis," *Proceedings, Business and Economics Section, American Statistical Association*, pp. 241–244.

Erickson, E. W., and Henry, W. H. (1979). "Economic Analysis of Potential Damages," Report submitted in the *Chicken Antitrust Litigation*.

Finkelstein, M. O. (1973). "Regression Models in Administrative Proceedings," *Harvard Law Review*, **86,** 1442; reprinted in Finkelstein (1978). *Quantitative Methods in Law*, New York: The Free Press, Chapter 7.

Finkelstein, M. O. (1980). "The Judicial Reception of Multiple Regression Studies in Race and Sex Discrimination Cases," *Columbia Law Review*, **80,** 737–754.

Fisher, F. M. (1980). "Multiple Regression in Legal Proceedings," *Columbia Law Review*, **80,** 702–736.

Harrison, J. L. (1980). "The Lost Profits Measure of Damages in Price Enhancement Cases," *Minnesota Law Review*, **64,** 751–788.

Intrilligator, M. D. (1978). *Econometric Models, Techniques, and Applications*, Englewood Cliffs, N.J.: Prentice-Hall.

Johnston, J. (1972). *Econometrics Methods*, 2nd ed., New York: McGraw-Hill.

Leitzinger, J. J. (1984). "Regression Analysis in Antitrust Cases: Opening the Black Box," *The Philadelphia Lawyer*, **20**(2), 1–8.

Levin, B., and Robbins, H., (1983). "Urn Models for Regression Analysis With Applications to Employment Discrimination Studies," *Law and Contemporary Problems*, **46**(4), 247–267.

Parker, A. (1977). "Economics in the Courtroom: Proof of Damages in a Price-Fixing Case," *Antitrust Law and Economic Review*, **9**(4), 61.

Robbins, H., and Levin, B. (1983). "A Note on the 'Underadjustment Phenomenon,'" *Statistics and Probability Letters*, **1**, 137–139.

Rubinfeld, D. L., and Steiner, P. (1983). "Quantitative Methods in Antitrust Litigation," *Law and Contemporary Problems*, **46**(4), 69–141.

Regression Analyses in Employment Discrimination Cases

Delores A. Conway **Harry V. Roberts**
Graduate School of Business
University of Chicago
Chicago, Illinois

ABSTRACT

Regression analyses have been increasingly used in litigation over alleged employment discrimination to compare mean salaries of male and female employees after adjusting for legitimate differences in qualification between the two groups. The fundamental ideas behind regression analyses can be explained in terms of simple tabular comparisons that are more easily understood. Tabular comparisons motivate two different methods for comparing male and female salaries and qualifications to detect possible employment discrimination. One viewpoint (direct regression) compares male and female mean salaries for a given level of qualifications, whereas a second viewpoint (reverse regression) compares male and female mean qualifications for a given salary. The two viewpoints can lead to different conclusions about possible discrimination and highlight the importance of the choice of dependent and independent variables in regression models for discrimination studies.

We illustrate some of the important legal and statistical issues that pertain to the formulation, checking, and interpretation of regression models with a data set from a recent legal case. Of particular concern are the potential sources of bias in regression results that arise from the use of noncompeting groups, imperfect measures of regression variables, and the omission of variables relevant to the employment process. Some of the special problems facing statisticians who serve as expert witnesses in discrimination cases, such as disclosure of methods and access to data bases, are also discussed.

1. Introduction

Litigation over alleged employment discrimination has seen increasing use of statistical analyses, both by defendants and plaintiffs. Initially, the analyses tended to emphasize the simplest comparisons, such as mean salaries of males against mean salaries of females. But litigants became aware that such comparisons leave out legitimate nondiscriminatory explanatory variables of salary differentials. For example, salaries may reflect seniority to some degree. White males may have both higher salaries and higher seniority. To make some allowance for such variables, tabular comparisons were increasingly used.

As long as the statistical analyses are confined to the tabular level, statistical and legal professionals meet on common ground. Reading a table does not demand extensive statistical training, and the interpretation of data in tables is reasonably straightforward. Lawyers as well as statisticians can contribute to the development of evidence.

When litigants increasingly recognized the importance of allowing for multiple explanatory variables, they discovered that the tables necessary to reflect fully these variables became too complicated to present compactly or to interpret easily. Statistical experts then introduced regression analyses, since regression can make some allowance for a larger number of explanatory variables than can be readily comprehended in tabular comparisons. Regression analysis reduces the complexity of tables to compact mathematical descriptions of the relationship of explanatory variables to salaries. The mathematical models underlying regression analysis typically specify simple relationships, even though there may be many explanatory variables. (Tables are more flexible exploratory tools for bringing out complex interaction effects and other nonlinear relationships.)

The potential gain from regression over simple tabular analysis is limited by the fact that regression models entail simplifying assumptions that may not adequately describe the actual employment process. Regression models are approximations to the employment practices of the firm. They typically capture some features of the process, but may miss others.

Moreover, the technical sophistication of regression methods tends to hide the underlying rationale from nonstatisticians. A great deal of formal statistical machinery underlies regression analyses, so that interpretation of results may require understanding of statistical concepts that are opaque to legal professionals. A judge without statistical training is faced with an awesome task in evaluating regression studies done by statisticians. The recent opinion in *Vuyanich v. Republic National Bank* (1980) points to the scale of this task. Judge Higginbotham delved deeply into the methodology of regression to evaluate conflicting statistical testimony.

The tightly written opinion fills 127 pages and addresses many technical statistical issues.

This paper is directed mainly to statisticians who may be asked to be expert witnesses in cases in which regression analysis is prominent. Exposition of regression results is an essential part of the role of the expert statistical witness. Lawyers, judges, and jurors typically need assistance to understand regression results. Some of the mystery surrounding regression analyses can be removed by relating regression models to simple tabular analysis. This approach helps statisticians to translate the main issues into terms that are more readily understood by legal professionals.

The main objective of the paper is more ambitious. Because regression models reduce the features of complex tables to condensed summaries, great care is required for correct interpretation of the results. Statisticians know that regression analyses may be misleading or invalid if they rest on faulty assumptions, but each area of application raises its own special questions about the assumptions. We hope to bring out some of the special questions that must be raised to interpret regression results in discrimination cases. To focus our discussion, we provide an analysis of data used in a recent employment discrimination case.

2. Regression Models of Employment Practices

Title VII of the Civil Rights Act of 1964 prohibits discrimination in employment practices based on race, color, religion, sex, or national origin. In a class action lawsuit under Title VII, the plaintiff challenges the employment practices of a company and argues that members of a protected class have been affected adversely with respect to such decisions as hiring, job placement, salaries, promotions, and terminations. Unfair treatment is alleged to result from discriminatory intent on the part of the employer. To establish a prima facie case of discrimination, the plaintiff provides evidence that the employment practices result in disparate treatment or impact on members in the protected class.

Once a prima facie case has been established, the burden of proof for rebutting the charge of discrimination shifts to the defendant. In cases where adverse impact has been demonstrated, the touchstone of the defense has been to show that the adverse impact may result from practices justified by "business necessity" (Fiss, 1979). Here, the defendant provides evidence to demonstrate that any apparent disparate impact to a protected class results from the use of legitimate nondiscriminatory criteria, essential to the safe and efficient conduct of a business. This defense is based on the intent of Title VII to prevent unequal treatment of individuals but not to infringe on employment decisions because of reasons of business efficiency, as long as the employer does not take the group status of the individual into consideration ("Business Necessity Under Title VII . . . ," 1974).

Statistical evidence is often used to establish or rebut a prima facie case of employment discrimination. Data from employee records are examined to determine whether the company's employment practices lead to substantial and systematic differences in salaries or rates of hiring, promotion, and termination for the protected class from those of other employees (Baldus and Cole, 1980). For at least two reasons, statistical methods are important tools for measuring differences and assessing whether they can be attributed to discrimination. First, statistical methods help to assess whether the observed differences are systematic or merely the result of chance variation in the data. Second, statistical methods permit allowance or adjustment for factors in the employment process, such as employee job qualifications, that may legitimately influence employment decisions.

For example, in legal cases involving allegations of salary discrimination, statistical analyses have emphasized comparisons of male and female salaries (or salaries of whites and minorities). The courts have recognized that differences in mean salaries do not by themselves establish a prima facie case of discrimination, since these differences could have arisen from differences in job qualifications or other job-related factors among employees. In the case of *Vuyanich v. Republic National Bank* (1980), Judge Higginbotham ruled,

> Not surprisingly, where an employer has employees in differing occupations and of different backgrounds, a simple comparison of all average wage of all white employees and all black employees (or all male employees and all female employees) will not be enough to prove salary discrimination. . . . we reject use of "raw" average pay differentials as the basis for a prima facie compensation case where the pay is averaged across a wide range of jobs and backgrounds.

Economists define salary discrimination in terms of pay differences between groups that do not correspond to productivity differences between the groups (Becker, 1957). This definition formalizes the commonsense view that an individual with more education and experience than another may contribute more and thus be entitled to higher pay. Title VII specifically mentions the use of seniority, employee merit, quantity of work, and quality of production as valid reasons for disparate treatment of employees (Gwartney et al., 1979).

Statistical methods are used to compare the salaries and job qualifications of employees in the protected class with those of other employees to determine whether or not a group has been treated unfairly. In order to identify unexplained salary or qualification differences, statistical comparisons must reflect an accurate representation of the relationship between salaries and job qualifications. Although the statistical task may appear straightforward, the task effectively requires modeling the company's employment practices. Because the process by which salaries are established is often complex, it is not surprising that sophisticated statistical methods are necessary to evaluate evidence.

Tabular Comparisons of Employee Salaries and Qualifications

For concreteness, we examine data used in the case of *United States Department of the Treasury v. Harris Trust and Savings Bank* (1981). A major charge by the plaintiff alleged salary discrimination against female employees of the bank. Legal considerations led to the analysis of employment records for all full-time employees on the bank's payroll as of 31 March 1977 who were hired between 1965 and 1975. To simplify discussion, we consider the subgroup of white employees hired between 1969 and 1971. Within this subgroup, or cohort of employees, there were 176 white males and 194 white females.

A naive comparison of the 1977 annual salaries shows that the mean salary for white females in the cohort is $10,700, whereas that for white males is $17,800. Expressed another way, the mean female salary is approximately 60% of that for males. The plaintiff did not allege that the entire 40% female shortfall was due to discrimination. Statistical analysis was needed to determine how much, if any, of the difference could be ascribed to legitimate factors affecting job performance.

Formal education of employees was recognized by both sides as a job-related qualification that is relevant to salaries in the bank. Different jobs require different levels of education for effective performance. Expected educational requirements range from a high school degree to advanced graduate degrees, depending on the specific job and level of expertise required. Table 1 examines the relationship between salary and education and displays the joint frequency distribution of male and female employees in the cohort according to the employee's salary and educational level. The frequency distribution of salary shows that more than half the females, but fewer than 5% of the males, earn under $10,000. Similarly, the frequency distribution of education shows that 27% of the females have additional formal education beyond a high school degree, whereas the corresponding proportion for males is 74%.

The joint frequency distribution of education and salary reveals that higher levels of education tend to be associated with higher salaries. Hence, part of the overall mean salary difference between males and

TABLE 1. **Distribution of White Male and Female Employees Hired at Harris Bank in 1969–1971 by Salary and Educational Level**

Salary	Educational Level: High School or Less	Some College	College Degree	Graduate Degree	Number	Mean Years of Education Given Salary
(A) Male Employees						
Over $18,000	—	7	29	36	72	17.4
$ 16– 18,000	2	6	6	—	14	15.0
$ 14– 16,000	4	12	—	1	17	14.4
$ 12– 14,000	18	22	2	—	42	13.3
$ 10– 12,000	18	21	—	—	39	13.2
Under $10,000	8	2	—	—	10	11.4
Number	50	70	37	37	194	
Mean salary given education	$11,735	$13,735	$23,028	$28,407		
(B) Female Employees						
Over $18,000	—	—	7	—	7	16.0
$ 16– 18,000	1	—	3	—	4	15.0
$ 14– 16,000	2	2	5	—	9	14.9
$ 12– 14,000	9	5	6	—	20	13.8
$ 10– 12,000	30	9	1	—	40	12.4
Under $10,000	86	9	1	—	96	11.4
Number	128	25	23	—	176	
Mean salary given education	$ 9,600	$10,942	$16,428	—		

females may be related statistically to the fact that males tend to have higher educational levels than females.

One statistical viewpoint of salary discrimination is that members of a protected class with similar qualifications tend to be paid less (see, e.g., Weisberg and Tomberlin, 1983). To adjust, or make statistical allowance for, the effects of educational level on salaries, we would compare mean salaries for males and females within each education group. These mean salaries are displayed in the last row of Tables 1A and 1B and show that female mean salaries are less than the corresponding mean salaries of males for each educational level. The mean salary for females is approximately 80% of the mean salary for males in the first two education groups and 70% in the third. It is natural to say that educational level "explains" part of the overall salary difference, since female salaries within educational groups are a higher proportion of male salaries than the overall proportion of 60%. (We offset "explains" with quotation marks to emphasize that a statistical association like this one is not necessarily a causal one.)

The comparison just made summarizes one feature of the underlying data in the table, namely, the mean salaries of males and females in each of the educational groups. This particular comparison highlights the lower mean salaries of females within each educational group, but does not show the comparison previously noted, that females have less formal education overall than males.

Another perspective on salary discrimination compares the qualifications of employees at the same salary level. Female employees earning the same salary as their male counterparts may be substantially better qualified on average. Mean qualification differences favoring females over males at the same salary level also suggest disparate treatment of females. From this perspective, comparisons begin with salaries (in statistical terms, we "condition on salaries") and contrast mean educational levels of males and females. The numbers in the far right column of Table 1 show that males and females have about the same mean education in the six salary groups.

The second comparison summarizes another aspect of the table, namely, the mean educational characteristics of males and females within each of the salary groups. This comparison shows comparable educational levels for males and females within salary groups, but does not show that females have lower overall mean salaries than males.

The two comparisons summarize different aspects of Table 1 and the underlying data. In the first, we condition on education and compare mean salary; in the second, we condition on salary and compare mean education. The first comparison suggests the possibility of discrimination against females, whereas the second does not. Both comparisons are statistically meaningful. The problem is to decide what importance is to be attached to each in assessing the import of the evidence. Moreover, we have seen that there is other relevant information: males tend to have higher salaries than females without regard to educational groups, and males also tend to have higher educational levels than females without regard to salary groups.

Furthermore, the table is itself only a summary of part of the information in the underlying employment data. Even if we confine attention to comparisons of education and salary, the table is incomplete, since it displays only four educational groups and six salary ranges. As a result, the comparisons concern broad groups of employees and may fail to bring out relevant details about the relationship between education and salary. The statistician's objective is to exploit as fully as possible any relevant information in the data that permits precise comparisons of educational levels and salaries across individual employees. More detailed information is available from employee records: the 1977 annual salaries are recorded to the nearest dollar and years of formal education are recorded on a scale from 8 to 20, with 12 indicating a high school education, 16 a college degree, and so forth.

Direct and Reverse Simple Regression Models

When we examine individual educational levels and salaries, as opposed to the broad groupings above, tabular analysis becomes unwieldly. In-

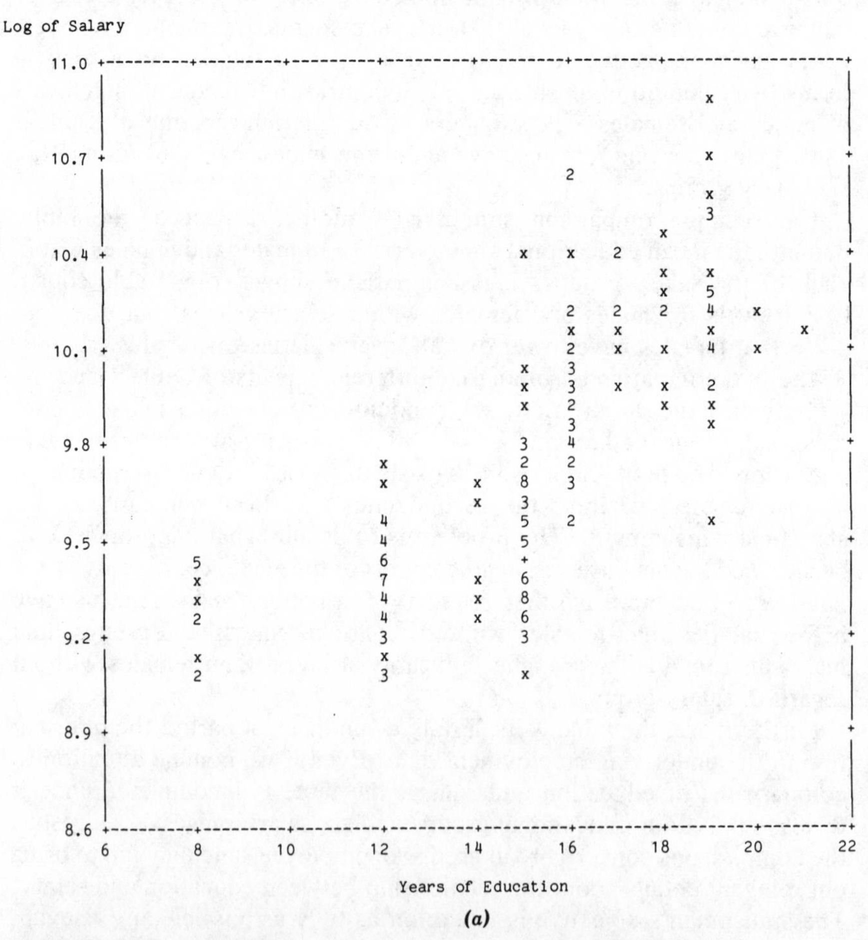

(a)

Figure 1. Distribution of employees in bank cohort by log salary and educational level. (*a*) White male employees. (*b*) White female employees.

stead, a graph of employees' salaries by their corresponding levels of education provides a compact but full display of the data. Separate scatterplots for male and female employees appear in Figure 1. On these scatterplots, salary has been rescaled by transforming each dollar salary into its corresponding logarithm.

More specifically, the scatterplot of Figure 1 graphs log salary versus years of education. On a log scale, the difference between two log salaries corresponds to a ratio of the salaries. Moreover, if the ratio is close to one, the difference between two log salaries times 100 can be interpreted directly as the approximate percentage difference. For example, if the

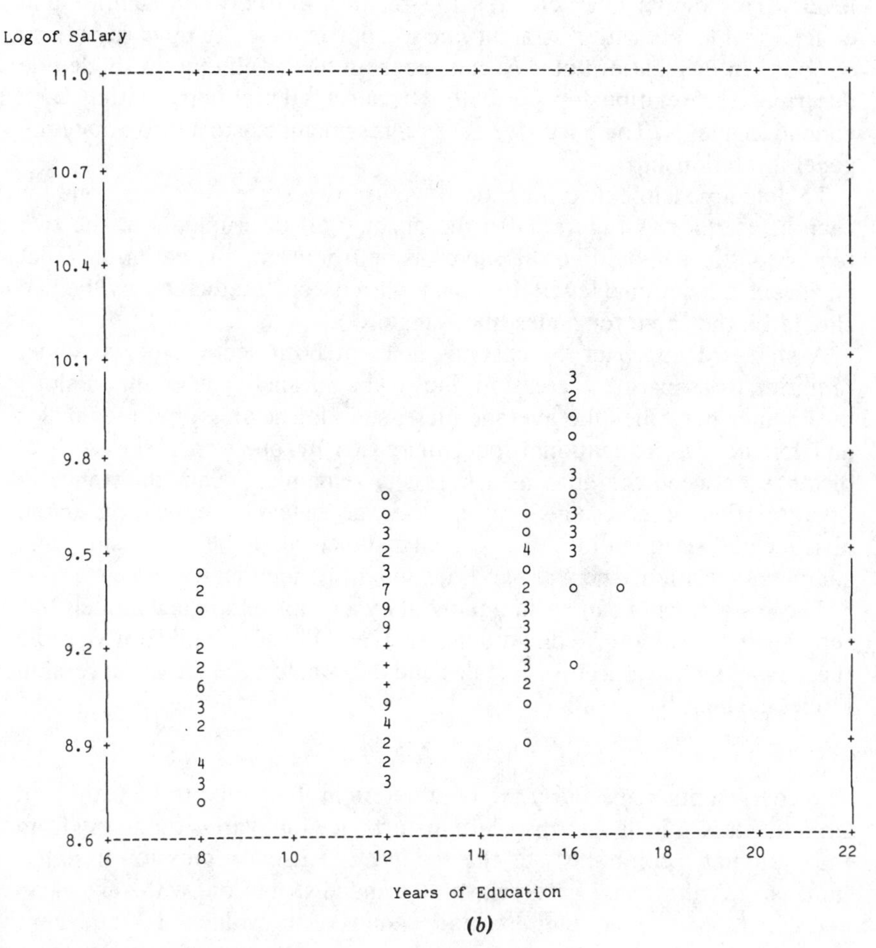

(b)

Figure 1. (Continued).

difference between a female and male employee's log salary is -0.10, we would say that the female's salary is approximately 10 percent less than that of the male. The same difference in log salaries could apply to a male salary of $10,000, $20,000, or $50,000.

The scatterplots provide more detail about the information summarized in Table 1 but verify the same general conclusions. Male employees tend to have higher levels of education and higher salaries, whereas female employees tend to have lower levels of education and lower salaries. However, comparison of salaries for males and females at the same educational level, or of educational levels for given salary, can be made by regression analysis.

The direct regression model compares salaries at given educational levels. This model characterizes the relationship between salaries and educational levels with a straight line to approximate the upward pattern of points in the scatterplot. We can see from the graph that no single line describes the relationship perfectly, because salaries vary within each educational level. The purpose of the regression line is to approximate the general relationship.

To detect possible discrimination, we use two regression lines, one for each scatterplot in Figure 1. In the absence of discrimination, the two regression lines should be the same, aside from sampling variation. That is, at each educational level, the salary "fitted" or "predicted" by the line should be the same for males and females.

A standard approach to analysis is to fit both scatterplots at once, requiring the separate regression lines to be parallel. The common slope of the lines describes the average increase in log salaries, both for males and females, as educational level increases by one year. The vertical distance between the lines measures any remaining salary difference in fitted salaries for males and females at any given level of education. In the absence of discrimination, this vertical distance would be zero, aside from sampling variation, and the two lines would be identical.

These statements can be summarized by a simple statistical model. If Y represents log salary, X educational level, and Sex is an indicator variable that assumes the value 1 for females and 0 for males, the direct regression model is given by

$$Y = \alpha + \beta X + \gamma \text{Sex} + \varepsilon. \tag{1}$$

The coefficients α, β, and γ are parameters in the model to be estimated from the data. The last term, ε, is a disturbance that varies randomly from the systematic component of the model. The usual regression assumptions specify that these errors have a normal distribution, with zero mean and constant variance, and are independent of the values of X and Sex.

The coefficients in (1) are estimated from the data by minimizing the sum of the squared errors between the observed and predicted values of log salaries. In terms of Figure 1, this leads to the choice of two parallel lines that minimize the sum of the squared vertical distances between the observed points and the lines. We write the estimated model as

$$\text{Fitted}(Y \mid X) = a + bX + c\text{Sex}, \tag{2}$$

where Fitted($Y \mid X$) refers to the expected salary for a given educational level X. We use a, b, and c rather than α, β, and γ to distinguish the sample estimates (now relevant) from the unknown parameters of the assumed regression model. The observed salaries can be expressed as $Y = \text{Fitted}(Y \mid X) + e$, where the residual e represents the discrepancy between the actual and predicted log salary for each individual. The two fitted regression lines are graphed in Figure 2.

The standard regression model assumes that the errors arise from sampling variability in the data. In some employment discrimination studies, however, the data encompass the entire workforce, rather than a random sample from the population. For these studies, the errors are often viewed as resulting from other deterministic, but inaccessible factors, that are not specified in the model. This interpretation has been considered by Dempster (1984) and Goldberger (1984) in the context of causal models of the employment process. Levin and Robbins (1983) consider an alternate randomization approach, where regression is viewed as an adjustment procedure and all statistical inferences are conditional on the observed data. In this paper, we shall regard the disturbance as resulting from other factors in the employment process that are not specified by the model.

Assuming that the model is correctly specified, the statistician draws inferences about the parameters from the estimated coefficients. The coefficient of the education variable estimates the common slope of the lines as 0.088 with a standard error of 0.0050. The interpretation is that fitted log salary increases an estimated 9% for each additional year of education. The t-statistic for education is 17.5, so that the slope differs significantly from zero.

The sex coefficient is -0.232 with a standard error of 0.0296. The t-statistic for this coefficient exceeds 7.8 in absolute value and is statistically significant. The sex coefficient corresponds to a female salary shortfall of -0.232 in log units or approximately 21% at any given educational level. This result parallels closely the conclusion from the tabular comparison in the previous section, where the female salary shortfalls within the education groups varied between 20% and 30%.

Two summary statistics appear in the last row of Table 2A and measure the goodness of fit of the model to the data. The adjusted multiple R^2

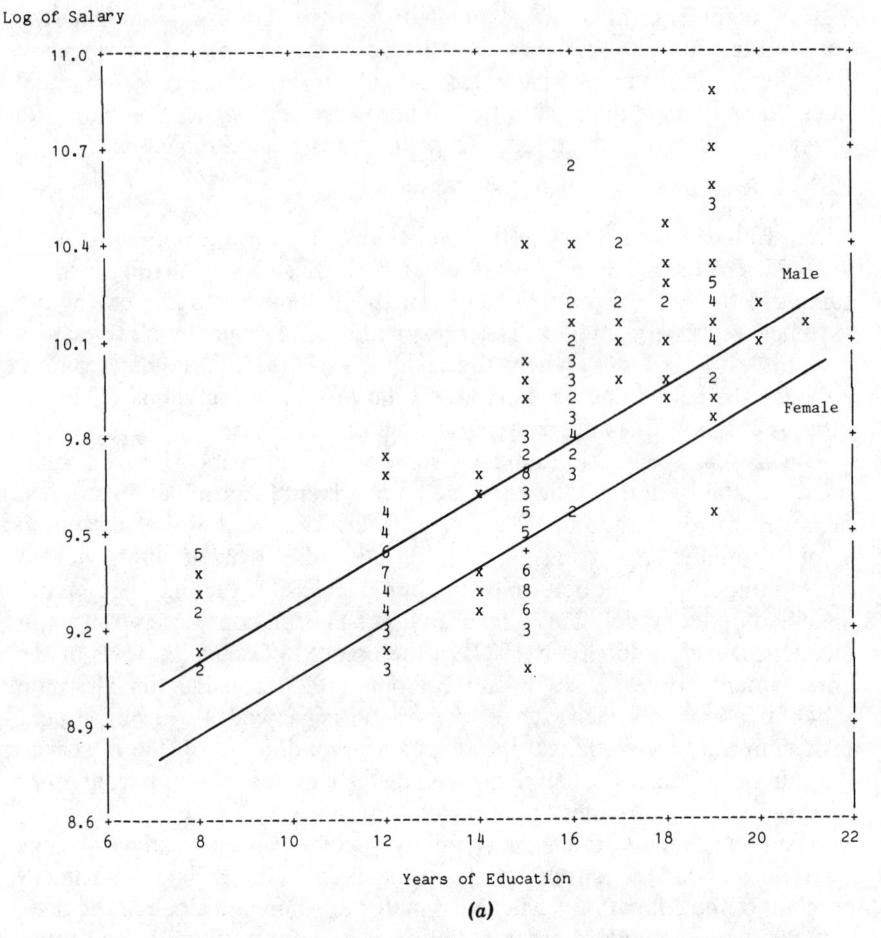

(a)

Figure 2. Simple direct regression fit of log salary by educational level for employees in bank cohort. (*a*) White male employees. (*b*) White female employees.

statistic is interpreted as the proportion of total variability in log salary attributed to the regression relationship. For the model just fitted, this proportion is 62%. The second goodness-of-fit measure, the *s*-statistic, is the standard deviation of residuals, a useful summary measure for comparing different models. If one model has a smaller standard deviation of residuals than a second model, less variability of salaries is attributed to factors excluded from the analysis.

To compare educational levels for given salaries, we introduce the reverse regression model. Again, two parallel lines are used to approxi-

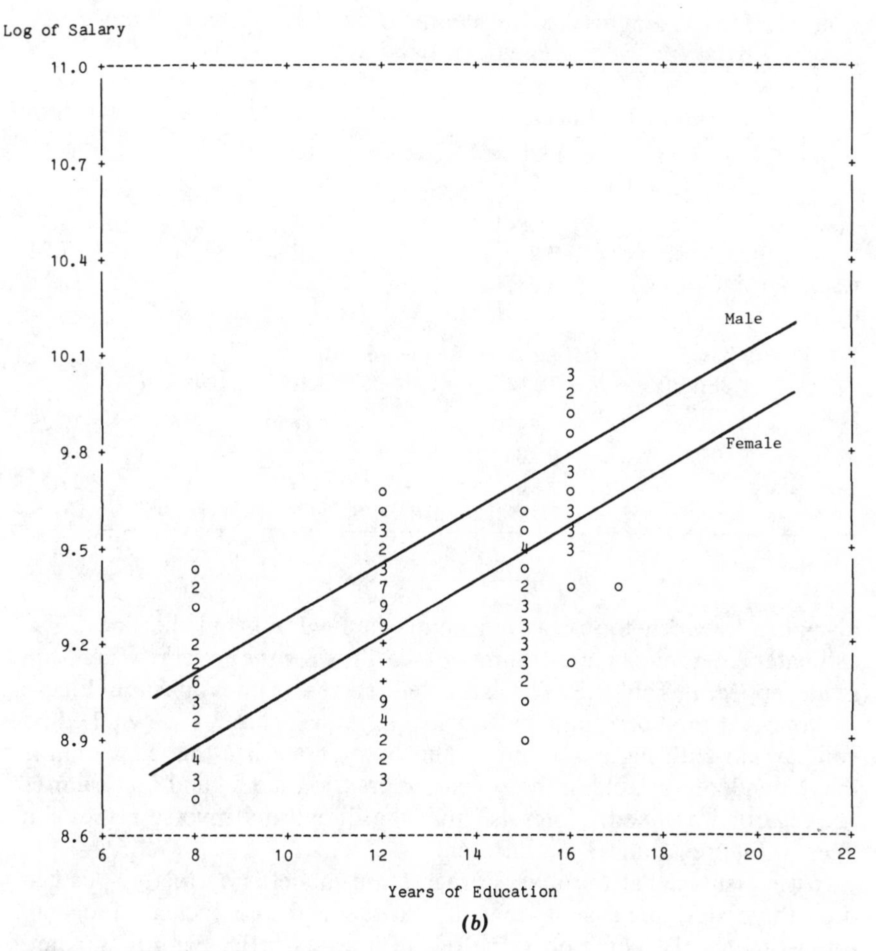

Figure 2. (Continued).

mate the relationship between salaries and educational levels for each sex. But now salaries are regarded as given and educational levels vary for each salary level. The reverse regression model is written

$$X = \alpha^* + \beta^* Y + \gamma^* \text{Sex} + \varepsilon^*. \tag{3}$$

The coefficients α^*, β^*, and γ^* are parameters in the model to be estimated from the data. Again, the standard regression assumptions require that the disturbance has a normal distribution, with mean zero and constant variance.

The coefficients are again estimated by least squares, which now selects two parallel lines to minimize the sum of the squared horizontal

TABLE 2. Simple Regression Results for Bank Employees Relating Salary to Educational Level and Sex

Variable	*B* Coefficient	Standard Error of *B*	*t* Statistic
(A) Direct Regression Model			
Log Salary = 8.384 − 0.232Sex + 0.088Education			
Constant	8.384	0.0773	108.49
Sex	−0.232	0.0296	−7.84
Educational level	0.0881	0.00503	17.52
	Adjusted R^2 = 0.621	s = 0.256	
(B) Reverse Regression Model			
Education = −35.197 − 0.206Sex + 5.168Log Salary			
Constant	−35.197	2.8650	−12.29
Sex	−0.206	0.2447	−0.84
Log salary	5.168	0.2950	17.52
	Adjusted R^2 = 0.558	s = 1.957	

distances between the observed points in Figure 1 and the lines. The estimated coefficients and summary measures for the reverse regression model appear in Table 2B. The estimated regression lines appear in Figure 3. Note that the horizontal and vertical axes in Figure 3 are switched by comparison with Figures 1 and 2. This is to emphasize that education is the dependent variable in the reverse regression model and that salaries are regarded as fixed. Note also that the direct and reverse regression lines in Figures 2 and 3 are different.

If we assume that the reverse regression model is correctly specified (i.e., the underlying assumptions are satisfied), the coefficient of log salary estimates the common slope of the lines as 5.168, with a standard error of 0.296. The t-statistic for salary is highly significant. The coefficient of sex is −0.206, with a standard error of 0.2447, and the t-statistic is not significant. The practical conclusion is that the data do not point to any substantial difference between male and female mean education at any given log salary. This result agrees with the tabular comparison in the previous section that showed roughly the same mean educational levels for males and females within each salary group. The adjusted R^2 statistic of 0.558 indicates that approximately 56% of the variation in educational levels is associated with the fitted relationship to salaries and sex.

It is helpful to write the estimated reverse regression model in a form that parallels the direct regression model. Let Fitted($X \mid Y$) represent the mean educational level for a given salary Y as given by the reverse regression model. Then, the fitted model can be written

$$Y = -(a^*/b^*) + (1/b^*)\text{Fitted}(X \mid Y) - (c^*/b^*)\text{Sex}. \tag{4}$$

Equations (2) and (4) have the same appearance, but Y in (4) does not equal Fitted($Y \mid X$) in (2), and the coefficients in the two equations will be different unless the data are fitted perfectly by both models (Dempster, 1984).

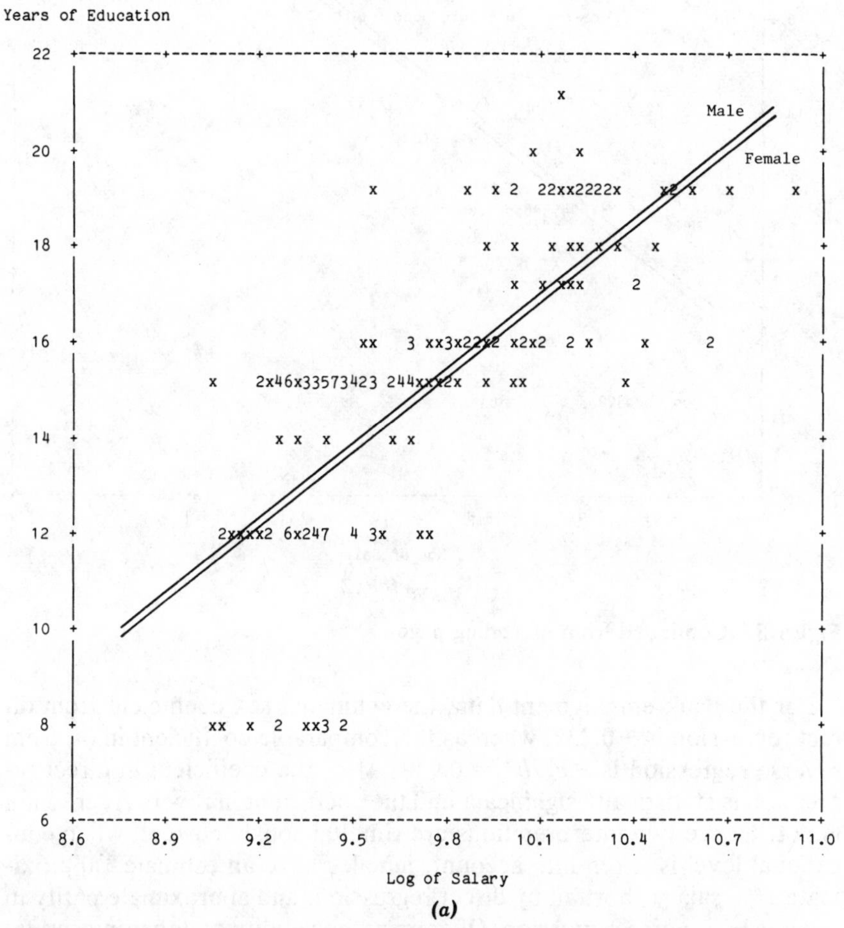

(a)

Figure 3. Simple reverse regression fit of educational level by log salary for employees in bank cohort. (*a*) White male employees. (*b*) White female employees.

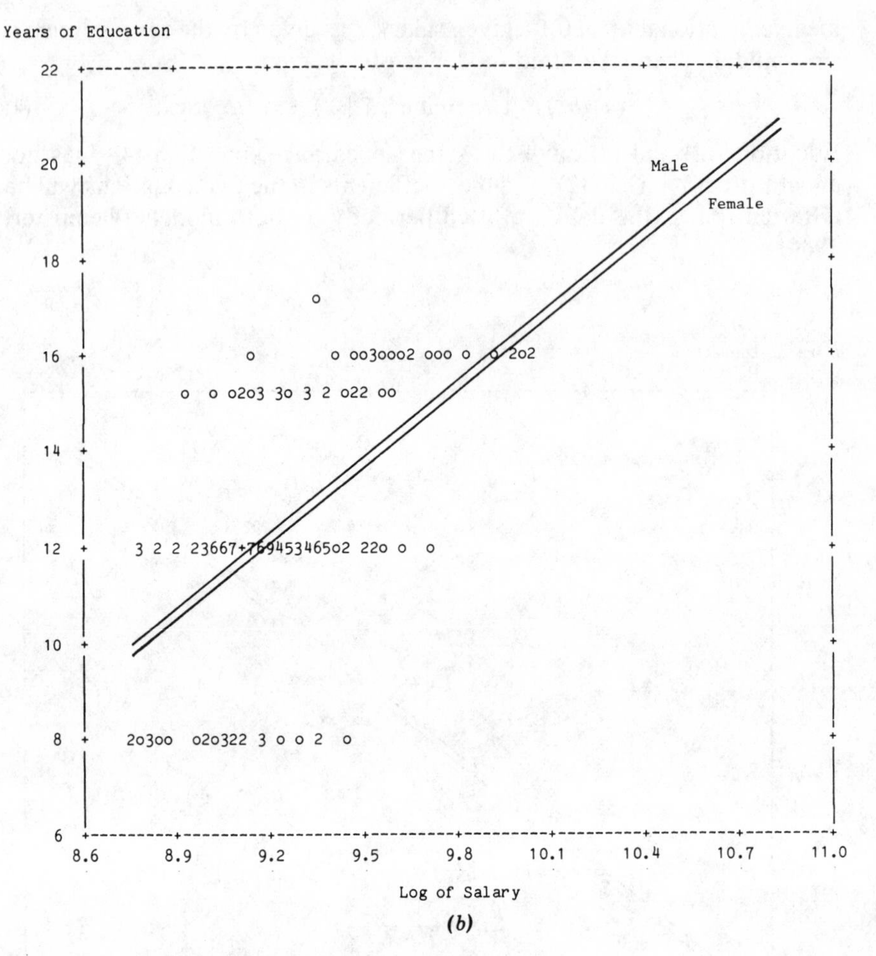

(b)

Figure 3. (Continued from preceding page).

For the bank employment data, the estimated sex coefficient from direct regression is -0.232, whereas the comparable coefficient in (4) from reverse regression is $-c^*/b^* = 0.040$. Also, the coefficient in direct regression is statistically significant and the coefficient in reverse regression is not. Hence two interpretations are simultaneously correct: when educational level is taken into account, females have an estimated approximate 23% salary shortfall by direct regression, and approximate parity in salaries by reverse regression. Of course, the validity of either interpretation depends on the correctness of specification of each model, that is, on the degree to which the underlying model assumptions are met.

Checks on Regression Model Assumptions

In order to demonstrate that the regression models accurately characterize the relationship of salaries with sex and educational levels, statisticians conduct numerical and graphic diagnostic checks on the model specification. For the bank employment data, the linearity assumption is questionable for both the direct and reverse regressions, as can be seen from Figures 2 and 3. The two parallel lines do not closely approximate the relationship between salaries and education. In Figure 2, actual salaries exceed the fitted salaries on the direct regression lines for low and high educational levels; for middle levels of education, the opposite is true. Similarly, in Figure 3, actual educational levels are higher than the fitted levels on the reverse regression lines for middle salaries and are lower than fitted for higher salaries. A good next step is to attempt to correct this shortcoming by a nonlinear fit, and we shall later explore a method of doing this.

The validity of the assumptions about disturbances can be partially checked by examining scatterplots of the residuals versus fitted values for the direct and reverse regression models; these appear in Figures 4a and 4b, respectively. The graphs show that the mean residuals do not have a zero mean at different fitted values. This reflects an inadequate approximation by the two lines to the nonlinear relationship between salary and education.

Inspection of the data in Figure 1 shows that only 10 levels of education are represented in the data, whereas log salary is on a continuous scale. The discreteness of education causes the appearance of lines and valleys in the graphs in Figure 4. The valleys are vertical for direct regression when education is the predictor variable and diagonal for reverse regression when education is the dependent variable. We expect these lines and valleys in the residual plots, whenever the regression model relates a discrete variable to a continuous one.

Although examination of Figure 4 suggests also a possible tendency for residual dispersion to increase with larger fitted values, the only serious problem exposed by diagnostic checks is the clear indication of nonlinearity in the relationship between education and log salary for both sex groups.

Direct and Reverse Regression Perspectives and the Regression Phenomenon

Apart from the technical problems of possible model misspecification, a more compelling question concerns the conflicting conclusions about ad-

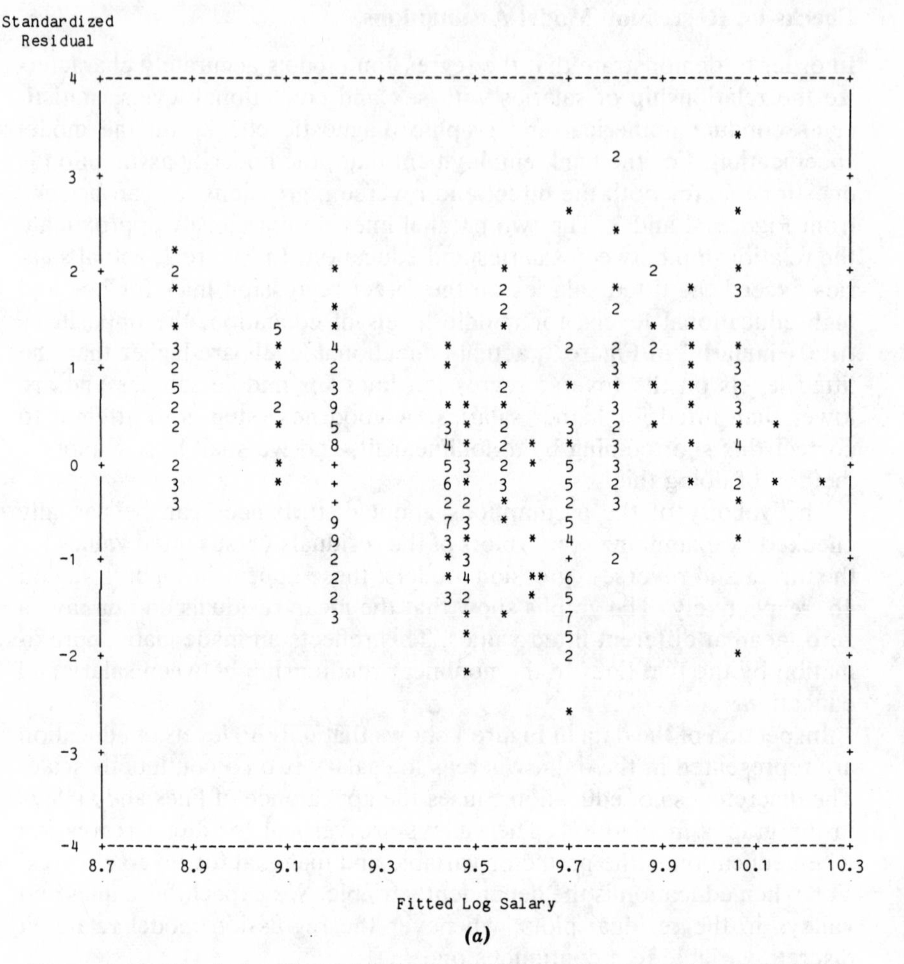

Figure 4. Residual plots from simple direct and reverse regressions in Table 2 for employees in bank cohort. (*a*) Direct regression model. (*b*) Reverse regression model.

justed salary differences between males and females obtained from direct and reverse regression. As a result of different directions of conditioning, direct and reverse regression generally lead to different conclusions about relationships among variables. Studies of alleged discrimination are not unique in this respect. To illustrate, we restrict attention to the white male employees in the bank cohort and consider the relationship of salary to education. The fitted direct regression model is given by

$$\text{Fitted(Salary} \mid \text{Education)} = 8.156 + 0.103\text{Education}$$

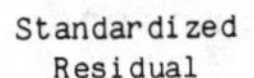

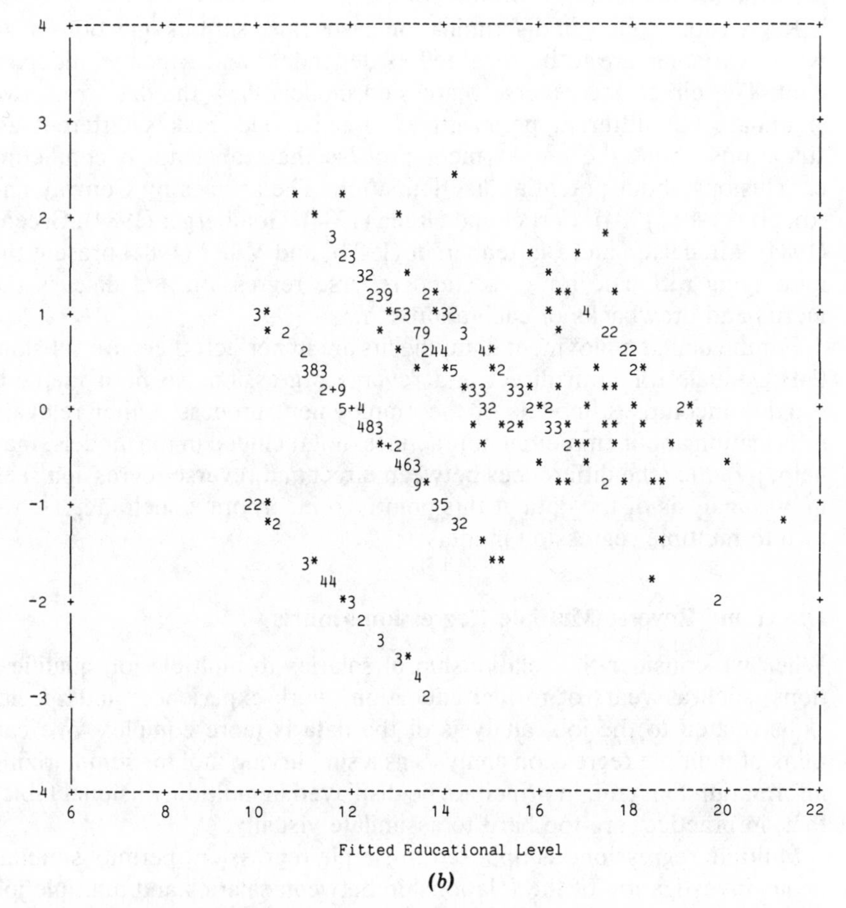

(b)

Figure 4. (Continued).

and the fitted reverse regression model can be written in the parallel form

$$\text{Salary} = 6.770 + 0.196\text{Fitted(Education} \mid \text{Salary)}.$$

Direct regression estimates the increase in salary for each additional year of education as 10%, whereas reverse regression estimates the increase as approximately 20%. The difference between the estimates from the two models is substantial. We are seeing here an illustration of the regression phenomenon, the tendency for the two regression relationships to diverge in the presence of imperfect correlation. If education were a perfect linear predictor of salary, then there would be no conflict between the direct and reverse regression viewpoints. Education would fully predict salaries and the estimated coefficients would coincide.

Regression studies in discrimination cases raise serious questions as to which variables are to be regarded as dependent and which as independent. The direct and reverse regression models view the data from two legitimate but different perspectives. Each model makes different assumptions about the employment process that can lead to conflicting conclusions about potential discrimination. The articles by Conway and Roberts (1983, 1984), Ferber and Green (1984), Goldberger (1984), Greene (1984), Michelson and Blattenberger (1984), and Miller (1984) present the underlying rationale for direct and reverse regression and discuss the merits and drawbacks of each approach.

For the bank employment data, the fits are imperfect; there are substantial residuals for both direct and reverse regression, so both methods provide incomplete models of the employment process. Other relevant information about employment practices, not included in the models, may help to reduce the differences between direct and reverse regression seen in our analysis of the data at this point. To incorporate such factors, we turn to multiple regression models.

Direct and Reverse Multiple Regression Models

When we consider the relationship of salaries to multiple job qualifications, such as years of formal education, work experience, and special skills related to the job, analysis of the data is more complex. We can think of multiple regression analysis as a simplifying tool for summarizing information that can, in principle, be displayed in multidimensional tables that, in practice, are too hard to assimilate visually.

Multiple regression, as opposed to simple regression, permits simultaneous investigation of the relationship between salaries and multiple job qualifications. As applied to discrimination studies, multiple regression models assume that salaries (or log salaries) are related to a simple index of job qualifications. For example, consider two qualifications, such as educational level X_1 and work experience X_2. One summary of the two qualifications is an index, or linear combination, of the form $\alpha + \beta_1 X_1 + \beta_2 X_2$. The coefficients β_1 and β_2 reflect the relative weights attached to education and work experience, respectively, in fitting salaries by means of the index. Estimates of the coefficients in the index are obtained by application of regression to observed data on education, work experience, and salaries for employees in the company.

A fitted index of qualifications now plays the role of the single qualification "education" used in simple regression. The major additional assumption of multiple regression models is that the qualification index adequately characterizes the relationship of salaries to multiple qualifications. Different indices, based on different sets of qualifications or linearizing transformations thereof, can lead to different conclusions about possible discrimination in salary practices.

As with salaries and a single qualification, there are two perspectives for comparing salaries and multiple qualifications. We can condition on qualifications and use direct regression to compare salaries, or we can condition on salaries and use reverse regression to compare job qualifications. So we shall speak to the direct and reverse regression approaches for multiple, as well as simple, regression models.

It is useful to think of the relevant job qualifications as a vector, $X = (X_1, X_2, \ldots, X_p)'$. The individual qualifications X_i might include education, months of work experience, age, and seniority level in the company. We define Sex as an indicator variable that assumes the value 1 for females and 0 for males. As before, Y represents salary or log salary.

To compare salaries for given job qualifications, we write the direct multiple regression model as

$$Y = \alpha + \beta_1 X_1 + \cdots + \beta_p X_p + \gamma \text{Sex} + \varepsilon.$$

This model is often written in more compact form using vector notation as follows:

$$Y = \alpha + \beta' X + \gamma \text{Sex} + \varepsilon. \tag{5}$$

The model states that salaries are related to a linear index of the X's and Sex, where the coefficient vector β and the coefficient γ determine the relative weights of the variables for predicting salaries. The coefficient γ is a particular interest, since it measures the mean difference in salaries between males and females for any values of job qualifications X. The coefficients β_i summarize the relative weights of individual job qualifications X_i in fitting salaries. It may be necessary to transform Y or the X's to achieve closer conformity to the linearity assumption of the model. The standard assumptions apply, including specification of the behavior of the disturbance ε. The coefficients in the model are estimated from the data using the same procedure as for simple regression, that is, minimizing the sum of the squared residuals between actual and fitted salaries. The fitted model is then written

$$\text{Fitted}(Y \mid X) = a + b' X + c \text{Sex}. \tag{6}$$

The second perspective for comparing salaries and multiple qualifications is the reverse regression approach. The objective of reverse regression is to compare qualifications at a given salary for males and females. Unlike simple regression, the dependent variable X is now a vector, consisting of p qualification variables X_i. We say that the reverse regression model is "multivariate" since the dependent variable X represents not a single variable, but a set of variables.

It is worth mentioning that, in reality, the direct regression model is also multivariate because of the multivariate nature of income. The dimensions of employee income typically include a salary and job position, and may include other variables as well, such as a promotion. Thus, one can view the employment process as one in which employees with multiple qualifications receive multiple income variables. In most litigation, income variables tend to be treated separately, and for the present, we shall follow this practice, using "salary" as a generic income variable.

That the dependent variable is a vector of multiple job qualifications introduces a new technical problem, and there are several methods for dealing with it. One method for making reverse regression comparisons is to perform p simple reverse regressions of each job qualification on Y and Sex. These regressions yield separate comparisons of each job qualification at given salaries for males and females.

Several authors have suggested other methods to account for these relationships and analyze the multiple qualifications simultaneously. Kamalich and Polachek (1982) propose a system of regression equations in which each qualification is regressed on salary, sex, and the remaining qualifications. Leamer (1978, Chapter 7) uses the system of p simple regressions of each qualification on Y and Sex to develop a more precise estimate of the qualification differential between males and females at fixed salaries. Birnbaum (1979a) suggests a very different approach where a panel of independent judges evaluates a set of job qualifications and assigns merit ratings. The average merit rating is then used as the dependent variable in the reverse regression model.

A simplifying approach that we have found useful is based on the estimated qualification index from the direct regression model in (6). In terms of causal models of the employment process, Goldberger (1984) delineates the assumptions about the disturbance that justify this approach. We define the estimated qualification index as

$$Q = a + b'X. \tag{7}$$

Intuitively, we can think of the index as an estimate of how the employer prices qualifications. Note that Q differs from the fitted values from direct regression, since it excludes the estimated sex coefficient. This removes the sex effect from the index, so that subsequent comparisons treat all employees as white males. The qualification index is now univariate, and the reverse regression model can be written

$$Q = \alpha^* + \beta^* Y + \gamma^* \text{Sex} + \varepsilon^*. \tag{8}$$

The coefficients β^* and γ^* have interpretations parallel to those for simple reverse regression.

Again, we can write the fitted reverse regression model in a form that parallels the fitted direct regression model in (6):

$$Y = -(a^*/b^*) + (1/b^*)\text{Fitted}(Q \mid Y) - (c^*/b^*)\text{Sex}. \tag{9}$$

In this form, the coefficient $-c^*/b^*$ is the reverse regression analogue to the sex coefficient c in equation (6). Fitted(Q | Y) represents the fitted values from reverse regression, and Y in (9) is actual salary, not Fitted($Y \mid X$) computed from the direct regression in (6). The two sex coefficients will be different unless the qualification index is a perfect linear predictor of salaries. To prevent a possible misunderstanding, we observe that this way of writing the model is intended solely to facilitate comparison between numerical results from direct and reverse regression.

3. Legal Considerations in the Development of Regression Models

The accuracy of multiple regression models for assessing possible employment discrimination depends critically on the data used in the analysis. The focus of litigation may concern the fairness of salary, promotion, or hiring practices by the employer. Litigants may present regression analyses to show how sex, age, education, work experience, or other job-related factors have bearing on these practices.

The courts consider carefully whether statistical data and analysis are properly tailored to the allegations presented, so that inferences about alleged discrimination are relevant and probative to the case. The presiding judge in *Rich v. Martin-Marietta* (1979) noted "in discrimination cases, statistical data should be closely related to the specific issues presented." The appropriateness of the data is often challenged by opposing litigants and these challenges can lead to debate and dispute in court.

Selection of the Employee Population

The legal requirement in Title VII cases that plaintiffs must file a charge within 180 days of the act of discrimination has important implications regarding the employee population selected for analysis. Morris (1979) cites two important Supreme Court decisions that restrict statistical analysis to employees within the time period of legal responsibility. In *Teamsters v. United States* (1977), the Supreme Court ruled that statistical evidence may be deficient if it is based on pre–Title VII employment practices. The Supreme Court introduced additional time-period restraints in *United Air Lines v. Evans* (1977) by equating claims of discrimination not pursued within the 180-day charging period to actions occurring before the passage of Title VII. The Court stated, "a discriminatory act which is not made the basis for a timely charge is the legal equivalent of a discriminatory act which occurred before [Title VII] was passed.'' If no timely charge is filed, an employer is ''entitled to treat the past act as lawful. . . . It is merely an unfortunate event in history which has no present legal consequences.''

However, there is considerable debate as to what constitutes the period of legal responsibility in specific cases. Finkelstein (1980) notes that the employer's practices during the prestatute of limitations period may be relevant to proof of compensation or promotion discrimination, if current employment practices merely perpetuate past inequities. He cites the Supreme Court ruling in *Hazelwood School District v. United States* (1977), stating, ''proof that an employer engaged in racial discrimination prior to the effective date of Title VII might in some circumstances support the inference that such discrimination continued, particularly where relevant aspects of the decision making process had undergone little change.''

The selection of the appropriate employee population for analysis is thus made with regard to the period of legal responsibility. This might entail the identification of a group, or ''cohort,'' of employees hired during the relevant time period and restriction of analysis to the cohort. When evidence is based on the entire work force, the analysis often excludes exempt employees or employees' experience with the company prior to the period of legal responsibility.

Statistical evidence may also be discredited if the employee population leads to inappropriate comparisons. Results from statistical analysis are viewed as suspect if the employee population includes groups that are not regarded as competing. For example, in *Pouncy v. Prudential Insurance* (1980), the court viewed the plaintiff's statistics on black promotion rates as inadequate, since they were based on the company's entire workforce rather than competing groups:

> Within defendant's own workforce, there will be, of necessity, definite groups of employees who are eligible for certain promotions, while other employees without the requisite training and experience will be ineligible for the same promotions. . . . Further, because the company would naturally be eager to utilize the experience obtained by an employee while he worked in a particular division, it is logical to assume that when a position in one division becomes open, those considered for promotion to that position will be persons within that same division. Therefore, as is more fully delineated hereinafter, the promotion awards in each level and area must be carefully analyzed in the context of the relevant promotional pool inasmuch as the question of whether a prima facie case has been established depends on

> "the demonstration of a disparity between a minority's presence in a pool qualified for a position and its representation in the group selected for that position."

The definition of competing groups varies from case to case, depending on the employer's workforce and the specific allegations of discrimination being contested.

The total number of employees also plays a role in determining the population of interest, since the reliability, validity, and usefulness of statistical results vary depending on the size of the population. Inferences based on small groups of employees may yield imprecise comparisons. Judge Sirica summarized the tensions that occur in identifying the appropriate reference population and maintaining an adequate population size.

> To be persuasive, statistical evidence must rest on data large enough to mirror the reality of the employment situation. If, on the one hand, the courts were to ignore broadly based statistical data, that would be manifestly unfair to Title VII complainants. But if, on the other hand, the courts were to rely heavily on statistics drawn from narrow samples, that would inevitably upset legitimate employment practices for reasons of appearance rather than substance. The courts must be astute to safeguard both of these conflicting interests. (*Dendy v. Washington Hospital Center,* 1977)

Statisticians must assess whether the total number of employees in the population is adequate to make reliable statistical comparisons of jobs, salaries, and qualifications.

Legitimate Variables for Regression Analyses

Once an appropriate employee population has been defined and identified, the courts also pay attention to the "legitimacy" of the variables used in statistical analyses. Litigants typically use employee biographical qualifications, such as years of education, work experience, age, advanced degrees, and seniority, to account for salary differences or differential rates in hiring or promotion among employees. These variables are generally not subject to challenge, as long as the qualifications are related to job performance and not subterfuges for indirect discrimination.

An example of a subterfuge appears in the Supreme Court ruling on *Dothard v. Rawlinson* (1977). The plaintiff presented statistical evidence showing that Alabama's requirement for prison guards to weigh at least 120 pounds and be at least 5 feet 2 inches tall excluded a disproportionate percentage of women: the requirement excluded 33% of the women and only 1% of the men in the United States between the ages of 18 and 79.

The Supreme Court ruled in favor of the plaintiff, since the defendant failed to demonstrate that the height and weight requirements were job-related.

Other potentially relevant variables in employment discrimination cases include job content factors, such as job position, level of responsibility, and number of employees supervised, and job performance measurements, such as productivity ratings, output measurements, and supervisor ratings. These variables are not always available; if they are available, the accuracy and interpretation of the measurements may be unreliable. Legal proceedings often exclude the use of positions or evaluations bestowed by the employer, when there is a possibility of "taint" (Finkelstein, 1980). A variable is tainted if there is some presumption that the employer can slant the definition or measurement to favor one group of employees over another.

For example, in the case of *EEOC v. Akron National Bank and Trust* (1980), the defendant used a regression study to explain different promotion rates between males and females in terms of seniority, type of college degree (business or otherwise), absenteeism, and previous part-time status of employees. The court rejected the inclusion of part-time status, deeming it to be tainted:

> Inasmuch as a large percentage of tellers employed by the bank were females, the failure to promote individuals who had previously worked part-time could have as readily been the result of their sex, not the fact that they had previously worked part-time.

Absenteeism was also rejected because of unreliability:

> The Court is concerned that inasmuch as the bank's female employees were assigned to lower level positions, the absentee records for those positions may have been recorded more accurately than for higher positions.

The judge ruled that the defendant had failed to carry the burden of proof.

Regression analyses may also be rejected if they fail to include legitimate variables that are deemed essential to employment practices. For example, in *Presseisen v. Swarthmore College* (1977), a class action suit was brought against Swarthmore alleging discrimination against women in hiring, promotion, and salary. The court rejected the regression studies introduced by both sides for failure to include such factors as scholarship, publications, teaching evaluations, quality of degree, career interruptions, administrative responsibilities, and committee work.

The legal cases just cited point to the difficulty of selecting all relevant variables for statistical analysis, quantifying them appropriately, and excluding irrelevant or tainted variables. Available information about employees from personnel files may also impose severe data limitations. In actual cases, the list of relevant variables is often far short of what is needed for a precise assessment of alleged discrimination practices. This limitation poses serious problems regarding statistical inferences from regression models.

Legal precedents offer some guidelines about which variables may be appropriate. However, legal precedents are often conflicting. What is acceptable as evidence in one case may be discredited in another. For cases involving the use of employment tests, the Equal Employment Opportunity Commission developed the *Uniform Guidelines on Employee Selection Procedures* (1978) to provide minimum, though complex, standards for test validation. The guidelines are restricted to test validation and fail to cover many other types of variables. Kaye (1982, p. 791) points out that the guidelines are also not necessarily definitive, since less demanding forms of validation are often accepted in court.

Because the requirements in specific cases vary, new proposals have been suggested to help ensure the probative value of statistical evidence prior to the collection of data and subsequent analysis. Finkelstein (1973) proposes a two-stage approach in which the court first delimits the data to be analyzed by the opposing parties, specifying the legitimate job-related factors and the population for study. The second stage then focuses on the analysis of the data so delimited. Implementation of such a proposal might sharpen and simplify the primary issues for the court to consider and help to avoid unnecessary costs of litigation.

Development of Multiple Regression Models

Once the relevant employment variables and population of employees have been identified, the statistician develops a multiple regression model to describe the relationships between salary and job qualifications, drawing both on statistical background and judgments about the employment process being studied. All models are to some degree tentative, and any model should always be diagnostically checked for adequacy by inspecting the data for departures from the assumptions. The model-building process entails analysis of the data in such a way that the adequacy of models are checked to suggest better models and bring to light unsuspected relationships or problems with the measurements themselves.

An example will illustrate the process. For the bank employees in Table 1, four basic qualification variables (education, age, previous work experience, and seniority) were regarded as important determinants of salaries. Regression models were used to quantify the relationship of salaries, the four basic qualifications, and sex.

The specification of multiple regression models requires consideration of possible nonlinear relationships between salary and the basic qualifications. We have already seen that diagnostic checks on the simple regressions in Table 2 suggested a nonlinear relationship between salary and education. Consider now the results from two multiple regression models that characterize this nonlinear relationship in different ways. The first model, referred to as Model 1, characterizes the relationship as a quadratic one and uses education and education squared as qualification variables in the direct regression. The second model, Model 2, uses five indicator variables to identify different levels of education, such as college degree or master's degree.

The direct and reverse regression results from Model 1 appear in the first two columns of Table 3. The fitted lines relating salaries to the qualification index appear in Figure 5. The direct regression results show that the estimated coefficients for education, age, and seniority are statisti-

TABLE 3. Multiple Regression Results for Bank Employees from Models 1 and 2 with Different Specifications for Education Effects

	Model 1		Model 2	
Variable	*B* Coefficient	*t* Statistic	*B* Coefficient	*t* Statistic
(A) Direct Regression Models				
Constant	10.126	41.20	9.219	98.04
Sex	−0.174	−5.94	−0.207	−7.85
Educational level	−0.215	−6.80	—	—
Education squared	0.0111	9.70	—	—
Some college work	—	—	0.100	3.48
College degree	—	—	0.539	16.24
Some graduate work	—	—	0.742	9.99
Master's degree	—	—	0.804	10.69
MBA or advanced degree	—	—	0.849	19.08
Age of employee	−0.00049	−3.02	−0.00057	−4.14
Work experience	0.00072	1.53	0.00111	2.70
Work experience squared	−0.0000011	−1.11	−0.0000021	−2.40
Seniority	0.00454	3.81	0.00436	4.13
Adjusted R^2	0.701		0.766	
s	0.227		0.201	
(B) Reverse Regression Models				
Constant	4.111	16.41	3.193	13.40
Sex	−0.0213	−0.99	0.0559	2.75
Log salary	0.576	22.33	0.671	27.35
Adjusted R^2	0.672		0.724	
s	0.171		0.163	

cally significant and the coefficient for previous work experience has borderline significance. The estimated sex coefficient from direct regression is significant, whereas the sex coefficient from reverse regression is not. The adjusted salary difference between females and males is estimated as -0.174, or -16%, from direct regression, and $0.0213/0.576 = 0.037$, or $+4\%$, from reverse regression.

The regression results from the second model appear in the last two columns of Table 3. Model 2 uses five categorical variables to identify specific levels of education and describe the nonlinear relationship with salary. The fitted qualification index and regression lines from Model 2 appear in Figure 6. The qualification index from Model 2 differs from the

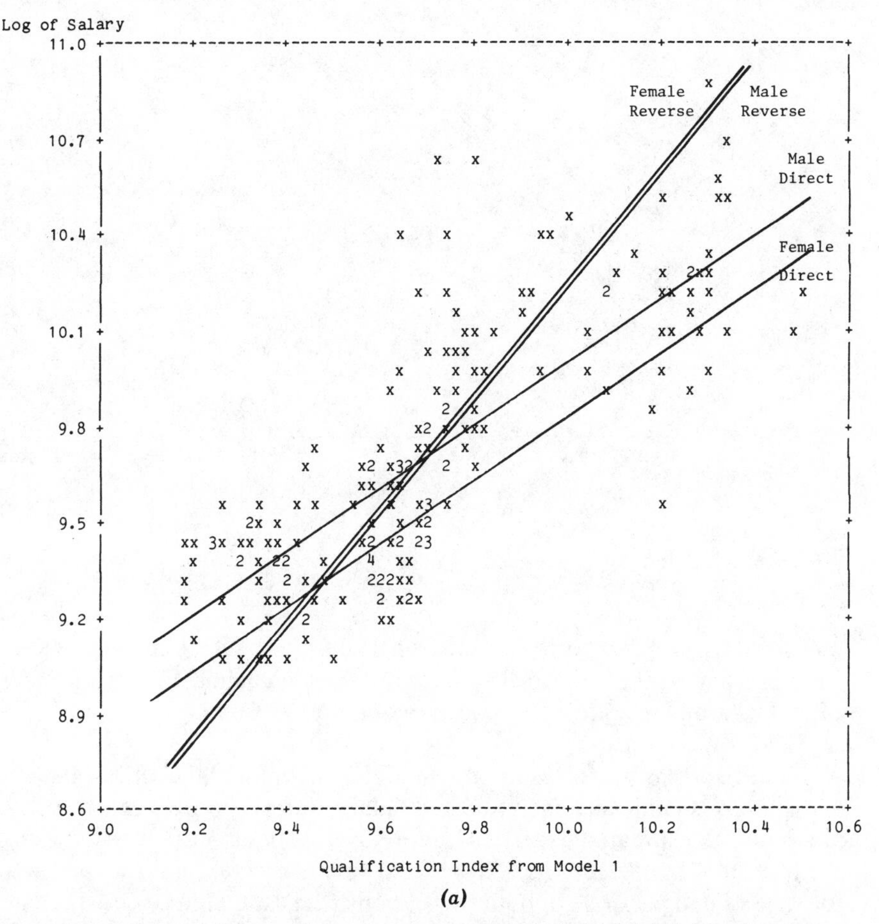

(a)

Figure 5. Multiple direct and reverse regression fits from Model 1 for employees in bank cohort. (*a*) White male employees. (*b*) White female employees.

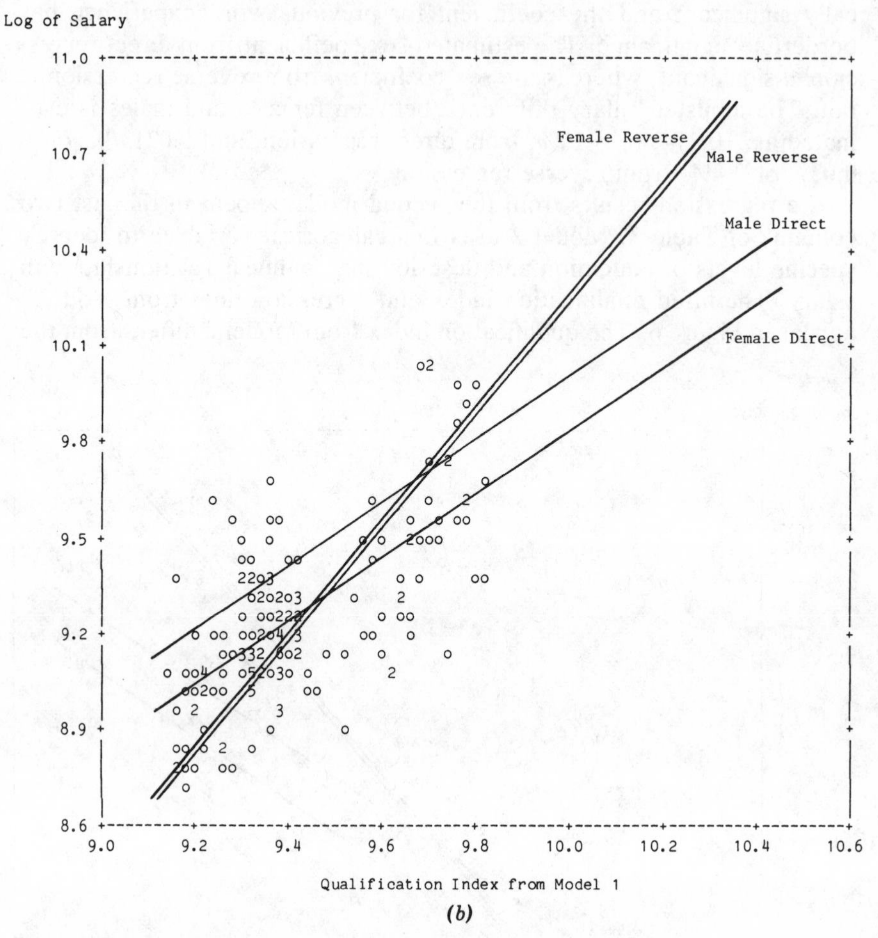

(b)

Figure 5. (Continued from preceding page).

one for Model 1, since it is based on a different set of qualifications. Figures 5 and 6 reflect these differences, since the qualification index from Model 2 results in a greater separation of employees at high salaries from those at low salaries.

The direct regression results for Model 2 show that all coefficients are statistically significant. The adjusted salary difference between females and males is estimated as -0.207, or -19%, from direct regression, and $-0.0559/0.671 = -0.083$, or -8%, from reverse regression, reexpressed for ease of comparison. In both regressions, the estimated sex coefficient is statistically significant.

The differences between the two models concern only the mathematical form of the relationship between salary and education; yet, the regression results are similar in certain respects and different in others. For example, the estimated coefficients for age, work experience, and seniority are approximately the same. Also, the results from both models indicate that salaries increase with educational levels. However, the conclusions about the effects of the sex variable are different. The direct regression sex coefficient depicts an adjusted salary shortfall of 16% from Model 1 and a shortfall of 19% from Model 2. In reverse regression Model 1 estimates an

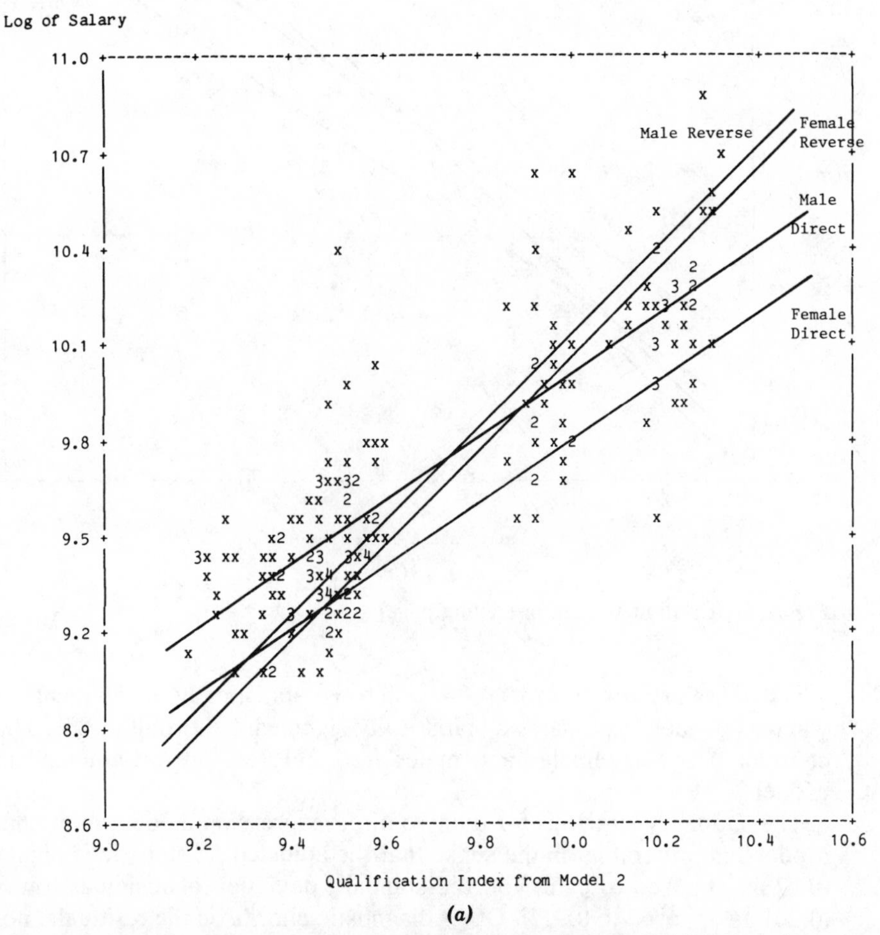

(a)

Figure 6. Multiple direct and reverse regression fits from Model 2 for employees in bank cohort. (*a*) White male employees. (*b*) White female employees.

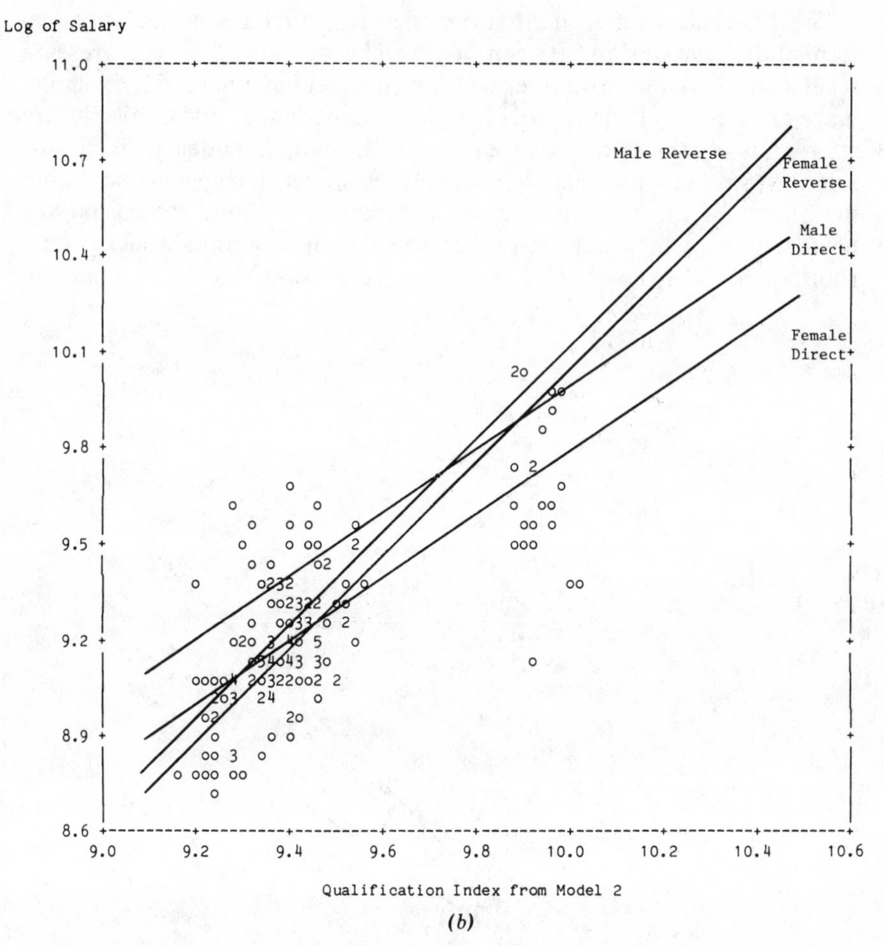

(b)

Figure 6. (Continued from preceding page).

adjusted salary excess of 4%—which is statistically insignificant—whereas Model 2 estimates a statistically significant shortfall of 8%. The contrast between females and males is therefore more pronounced in Model 2.

The summary statistics for the two direct regression models show that Model 2 is preferable, in the sense that the adjusted R^2 statistic is higher (0.77 as opposed to 0.70) and the standard deviation of residuals lower (0.201 as opposed to 0.227). Other diagnostic checks on the residuals, not presented here, suggest that Model 2 better satisfies the underlying assumptions of the regression model and is therefore better specified. We can also compare these summary statistics to the corresponding ones for the simple regression of salary on education and sex in Table 2. The simple direct regression model has an R^2 statistic of 0.62 and a standard deviation of residuals of 0.256. Inclusion of the additional qualifications, age, work experience, and seniority has led to somewhat more accurate salary predictions.

Both Models 1 and 2 approximate the nonlinear relationship between salary and education in reasonable ways according to standard statistical practice. Combinations of the two approaches could also have been used. The differences in the forms of the models are relatively minor and would be regarded as technical refinements; in many applications, such refinements have little effect on the salient conclusions. In this application, by contrast, conclusions about possible salary discrimination are sensitive to model specification. This sensitivity leads us to explore the data further and to consider a more complex relationship between salary, education, and sex in the data than the ones we have modelled so far. We explore this possibility in Section 4.

The goal of regression analyses in discrimination proceedings is to obtain an adequate approximation of the employment process. Many models that are tentatively explored during analysis are later dropped from consideration in favor of other models that provide better approximations. In this case, Model 2 displaces Model 1 since the goodness of fit to the data is better, suggesting a more accurate approximation of actual employment practices. But Model 2 is in turn tentative, and other models could displace it using the same criterion. Part of the statistician's job is to ensure that final models are well specified by accepted statistical standards and are relatively insensitive to mild perturbations in assumptions.

Underlying assumptions about the employment process are also critical in discrimination cases. Although a frequent objective in many regression applications is simply accurate predictions from a regression model, the objective in discrimination cases goes beyond prediction and addresses causal inferences about the effects of different factors in employment decisions. We explore these issues in Section 4.

4. Potential Problems with Regression Analyses

Regression analyses in legal discrimination cases examine employee data to detect possible differences in income–qualifications relationships for members of a protected class and the corresponding relationships for other employees. Statistical inferences from the regression models provide an assessment of whether the observed outcomes—which reflect employer's decisions—are fair, and whether they are consistent with nondiscriminatory employment practices. Note that fairness refers to the outcomes themselves, whereas discrimination refers to the employment process that leads to the outcomes. The two ideas are distinct, and both are involved in the statistical analysis of discrimination. We first consider fairness of observed outcomes.

Direct and reverse regression assess two types of fairness in employment practices. For legal cases in which the protected class refers to females, direct regression evaluates whether the conditional distributions of income given job qualifications are the same for both sexes, a concept we have called fairness-1. Reverse regression evaluates whether the conditional distributions of job qualifications given income are the same for both sexes, a concept we have called fairness-2. (For further discussion of fairness, see Conway and Roberts, 1983, 1984.)

Our illustrative data set has shown that the perspectives of direct and reverse regression can lead to different conclusions about the fairness of observed outcomes. For example, females may have lower mean salaries for given qualifications (unfairness-1), while males and females may have equal qualifications for given salaries (fairness-2). Either concept of fairness is defined with respect to a given group of employees that is the focus of litigation, not to some underlying process or population. The concept of fairness does not entail statistical inferences in the usual technical sense, unless the employees studied are a sample from a larger finite population of employees, which is not typical. Fairness is depicted by the descriptive conclusions of a regression analysis.

It follows that assessments of fairness from direct and reverse regression may change if additional qualifications are introduced into a statistical model. The assessments may also change for given qualifications if the form of the relationship between income and qualifications is changed by the statistical analyst. But fairness is always depicted by a statistical description of the outcomes reflected in the observed data.

When we consider modeling the employment process, inferences are needed, and the inferences cannot be based on the data alone. Rather, the data are used in conjunction with a model of employment practices to infer whether the observed outcomes are consistent with nondiscriminatory practices. The model often incorporates economic assumptions about employment practices in the absence of discrimination. As in the assessment of fairness, different models can lead to different conclusions about the employment process.

When regression models in discrimination cases are focused explicitly on the employment process itself, they are often referred to as causal models, since they attempt to draw inferences about employer behavior from observational data. If the regression model adequately describes actual employment practices, then inferences about possible discrimination are likewise accurate.

But there is a major difficulty. Causal models are most convincing when applied to data called "experimental," that is, data collected in a way that is designed to remove confounding factors. The data typically available in discrimination cases are called "observational" to make clear the contrast with experimental data. When we try to base causal models on observational data, we have much less assurance that the model captures the important aspects of the process being studied. Cochran (1965) and Wold (1956) have pointed out difficulties with causal models in observational studies and potential problems regarding statistical inferences from these models. Wold (1956, p. 29) noted

> . . . the theoretical model on which the statistical analysis is based can be specified with less exactness and detail when passing from experimental to observational data, and this leads to a partial reorientation of the statistical techniques. . . . The conclusion reached is that with observational data the statistical analysis becomes more dependent on (a) co-ordination with subject-matter theory, (b) large-sample methods, and (c) checks and tests against other evidence.

Although the courts have recognized the probative value of statistical evidence in employment discrimination cases, they have also proceeded cautiously in the interpretation of regression analyses as accurate descriptions of actual employer behavior. In *Vuyanich v. Republic National Bank* (1980), Judge Higginbotham summarized general misgivings about regression and econometric models by stating,

> Despite their recent recognition, the econometric techniques employed in this case are not discrimination CAT scanners—ready to detect alien discrimination in corporate bodies. It [an econometric model] may reveal shadows but its resolution is seldom more precise. Ultimately the findings of fact here are not numerical products and sums but a human judgment that the facts found are more likely true than not true.

In this section, we discuss several potential problems in drawing statistical inferences about alleged discrimination in employment practices from regression analyses. As before, we illustrate in the context of possible salary discrimination against females, although the discussion applies equally to analyses of alleged hiring and promotion discrimination and to protected classes other than females.

Noncompeting Groups and Disaggregation

In employment discrimination cases, the reference population of employees selected for analysis must be appropriate for study of the relations between income and job qualifications. This includes, but is not limited to, the considerations noted in Section 3, namely, appropriateness to the period of legal responsibility and the allegations being contested.

One aspect of choice of reference population concerns the degree to which all employees are grouped together (aggregated) or subdivided (disaggregated) into departments or job groups. If, for example, the data are aggregated into large groups in such a way that the employees so grouped cannot be regarded as competing for the same jobs, salaries, or promotions, regression comparisons between males and females may be confounded with differences in the range of jobs within a company.

This possibility can be examined by analysis of subgroups within the population. For example, studies that compare the qualifications of males and females within entering job groups can be used to investigate possible discrimination in job placement. Disaggregation also assesses whether a causal model developed for the population reflects the same features of the employment process when applied to subgroups.

We explore disaggregation in the Harris data. In cooperation with the government enforcement agency, the bank classified entering jobs into 20 homogeneous groups based on job titles and levels of responsibility. The results in Table 4 give the distribution of employees for three entering job groups—clerical, security guards, and professional—groups that might be regarded as noncompeting. These groups are consolidations of the 20 groups just mentioned.

The distribution of employees by entering job group show that 110, or 57%, of the males, and 166, or 94%, of the females, entered clerical jobs. There were no females in the bank cohort hired as security guards, and this group of employees was not deemed relevant to the study of alleged discrimination. The number of male employees entering professional jobs was 70, or 36% of the males, and the corresponding number of females was 10, or 6% of the females.

The distributions of employees by educational level are different for the clerical and professional groups. In the clerical group, the number of employees having a college or more advanced degree is 21 out of 276 total employees, or approximately 8%. The corresponding number in the professional group is 76 out of 80, or approximately 95%. The main difference between the education distributions of males and females within the clerical group is that 128, or 77%, of the females have a high school education or less, whereas the corresponding number of males is 36, or 33%. Precise comparison of the education distributions by sex in the professional group is precluded by the small number of females in that group. Finally, the mean years of education for males and females within the clerical and professional groups are roughly comparable, with male employees having a slightly higher average.

TABLE 4. **Distribution of Male and Female Employees in Bank Cohort by Job and Educational Level**

Educational Level	Job Status			
	Clerical	Security	Professional	Totals
	(A) Male Employees			
High school or less	36	14	—	50
Associate degree	5	—	—	5
Some college work	61	—	4	65
College degree	6	—	24	30
Some graduate work	—	—	7	7
Master's degree	1	—	7	8
MBA degree	1	—	25	26
Advanced degree	—	—	3	3
Number	110	14	70	194
Mean years of education given job status	13.9	10.3	17.5	
	(B) Female Employees			
High school or less	128	—	—	128
Associate degree	—	—	—	—
Some college work	25	—	—	25
College degree	12	—	10	22
Some graduate work	1	—	—	1
Master's degree	—	—	—	—
MBA degree	—	—	—	—
Advanced degree	—	—	—	—
Number	166	—	10	176
Mean years of education given job status	12.1	—	16.0	

Before exploring the disaggregated picture for clerical and professional employees, we must deal briefly with the possibility that the disaggregated data may reflect discriminatory employment practices. For example, the disaggregated data may conceal possible discriminatory hiring, such as excluding females from professional positions or shunting of females into lower jobs than qualifications would suggest. If necessary, this possibility would have to be resolved by considering additional data or other nonstatistical evidence. An important characteristic of legal applications is that statistical evidence is focused to address only those issues pertinent to the case. This brings out a sharp distinction with the use of statistical methods in other applications, where the statistician may address a comprehensive set of issues, rather than specific allegations.

Because of the substantially different educational requirements for clerical and professional positions in the bank, we might question whether there is a confounding effect due to entering job in the multiple regression comparisons for Models 1 and 2 discussed in Section 3. The residual plots for the direct and reverse regressions for Model 2 appear in Figure 7. Within each plot, we have identified employees by the three entering job groups. In both the direct and reverse regression residual plots, there is a

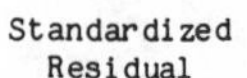

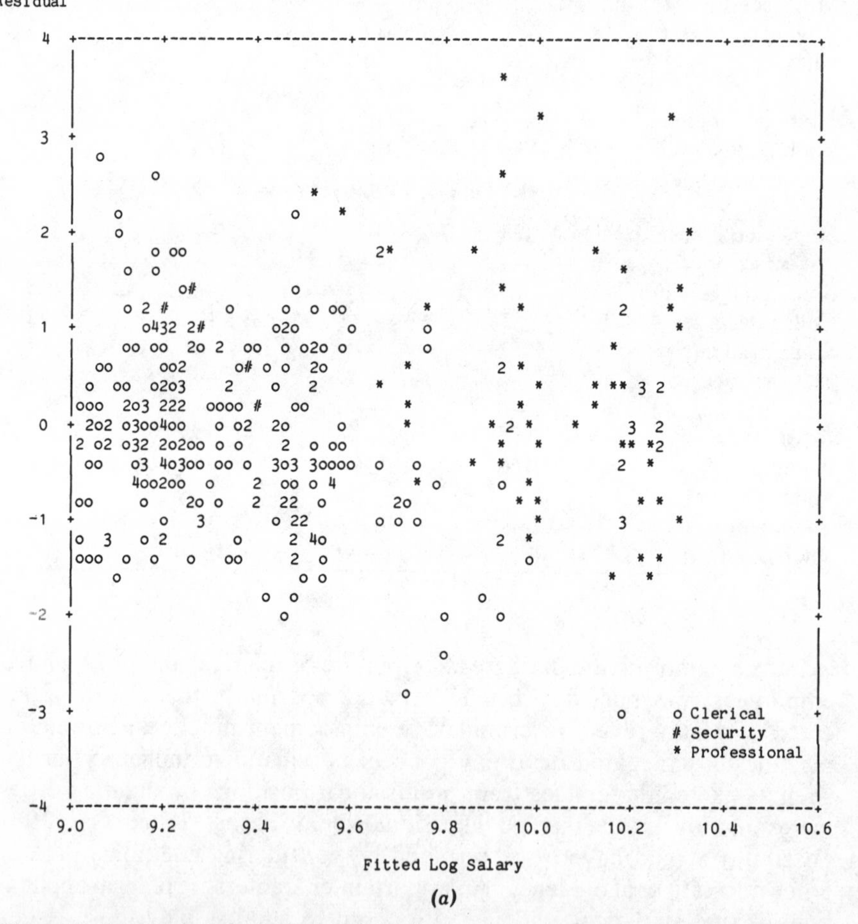

(a)

Figure 7. Residual plots from multiple direct and reverse regressions for Model 2 with employees identified by entering job group. (*a*) Direct regression model. (*b*) Reverse regression model.

clear separation of the clerical and professional employees. The clerical employees tend to have much lower job qualifications and salaries than the professional employees. This separation has a direct effect on the adjusted R^2 statistic from either regression, since the wide range of jobs tends to inflate the variability of the dependent variable (income or the qualification index) and also the variability of the fitted values. The result is a tendency for the regression lines to be steeper and the R^2 to be artificially high in comparison to the R^2 values associated with the separate regressions for each job group (see, e.g., Barrett, 1974). The residual plots for the direct and reverse regressions for Model 1 show the same separation and are not reproduced here.

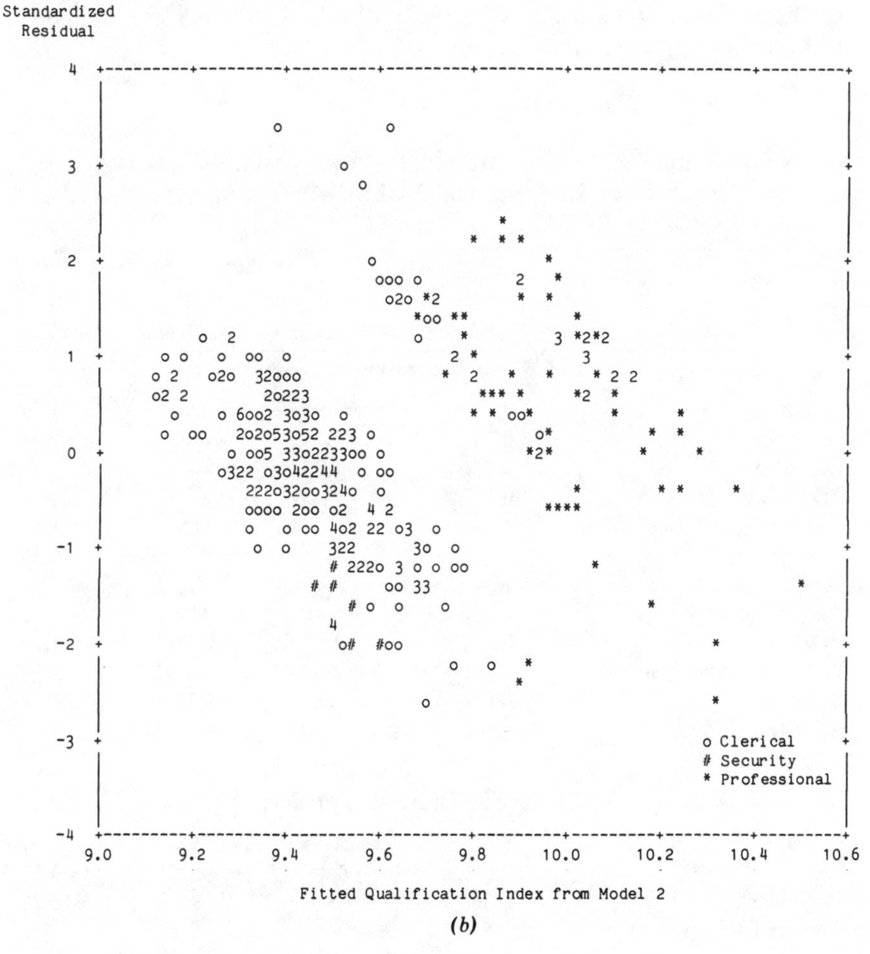

(b)

Figure 7. (Continued).

We can disaggregate, or stratify, the data, and fit separate regressions for the clerical and professional groups. The direct and reverse regression results for Models 1 and 2 for the clerical group appear in Table 5 and the corresponding results for the professional group appear in Table 6. In Model 2, the educational categories are slightly different from those in Table 4, because not all levels of education apply to the two groups.

Within both the clerical and professional groups, the estimated direct regression coefficients for age, work experience, and seniority in Models 1 and 2 are now in close agreement, and they differ from the corresponding coefficients for the entire cohort. Similarly, the reverse regression coefficients for salary more closely agree. The specific form of the nonlinear relationship between education and salary has much less of an effect on the reverse regression sex coefficient in the individual groups than it had on the aggregate comparisons in Table 4.

TABLE 5. **Multiple Regression Results for 276 Employees in Clerical Group from Models 1 and 2 with Different Specifications for Education Effects**

	Model 1		Model 2	
Variable	*B* Coefficient	*t* Statistic	*B* Coefficient	*t* Statistic
(A) Direct Regression Models				
Constant	9.772	35.98	9.228	98.45
Sex	−0.115	−4.29	−0.148	−5.52
Education	−0.136	−3.32	—	—
Education squared	0.00720	4.35	—	—
Some college work	—	—	0.115	4.34
College degree	—	—	0.370	8.25
Graduate work	—	—	0.507	4.84
Age of employee	−0.00062	−4.48	−0.00066	−5.06
Work experience	0.00125	2.72	0.00138	3.10
Work experience squared	−0.0000030	−2.87	−0.0000034	−3.38
Seniority	0.00418	3.82	0.00403	3.74
Adjusted R^2	0.495		0.515	
s	0.181		0.177	
(B) Reverse Regression Models				
Constant	6.015	21.85	5.752	20.59
Sex	−0.0423	−2.80	−0.0025	−0.16
Log salary	0.364	12.49	0.391	13.25
Adjusted R^2	0.479		0.453	
s	0.108		0.110	

The results concerning possible salary discrimination in the clerical and professional groups differ from those for the entire cohort. This is apparent from the fitted regression lines for each model in Figures 8 and 9. The direct regression lines are closer together for both the professional and clerical group, indicating a reduced estimated salary shortfall for females. The reverse regression lines show that male qualifications are either indistinguishable from or higher than those of females at the same salary.

For Model 1, the direct regression adjusted salary difference shows an approximate 16% shortfall for females in aggregate, and 11% and 14% shortfalls for female clericals and professionals, respectively. For Model 2, the estimated sex coefficient from direct regression shows an approximate 19% female salary shortfall in aggregate, and corresponding 14% and 12% shortfalls for female clericals and professionals. Stratification by job group has thus led to a reduction in the adjusted female salary shortfalls observed for the entire cohort.

TABLE 6. Multiple Regression Results for 80 Employees in Professional Group from Models 1 and 2 with Different Specifications for Education Effects

	Model 1		Model 2	
Variable	*B* Coefficient	*t* Statistic	*B* Coefficient	*t* Statistic
(A) Direct Regression Models				
Constant	−2.783	−0.63	9.060	26.58
Sex	−0.150	−1.85	−0.121	−1.42
Educational level	1.307	2.60	—	—
Education squared	−0.0355	−2.47	—	—
Some graduate work	—	—	0.188	1.98
Master's degree	—	—	0.161	1.63
MBA or advanced degree	—	—	0.231	3.60
Age of employee	0.00212	1.87	0.00207	1.85
Work experience	−0.00104	−0.71	−0.00107	−0.72
Work experience squared	−0.0000036	−0.76	−0.000004	−0.82
Seniority	0.00271	0.95	0.00250	0.85
Adjusted R^2	0.361		0.325	
s	0.214		0.220	
(B) Reverse Regression Models				
Constant	6.948	12.94	7.238	13.85
Sex	−0.0634	−1.49	−0.1011	−2.43
Log salary	0.315	5.95	0.286	5.56
Adjusted R^2	0.389		0.404	
s	0.116		0.113	

The estimated sex coefficients from reverse regression are consistently negative (i.e., consistent with lesser qualifications for given salary) for both models and both entering job groups. This translates as follows into statements that can be compared with direct regression. For Model 1, reverse regression shows an adjusted salary excess for females of approximately 4% in aggregate, 11% for clericals, and 19% for professionals. For Model 2, there is an adjusted salary shortfall for females of approximately 8% in aggregate. The disaggregated results, however, show an adjusted salary excess of approximately 1% for clerical and 30% for professional females.

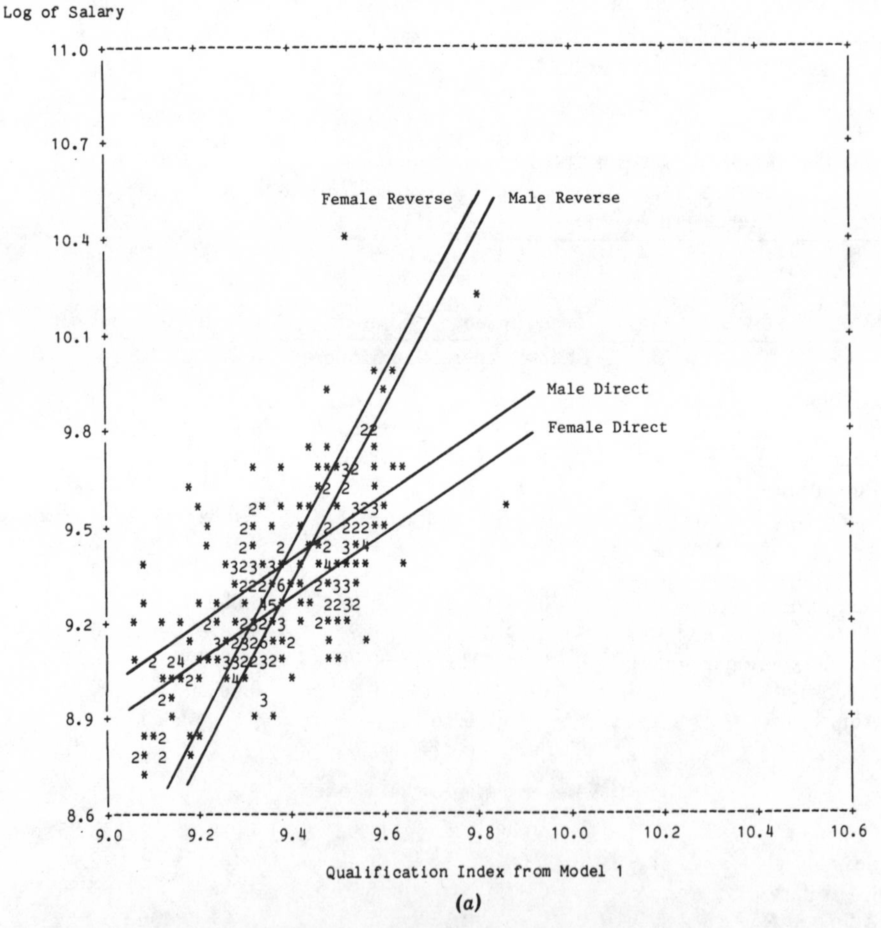

(*a*)

Figure 8. Multiple direct and reverse regression fits from Models 1 and 2 for clerical employees. (*a*) Model 1 regressions. (*b*) Model 2 regressions.

The standard deviation of the residuals for the direct regression of Models 1 and 2 are about 10% lower for the clerical group and about the same for the professional group as the corresponding statistics for the entire cohort. As a consequence, salary predictions from the disaggregated direct regressions have at least as high precision as those based on the aggregate regressions. The adjusted R^2 statistic drops sharply, however, for the disaggregated regressions. The direct regression results for Model 1 show that R^2 declines from 70% in aggregate to 50% for the clericals and 36% for the professionals. Similarly, for Model 2, the adjusted R^2 drops from 77% in aggregate to 52% for the clericals and 32% for

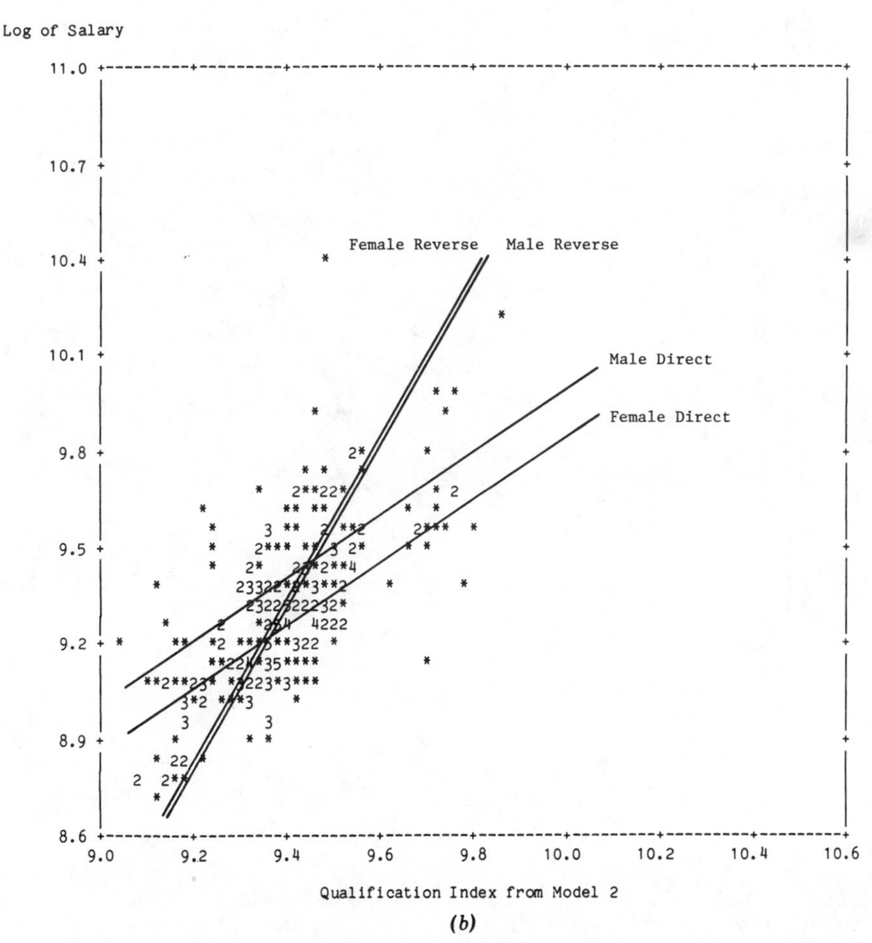

(b)

Figure 8. (Continued).

the professionals. On a disaggregated basis, a much smaller proportion of the variability of log salaries is "explained" by the regression.

Cochran (1965) and Cochran and Rubin (1974) point out the importance of disaggregation in observational studies for comparing treatment effects, since comparisons across homogeneous groups have less bias. In legal studies of discrimination, disaggregation is also motivated by economic reasons, since comparisons within groups regarded as competing for the same jobs, salaries, or promotions are generally more meaningful. The results given above suggest also that the disaggregated comparisons

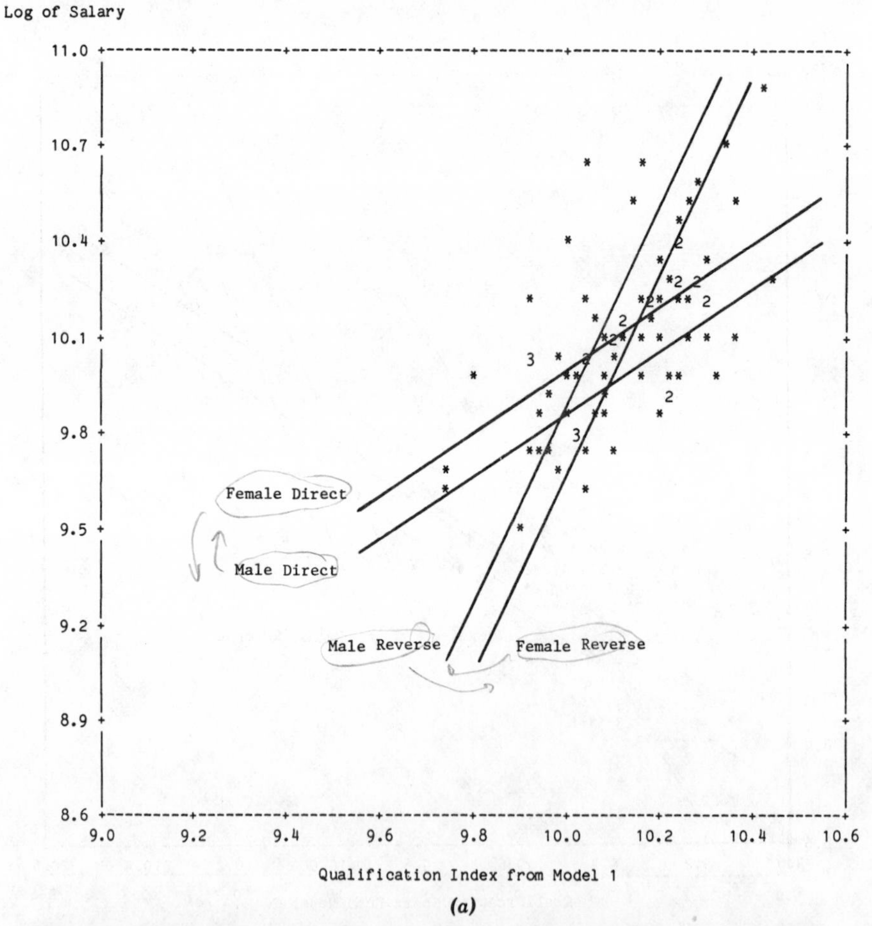

(a)

Figure 9. Multiple direct and reverse regression fits from Models 1 and 2 for professional employees. (*a*) Model 1 regressions. (*b*) Model 2 regressions.

of male and female salaries and qualifications may be more robust to perturbations in the model assumptions.

The usefulness of disaggregation for reducing potential bias in comparisons is counterbalanced by two other considerations. First, disaggregation may result in very small subgroups of employees, leading to comparisons that would not be regarded as statistically significant even if large true differences exist. Second, disaggregation may introduce what is known as a selection bias in estimating the multiple regression model (Heckman, 1979). This problem refers to bias in parameter estimates stemming from methods of disaggregation that may give truncated values

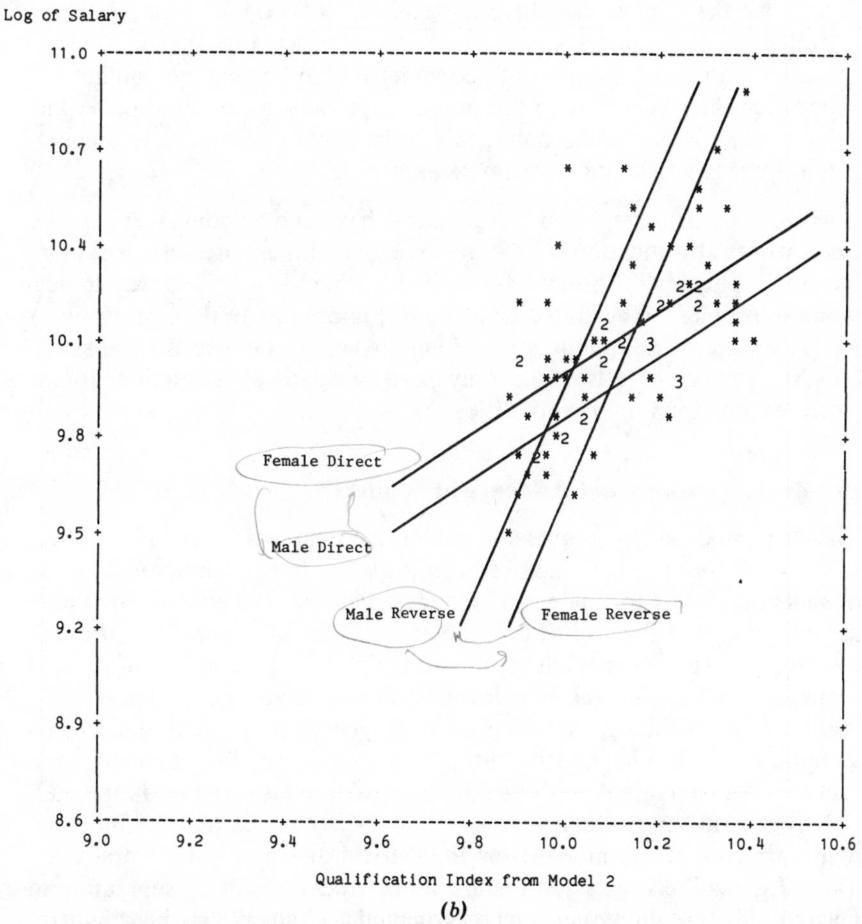

(b)

Figure 9. (Continued).

for the dependent or independent variables, by comparison with the full range of variation in the population.

The issue of selection bias was introduced in the case of *Vuyanich v. Republic National Bank* (1980). The defendant's expert statistician fitted separate regressions to entering clericals and professionals in the cohort of all employees hired by the bank in 1978. The plaintiffs' expert statisticians objected to the disaggregation because of the possibility of truncation bias in the separate regressions. The defendant's expert rebutted by stating the stratification was based on a variable external to those used in the multiple regression models and unlikely to create such a bias. After hearing the various objections, Judge Higginbotham ruled,

> Because the controversy here appears to center on an issue on the frontier of econometrics, and there seems, at least to a court unschooled in the intricacies of econometrics, to be genuine conflict between the experts as to the proper approach, we do not decide the issue. . . . More to the point, this court is not institutionally competent to resolve such a mathematical dispute.

The use of disaggregation in legal cases has become controversial. For cases where stratification is both meaningful and there are sufficient numbers of employees for precise comparisons, there is a natural way to deal with the problem. The statistician can fit regressions to the aggregate, as well as disaggregated, population of employees. If the results are dissimilar, then further investigation, only partly statistical, is required to explore reasons for the dissimilarities.

Parallel Regressions and Influence of Outliers

The direct and reverse regression models in equations (6) and (8), respectively, both assume that salaries have the same linear relationship to the qualification index for males and females. Possible sex effects in the data are reflected in a fixed constant, the sex coefficient, rather than through interactions with other qualification variables. The assumption of parallel regressions for males and females may also be challenged in court.

If there are sufficient numbers of employees, the parallel-relations assumption can be checked by fitting separate regressions to males and females. An equivalent approach includes additional variables in the multiple regression that describe interactions between sex and the independent variables of the model. For the clerical group of bank employees, Table 7 gives the regression results when Model 2 is fitted separately to male and female employees. The estimated coefficients in the direct regression model for education and seniority are very close for males and females. The sign and magnitude of the age and work experience variables

TABLE 7. **Multiple Regression Results for Clerical Group When Model 2 Is Fit Separately to Male and Female Employees**

Variable	Males *B* Coefficient	Males *t* Statistic	Females *B* Coefficient	Females *t* Statistic
(A) Direct Regression Models				
Constant	9.252	46.81	9.067	79.03
Some college work	0.110	2.64	0.102	2.74
College degree	0.337	3.74	0.377	7.26
Graduate degree	0.533	3.68	0.448	2.50
Age of employee	−0.00048	−0.86	−0.00080	−5.81
Work experience	0.00023	0.24	0.00239	4.10
Work experience squared	−0.00000097	−0.60	−0.0000060	−4.05
Seniority	0.00368	2.08	0.00460	3.28
Adjusted R^2	0.285		0.433	
s	0.192		0.168	
(B) Reverse Regression Models				
Constant	6.321	14.77	4.998	13.96
Log salary	0.331	7.31	0.457	11.75
Adjusted R^2	0.325		0.454	
s	0.107		0.111	
Sample size	110		166	

are roughly the same. The standard deviation of the residuals for direct regression is lower for females, suggesting more accurate predictions of salaries. Finally, the estimated coefficients for log salary in the reverse regressions are comparable. On balance, there are no serious violations of the parallel-relations assumption for either direct or reverse regression.

In applications where the parallel-relations assumption is violated, there can be serious difficulties of legal interpretation. As noted by McCabe (1980), separate lines describing the relationship between salary and the qualification index might cross, as in Figure 10. Consequently, females with relatively low qualifications would appear to be underpaid relative to males, while females with relatively high qualifications would appear to be overpaid relative to males. Of course, this problem is related to the issue of disaggregation discussed above.

Failure to model nonlinear relationships adequately may lead to a violation of the parallel regressions assumption. There is a need for diagnostic checks of models at all stages of the analysis, with special alertness to possible nonlinear relationships.

Outliers, or values that would be regarded as atypical if the model were

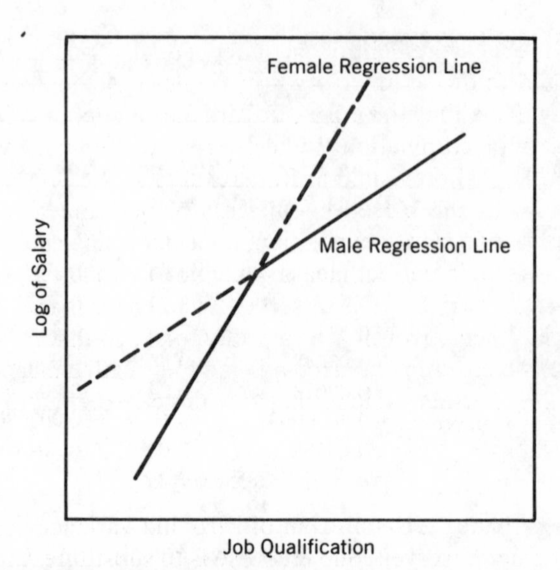

Figure 10. Illustration of separate regression lines for males and females relating salary to a single job qualification.

correct, may also lead to violation of the parallel-relations assumption, as well as other regression assumptions. Outliers should always be checked for clerical errors and, if necessary, corrected as part of the data cleaning process. If the outliers represent legitimate values, however, then the model should be checked to determine the sensitivity of the results to the outliers. The final regression models can be fitted with and without the outlying observations to see if the substantive conclusions depend critically on the outliers. We have also found it helpful to check additional information in employee personnel folders for individuals with outlying values for a possible explanation of the atypical values. This investigation often results in a better understanding of the employment process.

Measurement Errors and Underadjustment

The job qualifications used in regression analyses of employer practices sometimes approximate other measurements that are not directly available. For example, years of schooling might serve as a proxy for level of knowledge, which may have a more direct bearing on job performance. Absenteeism data may be used to indicate employee dependability. Years of additional training may reflect employee motivation and interest in the job.

Cochran (1968), Lord (1967), and Smith (1957) pointed out that the estimate of a group treatment effect from regression can be biased if there are errors in the measurement of the independent variables. Econometricians often refer to this as an errors-in-variables problem (see, e.g., Johnston, 1972). In explaining the errors-in-variables model to nonstatisticians in the context of discrimination studies, we have often used variations on a simple hypothetical presented in Roberts (1980). We present a theoretical explanation for the biased coefficients in the context of the direct regression model. To clarify exposition, consider a single job qualification variable X that is an imperfect measurement of a variable P, reflecting job performance. The variable X is observed, whereas P is not. Let η represent the measurement error in X as a proxy for P, so that $P = X + \eta$. We assume that the measurement error η is random, with a mean of zero and variance σ_η^2. We further assume that P is directly related to salary Y by the simple regression

$$Y = \alpha + \beta P + \gamma \text{Sex} + \varepsilon, \tag{10}$$

where the disturbance ε has a mean of zero and variance σ_ε^2.

Because P is not observed, one recourse is to substitute X for P and use direct regression on the observed data to obtain estimated coefficients a, b, and c for the parameters in equation (10). This substitution results in a biased estimate of the slope coefficient β. Assuming that X and ε are uncorrelated, Cochran (1968) shows that the estimated coefficient b has an expected value of $E(b) = \beta r$, where r is defined by $r = \sigma_\varepsilon^2/(\sigma_\varepsilon^2 + \sigma_\eta^2)$. The reliability coefficient r measures the accuracy of using X as a proxy for P in the regression model. The bias in the estimated slope coefficient is then directly proportional to the reliability of X as a proxy for P. When there is no measurement error, r equals 1, and the estimated coefficient b is an unbiased estimate of the slope β.

Even more relevant to regression studies of employment discrimination is the fact that the estimated sex coefficient is likewise biased. The estimated sex coefficient c has an expected value of

$$E(c) = \gamma + \beta(X_f - X_m)(1 - r), \tag{11}$$

where X_f and X_m refer to the observed mean X for females and males, respectively. Because of measurement error, direct regression fails to adjust completely for observed differences in X between males and females, and hence this problem is referred to as one of "underadjustment." If X is not a very reliable measure of P, so that r is small, the underadjustment bias can be substantial.

When the female mean of X is less than the corresponding male mean, $X_f - X_m$ is negative and the result is a downward bias in the estimate of the sex coefficient from (11). The bias works to create the appearance of discrimination if there is no discrimination and to intensify the appearance of discrimination if true discrimination exists. Unreliable measures of job performance accentuate the apparent discrimination when measured by direct regression, even if no discrimination exists. The extreme case attributes the overall difference in male and female salaries completely to discrimination.

An early rationale for the use of reverse regression in discrimination cases was that reverse regression is a method for dealing with the underadjustment problem (Roberts, 1980). Because income variables, such as salary and jobs, can be measured without error, reverse regression corrects for the underadjustment bias due to random measurement error in the qualifications. The measurement error becomes part of the disturbance in the reverse regression model and, consequently, the estimated coefficients are unbiased. If the measurement error is not random, then both direct and reverse regression may lead to biased assessments (Goldberger, 1984).

A number of empirical studies suggest that underadjustment bias is present in employment discrimination studies. Borjas (1978), using direct regression models, showed large wage shortfalls for females and minorities after standardizing for occupation and qualifications of employees in the Department of Health, Education, and Welfare. The size of the sex and race adjusted wage differentials are similar to those found by Oaxaca (1973) and Mincer and Polachek (1974) for the economy as a whole. Birnbaum (1979b) and Kamalich and Polachek (1982) used both direct and reverse regression to examine wage differences for females and minorities in other data sets. Their results showed large salary shortfalls for the protected class from direct regression and approximate parity from reverse regression.

Similar findings have arisen in studies of organizations that have been defendants in employment discrimination cases. A controversial question is whether such differentials should be attributed to discrimination or to limitations of statistical methodology common to these applications.

The courts are beginning to consider the problems of measurement error and underadjustment bias that may affect statistical evidence in legal discrimination cases. In *Vuyanich v. Republic National Bank* (1980), the plaintiffs' expert statisticians used age as a proxy for work experience in their direct regression models assessing compensation practices of the bank. The defendant's statisticians challenged the regression results on the basis that age tends to inflate the actual work experience of females and presented data showing that the proxy overstates female experience by approximately 30% on average. Commenting on these results, Judge Higginbotham stated:

> Given the seriousness of this proxy error, the plaintiffs should have used actual measures of general experience, or the plaintiffs should have met the burden of showing that such a serious proxy error did

not sufficiently affect their assertions of discrimination. Plaintiffs did neither. . . . The plaintiffs have failed to establish a prima facie case of compensation discrimination for females, either exempt or nonexempt, for any period of time.

Omitted Variables

In many regression studies of discrimination, not all pertinent job qualifications are available to the statistician. Indeed, the job qualifications actually available typically comprise a very incomplete listing of pertinent qualifications for different jobs. Rarely, for example, is any measure of performance included among available qualifications. In many employment discrimination cases, litigants challenge the results of regression analyses for failure to include important qualifications that should be used in statistical adjustment (Finkelstein, 1980).

Models of the employment process are generally crude approximations to the factors that enter into actual employment decisions. There is always concern that important factors used by the employer to arrive at salary and job decisions may be omitted from the model.

To demonstrate the problems created by such omission, we consider a different perspective of the data for the bank employees. We restrict attention to male employees only, to remove any effect due to possible sex bias in salaries. The data show that the difference in log salaries between male clerical and professional employees is $9.450 - 10.143 = -0.693$, yielding an approximate 50% shortfall. We might consider how much of the difference can be explained by the same observed qualifications used to account for salary differences between males and females. This should give an idea as to whether the observed qualifications reflect all relevant factors in employer decisions.

Table 8 gives the direct and reverse regression results from Model 2 when the clerical group is regarded as a "protected class" for males in the cohort. After adjusting for education, age, work experience, and seniority, we see that there are substantial salary shortfalls from direct regression and qualification shortfalls from reverse regression. Direct regression estimates the adjusted salary shortfall for clericals as approximately 28%, whereas reverse regression estimates an adjusted salary excess of approximately 95% for clericals. It is important to note that the results may be biased due to underadjustment. Nonetheless, the observed qualifications fail to explain a substantial portion of the observed salary differences between clerical and professional male employees. The results highlight the importance of including all relevant qualifications in regression models of employment practices.

The omission of key factors affecting actual employment decisions raises the possibility that statistical inferences about possible employment

TABLE 8. Multiple Regression Results for Male Employees from Model 2 Regarding the Clerical Group as a Protected Class

Variable	B Coefficient	Standard Error of B	t Statistic
(A) Direct Regression Model			
Constant	9.345	0.1750	53.40
Clerical job	−0.328	0.0638	−5.14
Some college work	0.103	0.0410	2.52
College degree	0.308	0.0736	4.19
Some graduate work	0.481	0.1085	4.44
Master's degree	0.532	0.0988	5.39
MBA or advanced degree	0.541	0.0824	6.57
Age of employee	0.000028	0.00042	0.07
Work experience	0.0000098	0.00068	0.01
Work experience squared	−0.00000043	0.0000011	−0.39
Seniority	0.00440	0.00144	3.05
	Adjusted $R^2 = 0.753$		$s = 0.202$
(B) Reverse Regression Model			
Qual. Index = 7.413 − 0.180Clerical + 0.269Log Salary			
Constant	7.413	0.3257	22.76
Clerical job	−0.180	0.0270	−6.64
Log salary	0.269	0.0321	8.39
	Adjusted $R^2 = 0.769$		$s = 0.102$

discrimination can change or be reversed by further evidence at any stage of the study. Inferences about possible discrimination from regression analyses depend directly on the particular set of qualifications used in the model. If important factors have been omitted, adjusted salary differences from the model may not accurately assess the degree to which discrimination is present in (or absent from) salary practices.

We illustrate the effects that different qualifications have on assessments of fairness by introducing the qualifications from Model 2 one at a time into the regression models. Tables 9 and 10 give the direct and reverse regression results for the clerical and professional employees, respectively, from a hierarchy of models in which each qualification in Model 2 is added to the previous set in succession. For example, the first model in Table 9 uses an indicator for some college; the second uses indicators for some college and college degree; the third uses indicators for some college, college degree, and graduate degree, and so forth. The last model coincides with Model 2, since it includes all the job qualifications used in that model.

TABLE 9. Direct and Reverse Regression Results for Clerical Group When Qualification Variables for Model 2 Are Introduced Successively

Variable Introduced	Sex Coefficient	t	Salary Coefficient	t	Adjusted Salary Difference	Adjusted R^2	s
(A) Direct Regression Models							
Some college	−0.214	−7.08	—	—	−19.3%	0.239	0.222
College degree	−0.204	−7.48	—	—	−18.4	0.380	0.201
Graduate degree	−0.194	−7.24	—	—	−17.6	0.409	0.196
Age of employee	−0.179	−7.06	—	—	−16.4	0.477	0.184
Work experience	−0.181	−6.80	—	—	−16.5	0.475	0.184
Work squared	−0.159	−5.82	—	—	−14.7	0.491	0.182
Seniority	−0.148	−5.52	—	—	−13.8	0.515	0.177
(B) Reverse Regression Models							
Some college	−0.027	−5.73	0.023	2.56	—%	0.196	0.034
College degree	0.008	0.65	0.208	8.45	−4.0	0.234	0.091
Graduate degree	0.008	0.59	0.247	9.45	−3.2	0.280	0.097
Age of employee	0.015	1.00	0.336	11.76	−4.4	0.373	0.106
Work experience	0.017	1.17	0.337	11.77	−5.1	0.371	0.106
Work squared	0.000	0.03	0.359	12.38	−0.1	0.416	0.108
Seniority	−0.002	−0.16	0.391	13.25	0.6	0.453	0.110

TABLE 10. Direct and Reverse Regression Results for Professional Group When Qualification Variables for Model 2 Are Introduced Successively

Variable Introduced	Sex Coefficient	t	Salary Coefficient	t	Adjusted Salary Difference	Adjusted R^2	s
(A) Direct Regression Models							
Some graduate work	−0.304	−3.58	—	—	−26.2%	0.134	0.250
Master's degree	−0.297	−3.45	—	—	−25.7	0.127	0.251
MBA or advanced degree	−0.167	−2.01	—	—	−15.4	0.289	0.226
Age of employee	−0.158	−1.89	—	—	−14.6	0.289	0.226
Work experience	−0.153	−1.88	—	—	−14.2	0.329	0.220
Work squared	−0.135	−1.60	—	—	−12.7	0.327	0.220
Seniority	−0.121	−1.42	—	—	−11.4	0.325	0.220
(B) Reverse Regression Models							
Some graduate work	−0.005	−0.67	0.007	0.75	—%	0.000	0.021
Master's degree	−0.011	−1.04	0.013	1.01	—	0.018	0.028
MBA or advanced degree	−0.080	−2.15	0.207	4.48	47.1	0.311	0.102
Age of employee	−0.086	−2.26	0.217	4.63	48.2	0.329	0.103
Work experience	−0.074	−1.82	0.271	5.35	31.5	0.360	0.111
Work squared	−0.090	−2.19	0.279	5.46	38.2	0.385	0.112
Seniority	−0.101	−2.43	0.286	5.56	42.4	0.404	0.113

The results for the clerical employees in Table 9 shows that the direct regression sex coefficient declines from a female salary shortfall of 19% in the first model to a shortfall of 14% in the last model. The reverse regression sex coefficient is close to zero for all models except the first. For both regressions, as successive qualifications are introduced, the fit of the model to the data improves.

The results for the professional employees in Table 10 are somewhat different. The estimated sex coefficient from direct regression declines from a female salary shortfall of 26% in the first model to 11% in the last. Also, the estimated coefficient in the last two models would not be regarded as statistically significant. The estimated sex coefficient from reverse regression is negative and significant once the indicator for graduate degree has been introduced. The corresponding adjusted salary difference is therefore positive. Note that the indicator of MBA degree is a key predictor variable, since the estimated coefficients from both models and the overall fit change substantially when this variable is introduced, and the results do not change much afterwards.

For the initial models in the table, the estimated salary coefficient in reverse regression measures the strength of association between salary and the set of qualifications. It is equivalent to the within-group sample multiple correlation coefficient (Conway and Roberts, 1984). We can see from the results for the first model in Table 9 that the estimated salary coefficient is 0.023, suggesting a very weak relationship between salary and the indicator for some college. Consequently, the estimated coefficients from this model are subject to large sampling variability and the adjusted salary difference is not reliable. Similarly, the within-group multiple correlations for the first two models in Table 10 are too low to obtain reliable estimates of the adjusted salary difference.

The introduction of additional job qualifications can affect the estimated sex coefficient from direct or reverse regression, as demonstrated by the above results. Goldberger (1984) shows that different assumptions about the distribution of an omitted variable and its relationship to salary can lead to bias in the estimated coefficients from the observed data. The extent of the bias depends on the additional information about salaries provided by the omitted qualification and its correlation with sex given the included qualifications.

The magnitude and direction of the effects that omitted qualifications may have on the assessment of fairness from a regression model are difficult to predict in advance. Assumptions about the distribution of the omitted qualification and its relevance to salary decisions are not to be presumed casually and must be supported by evidence in other data bases or empirical studies. The courts have generally ruled that such assumptions must demonstrate that a real effect has been omitted by excluding the variable and that exclusion systematically biases conclusions about fairness based on the observed data (see, e.g., *Vuyanich v. Republic National Bank,* 1980). Such requirements are reasonable since the effects of omitted variables are often difficult to assess in the absence of direct evidence.

5. The Role of Statisticians as Expert Witnesses

We now focus explicitly on the role of statisticians serving as expert witnesses in discrimination cases. The environment of litigation poses new challenges to the statistician in the role of expert witness. As mentioned earlier, statistical testimony has a substantial tutorial component. The statistician is often faced with a need to communicate statistical results quickly and with a clear interpretation. For this purpose, tables are especially helpful in explaining results from complex models. Along with the need for clear presentation of statistical evidence, the statistician uses professional judgment in the development of evidence.

Judgment in Model Development

Statisticians must exercise professional judgment in deciding how to carry out regression analyses to be offered as evidence. They may conduct tentative or exploratory analyses with various combinations of variables and relationships. Many of these exploratory analyses are superseded by later analyses. Exploration of models continues until the statistician is satisfied with one or more models that appear to bring out the message of the data; for simplicity we call any such a model a "preferred model," meaning that it is preferred in the light of the statistician's professional judgment. Each preferred analysis embraces a unique specification of variables to be included and a mathematical relationship of the variables, together with assumptions about the behavior of residuals from fitted models. The task for the statistician is to develop, explain, display, and defend preferred models.

Alternative regression analyses, even competent and disinterested implementations, can result in substantially different estimated coefficients for critical variables. Under the pressure of adversary proceedings, there may be a temptation to slant estimates in one direction or another. Slanting of estimates can be accomplished without obvious abuse of the regression methodology, and can be virtually undetected unless details of the regression analysis are accessible to the adversary in litigation. Full disclosure is vital, but the meaning of full disclosure is seldom self-evident.

The urgency of disclosure is suggested by the danger of overfitting, in which data exploration is carried to such extremes that the data fit not only the systematic component of the model but part of the random component as well. Further, data exploration can be tendentious in the sense that one can search the data until one finds a model that is especially favorable to the conclusions argued by the adversary who hires the statistician. Even though the preferred model may be the truth, it is not the whole truth.

Competent statisticians who work independently on the same data base and with the same general objectives typically arrive at somewhat different preferred regression models, although the broad conclusions from the models may be similar. The added pressures of litigation often lead to preferred regression models with sharply different conclusions. This may occur even if two competent statisticians, representing the adversaries in a legal case, work independently on the same data base.

Many of the sources of disagreement must be disputed in trials and resolved by judges or juries. But some sources of disagreement turn on factual disputes about the quality of the data and the details of how analyses were actually conducted. These disputes can be safely reduced by adequate disclosure of statistical investigations. Full disclosure helps judges and juries to resolve the disagreements between opposing statisticians, because the statisticians may have less to disagree about. Full disclosure also helps each statistician to criticize the opponent, and these criticisms can lead to clearer perceptions of why the conclusions diverge.

The Disclosure Problem

Disclosure has two aspects: (a) disclosure of methodology and (b) disclosure of preferred models. Disclosure of methodology is often called documentation. Good documentation describes the study in sufficient detail so that a competent statistician would be able to replicate the major numerical results. A pervasive obstacle for documentation, in legal work and elsewhere, is the reluctance to take time to document one's methodology when it seems that the time could be better used in further applications of the methodology and when documentation is often tedious and boring. But good documentation can be achieved if proper incentives exist. There is no conceptual or intellectual difficulty: one simply reports fully what was done.

Documentation of the preferred models in a regression study is intrinsically more difficult. Uniquely preferable regression models are not an automatic by-product of expertise in statistics and the area of application. In the process of data exploration, many models may be fitted and discarded for each one that survives. The only way to avoid trial and error would be to have greater a priori wisdom than most practicing statisticians would claim.

With typical data bases, an enormous amount of exploration is possible. The number of possible regression models is, genuinely, astronomical. The process of data exploration typically leaves behind many superseded models, raising the possibility that a superseded model may suggest different conclusions than those suggested by the final, preferred models. In adversary proceedings, there is the inevitable suspicion that superseded models may have been discarded because their conclusions were unfavorable, rather than that they were statistically unsound. Cynics may think that a defensible model can be found to support any desired conclusion. This view is extreme, but it reflects a genuine hazard.

One possible solution to the problem of superseded models is the retention of computer printouts on all models. Unfortunately, this requirement can lead to a mountain of computer output, even if one exempts output that has been made meaningless by errors of programming or incorrect use of a statistical package. The time required for the opposing statistician to wade through this mountain of printout can be enormous.

A further problem arises because of the increasing use of video display devices in carrying out interactive regression analyses. The statistical results are viewed on a screen, and no printed record may be made unless the model is a preferred one. Hence it is possible to examine and reject models without leaving a trace.

Even though a rigid requirement could be cumbersome and disruptive, the goal should be to make all computer printout available, on request, for examination by the opposing statistician.

Guidelines for Statistical Analysis

One way to avoid the danger of nondisclosure is to formulate rules that guide the data exploration so that tentative models unfavorable to one side of the case cannot be discarded simply because they are unfavorable. There is a measure of agreement among statisticians about desirable rules. A preferred regression model should fit the data well in the sense of small aggregate discrepancies, or residuals, between the values of the dependent variable Y and the fitted values from the regression model. One common measure is the standard deviation of residuals. (Such a criterion cannot be completely unambiguous, however, unless there is a specification of which variables and transformations are to be included.) Moreover, the fit should be adequate in the sense that the residuals are compatible with the assumptions of the regression model. Adequacy of fit can be verified by statistical and diagnostic checks on the residuals and fitted values.

Improved goodness and adequacy of fit in this sense does not predispose toward results that favor one side or the other. In regression studies of alleged discrimination, an improved overall fit of the model has no implication for the sign and magnitude of the sex coefficient that measures the adjusted difference in mean salaries of males and females, but it is reasonable to suppose that models with improved fits may provide better approximations to employer practices.

In addition to goodness and adequacy of fit, parsimony of the model specification also plays a role. Employment practices are complex, and there is much that is not understood about the specific determinations of salary and job decisions, especially at the level of firm. When relevant economic theory and other outside information do not yield precise forms for relationships among economic variables, simple forms of relationships tend to provide better approximations (Zellner, 1979). Parsimony in model specification helps also to avoid the danger of overfitting. Considerations of parsimony, however, should not not preclude consideration of all relevant qualification variables that may have a bearing on salary practices.

Often there will be variations on reported preferred models that could conceivably make an important difference. For example, it is conceivable that use of hourly wage rates instead of total annual salary could lead to different conclusions because of the differential availability of overtime work. One expert statistician may criticize the choice of earnings variable made by the other statistician. Often it is possible to resolve such a dispute by carrying out a regression analysis with the alternative earnings measure, and there is a presumption that, when possible, one should answer directly the "what if?" question by the appropriate statistical analysis, rather than speculate about it.

Reanalysis to Ensure Disclosure

A simple proposal, which may be feasible in some proceedings, would help to ensure adequate disclosure of regression results. Each statistician could be given the opportunity to analyze the other's data base after receiving a preliminary report on the other's regression analysis. (Sometimes it might be procedurally simpler to agree that a certain number of requests for analyses of the other side's data base be honored at an agreed-on cost. This could save an expert from having to learn new computer hardware and software, or from having to go through a painful transfer of a data base from one system to another.)

If there were agreement on a shared data base, reanalysis could be conducted automatically. Assurance of proper disclosure is an important addition to the other potential advantages of a shared data base. But since there is often strong resistance to sharing of data bases, the reanalysis proposal suggested here may often be the best attainable substitute from the standpoint of disclosure.

The attractiveness of the proposal is that armchair criticisms of statistical analyses tend to be unconvincing. All statistical analyses have limitations. A recital of these limitations does not get to the essential question of whether the limitations have a material effect on the conclusions of the analysis.

An active statistical reanalysis, as opposed to passive armchair criticism, can show concretely how the conclusions may change if the disputed data base is analyzed from a different perspective. Reanalysis may isolate the key differences in assumptions and implementation of regression that are responsible for major differences in conclusions by opposing adversary statisticians. The tedium of abstract criticism is replaced by the active examination and comparison of alternative conclusions and by attempts to isolate factors that contribute to the differences.

One important part of reanalysis would be an attempt to replicate the original findings. If replication is not successful, the failure suggests a failure of completeness or clarity of the original disclosure, and a statistician faced with this criticism is obliged to clarify the discrepancy.

Furthermore, the reanalysis offers the potential to develop alternative models from the perspective of the other side in the case. The differences between these models and the original models may help to delineate the essential differences between the two sides. The knowledge that one's data base may be analyzed by an expert on the other side not only disciplines the studies made by statistical expert witnesses, but almost automatically enforces the kind of documentation that is needed for adequate disclosure.

References

Baldus, D. C., and Cole, J. W. (1980). *Statistical Proof of Discrimination,* New York: McGraw-Hill.

Barrett, J. P. (1974). "The Coefficient of Determination—Some Limitations," *The American Statistician,* **28,** 19–20.

Becker, G. S. (1957). *The Economics of Discrimination,* (1st ed.), Chicago: University of Chicago Press.

Birnbaum, M. H. (1979a). "Is There Sex Bias in Salaries of Psychologists?" *American Psychologist,* **34,** 719–720.

Birnbaum, M. H. (1979b). Procedures for the detection and correction of salary inequities. In Thomas R. Pezzullo and Barbara E. Brittingham (eds.), *Salary Equity: Detecting Sex Bias in Salaries among College and University Professors,* Lexington, Mass.: Heath and Co., Chapter 10.

Borjas, G. J. (1978). "Discrimination in HEW: Is the Doctor Sick or Are the Patients Healthy?" *Journal of Law and Economics,* **21,** 97–110.

"Business Necessity Under Title VII of the Civil Rights Act of 1964: A No Alternative Approach," Note (1974). *Yale Law Journal,* **84,** 103.

Cochran, W. G. (1965). "The Planning of Observational Studies of Human Populations," *Journal of the Royal Statistical Society,* **128,** 234–265.

Cochran, W. G. (1968). "Errors of Measurement in Statistics," *Technometrics,* **10,** 637–666.

Cochran, W. G., and Rubin, D. B. (1974). "Controlling Bias in Observational Studies: A Review," *Sankhya A,* **35,** 417–446.

Conway, D. A., and Roberts, H. V. (1983). "Reverse Regression, Fairness and Employment Discrimination," *Journal of Business and Economic Statistics,* **1,** 75–85.

Conway, D. A., and Roberts, H. V. (1984). "Rejoinder to Comments on 'Reverse Regression, Fairness and Employment Discrimination,'" *Journal of Business and Economic Statistics,* **2,** 126–139.

Dempster, A. P. (1984). Alternative models for inferring employment discrimination from statistical data. In P. S. R. S. Rao and J. Sedransk (eds.) *W. G. Cochran's Impact on Statistics,* New York: John Wiley & Sons, 309–330.

Dendy v. Washington Hospital Center, 14 FEP Cases 1773, (D. D.C. 1977).

Dothard v. Rawlinson, 433 U.S. 321, 15 FEP Cases 10 (1977).

EEOC v. Akron National Bank and Trust, 22 FEP Cases 1665 (N.D. Ohio 1980).

Ferber, M., and Green, C. (1984). "What Kind of Fairness is Fair? A Comment on Conway and Roberts," *Journal of Business and Economic Statistics,* **2,** 111–113.

Finkelstein, M. O. (1973). "Regression Models in Administrative Proceedings," *Harvard Law Review,* **86,** 1442–1475.

Finkelstein, M. O. (1980). "The Judicial Reception of Multiple Regression Studies in Race and Sex Discrimination Cases," *Columbia Law Review,* **80,** 737–754.

Fiss, R. (1979). "A Theory of Fair Employment Laws," *University of Chicago Law Review,* **235,** 237–238.

Goldberger, A. S. (1984). "Redirecting Reverse Regression," *Journal of Business and Economic Statistics,* **2,** 114–116.

Greene, W. H. (1984). "Reverse Regression: The Algebra of Discrimination," *Journal of Business and Economic Statistics,* **2,** 117–120.

Gwartney, J., Asher, E., Haworth, C., and Haworth, J. (1979). "Statistics, the Law and Title VII: An Economist's View," *Notre Dame Lawyer,* **54,** 633–660.

Hazelwood School District v. United States, 433 U.S. 299 (1977).

Heckman, J. J. (1979). "Sample Selection Bias as a Specification Error," *Econometrica,* **47,** 153–161.

Johnston, J. (1972). *Econometric Methods,* 2nd ed., New York: McGraw-Hill.

Kamalich, R., and Polachek, S. (1982). "Discrimination: Fact or Fiction? An

Examination of an Alternative Approach,'' *Southern Economic Journal,* **49,** 450–461.

International Brotherhood of Teamsters v. United States, 431 U.S. 324, 14 FEP Cases 1514 (1977).

Kaye, D. H. (1982). ''Statistical Evidence of Discrimination,'' *Journal of the American Statistical Association,* **77,** 773–792.

Leamer, E. E. (1978). *Specification Searches: Ad Hoc Inference with Nonexperimental Data,* New York: John Wiley & Sons.

Levin, B., and Robbins, H. (1983). ''Urn Models for Regression Analysis, with Applications to Employment Discrimination Studies,'' *Law and Contemporary Problems,* **46,** 247–267.

Lord, F. M. (1967). ''A Paradox in the Interpretation of Group Comparisons,'' *Psychological Bulletin,* **68,** 304–305.

McCabe, G. P. (1980). ''The Interpretation of Regression Analysis Results in Sex and Race Discrimination Problems,'' *The American Statistician,* **34,** 212–215.

Michelson, S., and Blattenberger, G. (1984). ''Reverse Regression and Employment Discrimination,'' *Journal of Business and Economic Statistics,* **2,** 121–122.

Miller, J. J. (1984). ''Some Observations, a Suggestion, and Some Comments on the Conway-Roberts Article,'' *Journal of Business and Economic Statistics,* **2,** 123–125.

Mincer, J., and Polachek, S. (1974). ''Family Investments in Human Capital: Earnings of Women,'' *Journal of Political Economy,* **82,** S76–S108.

Morris, F. C. (1979). *Current Trends in the Use (and Misuse) of Statistics in Employment Discrimination Litigation,* Washington, D.C.: Equal Employment Advisory Council.

Oaxaca, R. L. (1973). Sex discrimination in wages. In Orley Ashenfelter and Albert Rees (eds.), *Discrimination in Labor Markets,* Princeton, N.J.: Princeton University Press.

Pouncy v. Prudential Insurance Co., 23 FEP Cases 1349 (S.D. Tex. 1980).

Presseisen v. Swarthmore College, 442 F.Supp. 593, 15 FEP Cases 1466 (E.D. Pa. 1977).

Rich v. Martin-Maretta, 22 FEB Cases 409 (D. Color. 1979).

Roberts, H. V. (1980). Statistical biases in the measurement of employment discrimination. In E. Robert Livernash (ed.) *Comparable Worth: Issues and Alternatives,* Washington, D.C.: Equal Employment Advisory Council.

Smith, H. F. (1957). ''Interpretation of Adjusted Treatment Means and Regressions in Analysis of Covariance,'' *Biometrics,* **13,** 282–308.

Uniform Guidelines on Employee Selection Procedures, 43 FR 38290-38315 (25 August 1978).

United Air Lines v. Evans, 431 U.S. 553, 14 FEP Cases 1510 (1977).

United States Department of Treasury v. Harris Trust and Savings Bank, Case No. 78-OFCCP-2 (1981), pending.

Vuyanich v. Republic National Bank, 24 FEP Cases 128 (N.D. Tex. 1980).

Weisberg, H. I., and Tomberlin, T. J. (1983). "Statistical Evidence in Employment Discrimination Cases," *Sociological Methods and Research,* **11,** 381–406.

Wold, H. (1956). "Causal Inference from Observational Data," *Journal of the Royal Statistical Society, Series A,* **119,** 28–61.

Zellner, A. (1979). "Statistical Analysis of Econometric Models," *Journal of the American Statistical Association,* **74,** 628–643.

Comment

Stephan Michelson
Econometric Research, Inc.
Washington, D.C.

1. Introduction

The Office of Federal Contract Compliance Programs (OFCCP), is mandated to enforce government contractor provisions for equal employment opportunity. That authority was originally "subcontracted" to eight federal agencies, each assigned to monitor related industry groups. Under a reorganization during the Carter administration, all oversight was centralized in the Department of Labor in 1978. This reconstituted OFCCP found itself engaged in a dispute with Harris Bank in Chicago, instituted by the Department of the Treasury, and headed for a hearing before an administrative law judge (ALJ).

OFCCP alleged failure to promote and pay blacks and females equally with whites and males. It proposed that Harris be debarred—prevented from being a federal funds repository—until it paid damages and changed its ways. The bank denied the allegations. I was retained to be the government's primary statistical expert, and Harry Roberts was retained in a similar capacity by the bank.

At the hearing in August and September 1979, I presented cross-section regressions. Observations were active full-time employees at a particular time. The same specification provided similar results when replicated on data from different dates. The dependent variable was the salary rate, and explanatory variables included typical personnel descriptors, as well as race, sex, and race–sex interaction.

For refusing to turn over studies of experts previously hired by the bank, in violation of his order, the ALJ said the bank could not put on a statistical defense. The bank could, however, rebut my testimony. Thus Roberts and others did in fact testify. They argued that my regressions were incorrect because they were traditional, direct regressions, whereas reverse regressions were called for. The ALJ did not accept that criticism, and found for plaintiffs (*Harris,* 1981).

The decision of the ALJ is only a recommendation to the Secretary of Labor, who remanded the case back to the ALJ with instructions to hear an affirmative defense (*Harris,* 1983). The remand in effect calls for a new trial for which, as I write (November 1984), the parties are preparing. I am not participating in these proceedings.

2. Why Regression?

I begin with the distinction between an *event* and a *situation*. Promotion to officer of a corporation, for example, is an event; being an officer of a corporation is a situation. Where or what people are is a situation; getting there requires events. The difference between an event and a situation has important legal and statistical implications.

Discrimination, a legal term, is defined by events, or a summary of events. To discriminate management must *do* something, not just *be* something. And it must do that thing in a recent time period—exactly what period being a matter of dispute in most litigation. Whatever the situation may be, discrimination is found if the defendant "committed a legally cognizable wrong" (*Ste. Marie,* 1981). The focus of statistical analysis should be "actions taken by the employer" (*Furnco,* 1978 at 580). The problem with situation analyses is that in them we do not know whose behavior we are observing or when it occurred. This problem is similar to identifying structural parameters of a model from a reduced form equation (Abowd, 1984).

Analysis Population

Conway and Roberts attribute to "legal considerations" their restriction of Harris Bank data to "all full-time employees on the bank's payroll as of March 31, 1977 who were hired between 1965 and 1977." There are no such legal considerations. "To simplify discussion" they restrict analysis to a "cohort" of (white) employees hired between 1969 and 1971. The limitation to a cohort is not a simplification. It is a method of statistical control for initial job. The relevance of that control, as well as of regression itself, is questionable.

Most legal theories hold that there is a date before which actions of the firm—events—are not at issue. I know of no compensation case, and certainly *Harris* is not one, that excludes persons from the class because they were hired before a certain date, even though they were employees at the time to which the complaint refers. In their situation analysis, Conway and Roberts cannot distinguish events preceding the critical data from events succeeding it, except by elimination of the individuals on whom the earlier events befell. However, those persons were eligible for and possibly harmed by subsequent employer actions. They have been eliminated from the data because regression is blind to the timing of events, not because the law is deaf to their complaints (see *Valentino,* 1982).

Similarly, defining the class to exclude persons hired after 1976 or persons not still employed on 31 March 1977 is a limitation of method, not of law. The cohort method requires that some time elapse between entrance and observation. Finally, we see in this study only one cohort, and no indication how Conway and Roberts would aggregate results from different cohorts to a summary picture to be presented to a judge.

Equal Pay Act Regression

In the Equal Pay Act of 1963 Congress defined the *paying* of wages as an event, if the wages were paid to males and females in the same job in the same establishment (physical location). The event occurs every pay day to active employees. Four exceptions that could justify unequal wages appear in the act [29 U.S.C. 206(d)]:

1. Seniority
2. Merit
3. Productivity based labor payment
4. "any other factor other than sex"

Once it is determined what jobs may be grouped together because they require "equal skills, effort and responsibility," and performance "under similar working conditions," and that workers receive wages, not piece-rate pay, Equal Pay Act allegations surely can be analyzed by regression (Fisher, 1980). The act even specifies a control variable, seniority, and it asks the analyst specifically to consider how to measure "merit." However, an Equal Pay Act regression will use, as observations, coinhabitants of a single occupation at a point in time, a population quite different from that studied by Conway and Roberts.

One might study the effects of seniority in general, and prior jobs (such as first job at the bank) in particular. A priori controls on such variables by stratification of the data set are uncalled for by the law and not helpful analytically. Not only are sample sizes reduced, but one cannot ask, with such a control, to what extent seniority and initial job are determinants of current salary.

If there are no Equal Pay Act (within job) violations, a direct analysis of events is called for. Initial assignment, the setting of wage rates for jobs, transfer, promotion, discipline, and discharge are events. Most analyses of discrimination in these events would not use regression.

Discrimination and Fairness

Conway and Roberts are careful to limit their analysis to allegations of "salary discrimination, although the discussion applies equally to analysis of alleged hiring and promotion discrimination. . . ." That limitation does not justify a situation analysis, and their discussion does not in fact apply to hiring and promotion. In most establishments, salary ranges are specified by job, grade, rank, or other subcategory of employee. The attainment of that subclassification is an event, and most judges want the expert's model of the employment process to distinguish those events from the setting of wages within subclasses (*Valentino,* 1982; *Eastland,* 1983; Baldus and Cole, 1984, Sec. 8.23). Following this principle, it is standard practice in university cases to hold faculty rank "constant," usually by estimating separate regressions at separate ranks (*Sobel,* 1983; *EEOC v. McCarthy,* 1983). Movement among ranks is analyzed separately.

This separation between the salary situation and the promotion (or initial assignment) event is made even in equal pay cases where only salary is at issue, because it correctly represents the nature of wage setting (salary is conditioned on rank) and correctly isolates events. Thus, for example, Borjas's (1978) satirical critique of regression fails as he estimates salaries in a federal agency from background characteristics. He studies neither how openings at high grades were filled, nor how people within a grade were paid.

Conway and Roberts seem to recognize the distinction between events and situations, calling the conclusion from an event analysis "discrimination" (or not), and from the situation analysis "fairness" (or not). Their language creates some difficulty. It is true that we assess the fairness of a coin (a situation), but we also require "fair" tosses (events). Fair employer actions can lead to apparently unequal situations because not all employees want the same jobs. Event analysis can control structurally for employee behavior, whereas situations are the results of employee and employer behavior. To call numerically unequal job or earnings distributions "unfair" when we do not know which employees sought which jobs, or were available to fill which openings when they occurred, is to pervert the meaning of the word. On the other hand, sex-biased acts of the employer that occurred before the legally cognizable time period may be unfair in the sense that most people would recognize, but not illegal. A distinction between fairness and discrimination can be made on the basis of when the events occurred, but both terms should describe events.

Other Regression Analyses

Regression is useful in a number of instances besides Equal Pay Act allegations, for example, analyzing the event "setting initial pay." In the so-called comparable worth case in Washington (*AFSCME,* 1983) I analyzed the state's setting wage scales per job by regression. For compensation, I used the wage of the lowest grade and step combination. Because each job's wage scale describes an action of management, I used jobs as observations. Because the law is about people, not jobs, I also provided regressions in which each job was weighted by its number of employees. I controlled for minimum required education both by variables and by a separate regression on jobs requiring minimal skills (high school or less).

I determined the percent female in a job category from active employees on a given date. The regression was therefore situation specific. It unambiguously described actions of management, however, because wages were set by a management committee. A separate analysis was required to determine whether these situations were so stable (whether a "female" job was consistently female, and whether a highly populated job was consistently highly populated) that one could interpret the results in terms of events. I had to know (and convince the court) that I was identifying the event of the setting of wages *with knowledge* of the typical characteristics of job holders.

Finally, regression can provide an effective "bottom line" presentation, even where events are at issue. If males and females are equally situated, conditional upon their qualifications, the idea that they have been unequally acted upon will be hard to accept (*Sobel,* 1983). Conway and Roberts show the opposite—that at Harris Bank males and females are not equally situated. A reasonable response to such a situation-based finding, in which direct regression indisputably gives significant sex coefficients, would be an events-based (selections) analysis showing that the disparate situation is not the result of illegal management actions (if it is not) (*Harrison,* 1982). *Vuyanich* (1980), often quoted by Conway and Roberts, represents a similar failure by defense experts to cure the fundamental flaw in plaintiffs' argument, reliance on situation analysis. (See also *Segar,* 1978, and *Trout,* 1981.)

3. Why Reverse Regression?

There is no question that there can be a fact situation in which direct regression produces estimates so biased that one incorrectly infers a negative relationship between salary and being female. We have reason to believe that such a situation is rare, and that typically reverse regression is biased against finding a difference between the sexes even when it exists (Ash, 1984; Blattenberger and Michelson, 1983). But it is possible (Goldberger, 1984). In such a situation, reverse regression may provide estimates with less bias and a more correct inference. How does one know which regression produces estimates that should form the basis of the expert's opinion?

A Reverse Regression Model

Consider the following model:

$$\text{SAL} = d_0 + d_1\text{PROD} + d_2\text{FEM} + d_e \tag{1}$$

$$\text{PROD} = b_0 + b_1\text{MEASURE} + b_2\text{FEM} + b_e \tag{2}$$

$$\text{MEASURE} = c_0 + c_1\text{FEM} + c_e \tag{3}$$

where

SAL = salary at hire
PROD = measured post-hire productivity
FEM = 1 if the employee is female, otherwise = 0, and
MEASURE = an indicator of future productivity measurable at hire, and admittedly used by management for initial salary determination.

Plaintiffs estimate the regression:

$$\text{SAL} = a_0 + a_1\text{MEASURE} + a_2\text{FEM} + a_e. \tag{4}$$

I adopt the notational convention that $a_0{}^*$ is the estimate of a_0, etc. Assuming $a_1 > 0$, then $a_2{}^*$ will be biased downward if $b_2 < 0$, biased upward if $b_2 > 0$, and will be unbiased only if $b_2 = 0$ (Goldberger, 1984).

Suppose plaintiffs estimate $a_2{}^* < 0$ from the regression in (4). To report this as his (her) estimate of the "true" relationship, the expert must believe that $b_2 = 0$, or that this is the proper assumption in the absence of a belief about the value of b_2. To offer reverse regression in rebuttal, defendant must have evidence that $b_2 \neq 0$, believe $b_2 \neq 0$, or hold that one should assume $b_2 \neq 0$ in the absence of evidence or belief. Let us take these alternatives in reverse order:

What to Hold in the Absence of Belief or Evidence. If you know nothing about the relationship between sex and productivity, after controlling for all available productivity indicators, I think you are bound to presume that there is no residual relationship in any other estimates (i.e., $b_2 = 0$). Conway and Roberts correctly indicate that courts will adopt this view in this circumstance.

Nonstatistical Bases of Belief. One may hold that, even if there is no statistical evidence, females appear to be "underadjusted" by measured variables. That opinion may be held in a specific case because of nonstatistical evidence. If that be so, the expert should relate the belief and its basis.

Statistical Evidence. If one had a measure of PROD, one could test equation (2) directly. The finding of a negative and "significant" b_2^* would make untenable plaintiff's claim that a_2^* was an unbiased estimate.

Note that although the relationship between MEASURE and FEM in equation (3) does not produce bias in direct regression coefficients, this relationship is not irrelevant to the legal issues. If females have a lower value of any such variable, use of that variable by employers will have a disparate impact. The appearance of a variable "explaining" some of the sex differential in wages is hardly the end of the statistical study. A measure with disparate impact must be justified (validated). If equation (2) can be estimated, then MEASURE can be tested for validity.

Thus an estimate of equation (2) is critical for a regression based defense. First, it may demonstrate that plaintiff's sex coefficient is biased. Second, it may demonstrate valid use of measures that have disparate impact. Since MEASURE is used to set salaries, we regress:

$$\text{SAL} = a_0 + a_1\text{MEASURE} + a_e \qquad (5)$$

and define a variable OTHER = a_e^* from equation (5). OTHER is the difference between salary predicted by a linear function of MEASURE and actual salary. It represents unmeasured or unrecorded information used by management to determine salary, as well as stochastic variation and subjective judgments.

Now we can regress:

$$\text{PROD} = b_0 + b_1\text{MEASURE} + b_2\text{FEM} + b_3\text{OTHER} + b_e. \qquad (6)$$

MEASURE and OTHER are uncorrelated variables that sum to salary, and test the salary determination process for validity against the criterion of productivity.[1] If $b_2^* > 0$, then the firm is relying on productivity indicators to the disadvantage of females. If $b_1^* = 0$ but $a_1^* \neq 0$ or, if $b_3^* = 0$, then the firm is using an invalid variable in setting salary.

[1] It is simple to show that b_2^* would have the same value if SAL replaced OTHER in the regression. Although the text exposition may seem roundabout, it will be far less confusing in court than the inclusion of both salary and one of its constituents as independent variables in the same regression. Courts do not like correlated independent variables (see *Sobel*, 1983).

Kouba v. Allstate

At Harris Bank there is no productivity measure for equation (2). In defending the bank by reverse regression, analysts must believe that the indicator variables underadjust, or they must hold that it is proper to assume underadjustment in the absence of evidence.

If a firm is to argue that it pays for expected productivity, it must be able to validate the indicators used to create that expectation. As shown above, even subjective judgments (OTHER) can be validated. The key is to have a criterion for setting salary. Here, I have considered that criterion to be a measure of productivity.[2] Without such a measure, reverse regression cannot be justified. With such a measure, reverse regression may be shown to be correct; but it is also unnecessary.

In *Kouba v. Allstate* (1984), in which I was the expert for defense, data were available to demonstrate that plaintiff's salary regression produced a biased coefficient for FEM. The appropriateness of reverse regression was not rejected by Goldberger's (1984) test, but the estimated FEM coefficient in the reverse estimate was not necessarily unbiased (because the model did not have the requisite error structure). After investigating sources of error, I concluded that the reverse regression gave a better picture of the relationship between sex and salary than did direct regression.

However, I also calculated the following validity equation:

$$\text{PROD} = -45.25 + \underset{(15.2)}{0.06\text{MEASURE}} - \underset{(4.8)}{23.02\text{FEM}} + \underset{(15.7)}{0.21\text{OTHER}} \qquad (7)$$

where *t*-statistics are in parentheses.[3] Both MEASURE (prior salary) and OTHER (the judgment of management to vary Allstate salary from a strict linear mapping of prior salary) are valid predictors of future sales ability. Using these predictors, management on average overpays females, in the sense that females have lower productivity than management is paying for at the rate management pays for productivity of males.

Although this approach may appear similar to reverse regression, the dependent variable is real productivity, not an indicator thereof. Indeed, the indicator is an independent variable. Equation (7) is a direct regres-

[2] Although Becker (1957) presented a productivity-based analysis of discrimination, many economists do not find productivity to be a usable or correct concept on which to base wage theories (see Blattenberger and Michelson, 1983). Commission sales jobs are a clear exception.

[3] For a number of technical reasons, the regression variables (except FEM) are covaried by region and year (that is, are defined as residuals from a regression on regions and years and, for PROD, the length of time over which productivity is measured). Regressions estimated on noncovaried variables were virtually identical, and would have been identical had FEM also been covaried.

sion, asking how well salary, which is management's prediction of productivity, does predict productivity. Allstate management has predicted productivity very well, using both objective and subjective criteria; but they have been unduly optimistic about productivity in setting salaries of females, even though female salaries, on average, are lower than male salaries. No similar statement can be made about Harris Bank salaries from the Conway–Roberts presentation.

4. Why Cohorts?

Regression analyses have been successful in litigation either because they were unopposed (Oaxaca, 1976), or because they have been correct (for example, *EEOC v. McCarthy,* 1983). They have also been successful when incorrect, either because the alternative presented to the court was even less convincing, (*Segar,* 1981; *Trout,* 1981) or because, as will happen, the court made a mistake (*Craik,* 1984). By and large, situation analyses will be successfully countered by sound events-based analyses (*Pouncy,* 1982; *Ste. Marie,* 1981).

The defense approach that most often fails to persuade judges in opposition to regression, besides alternative regressions, is cohort analysis. Why are cohort approaches typically rejected by courts?

Judge Robinson did not mince words (*Segar,* 1981, at 698):

> Cohort analysis is an untried method of statistical study, unsupported in any published statistical work or judicial decision.

The D.C. Circuit said that Judge Robinson "may have overstated a bit," but affirmed (*Segar,* 1984, at 57). In *Trout* (1981, at 651) Judge Greene also rejected the cohort analysis proffered in rebuttal of regression:

> The failure of the government's analysis to reject, in many instances, the hypothesis of no discrimination in promotions is in large measure attributable to the fact that it selected a statistical method with extraordinarily low power to detect sex-based disparities.

In another context: "By fragmenting the data into small sample groups, the statistical tests become less probative" wrote a Fifth Circuit panel, in a footnote referring to "the defendant's 'divide and conquer' technique" (*Capaci,* 1983). A Fourth Circuit panel expressed a similar sentiment in *Lilly* (1983, at footnote 7). See *EEOC v. IBM* (1984), however, for a court that accepted a highly disaggregated analysis.

Conway and Roberts's use of cohorts is clearly distinguishable from those of other analysts. Nonetheless, it shares some of the characteristics that have led most courts to reject cohort analyses, especially when submitted by defense. An expert would be well advised to cure the method's accepted defects (arbitrary elapsed cohort time and disaggregated statistics) before presenting a cohort approach in court.

5. Statistical Experts

One of the hazards encountered by experts is the temptation to play lawyer. On a practical level, the concept of shifting burdens is a mechanism for deciding a case, not for trying one. How and what burdens shift depend on whether the case is one of disparate treatment or disparate impact, and also on the circuit in which the trial is being held (*Segar,* 1984). Conway and Roberts are, in any case, incorrect: plaintiffs always bear the burden of proof (*Burdine,* 1981).

My intention here is to emphasize, as do Conway and Roberts, the importance of the judicial forum in determining what to analyze and how. One role of an expert in that forum is to demonstrate errors in the opposing expert's presentation. Technical errors may appear in data, method, presentation, or interpretation, and such errors should be pointed out to the court. For example, I would question the Conway–Roberts statement that the coefficient of salary "measures the strength of association between salary and the set of qualifications." I would comment on their use of years of school as if it were a cardinal variable in regression, although thankfully that use is subsequently modified. [Borjas (1978) makes a similar error, using federal pay grade as a dependent variable in regression.] But I would stress most of all the lack of correspondence between their statistics and the law.

Expert Procedures

If I were building a data base to analyze salary decisions at Harris Bank, I would try to obtain information on employees who were, or might have been, selected for positions above entry level. I would want to analyze selection from among pools of applicants (*Harrison,* 1982). Since the bank did not take applications for internal promotions, the challenge would be to find a way to assess which employees were available for each opening.

Data bases must be relevant as well as accurate (*McDowell,* 1983). Obviously the Conway–Roberts data base would be useless for the analysis I contend is relevant. Their suggestion that data bases be exchanged, often accomplished through the process of discovery, should not mask the difficulty created by different conceptions of what a proper data base would be.

To the Conway and Roberts suggestions for documentation and exchanges of data I would add the strong recommendation that experts exchange written reports (Special Committee on Empirical Data in Legal Decision Making, 1984). A report should articulate the issue as seen by the expert, sources and characteristics of data, analytic methods, and findings and interpretations (*McDowell,* 1983). For example, from my written report the defense expert was able to formulate his written response in *EEOC v. McCarthy* (1983). From my written rebuttal he was able to realize (as admitted at trial) that the errors were his, not mine. I cannot imagine this result being achieved through oral argument in trial! Indeed, in one case an expert used the population size rather than the sample size in calculating the standard deviation of the normal approximation to the binomial, making a difference of several standard deviations appear trivial. Throughout an oral exchange he steadfastly maintained his position. Surely later, after some reflection, he recognized his error. Written reports allow such reflection.

The Expert in Practice

In *EEOC v. IBM* (1984), defense expert offered 94 salary regressions, stratifying the data by grade and year. Stratification by grade seems proper for modeling. Salary level differences over time could have been handled by time shift variables. To account for lack of independence over time, one could have weighted observations from autoregressive procedures. The defense expert, however, offered the equations as equal weighted (regardless of number of observations) and independent. He reported that "only" 13 regressions contained "significant" negative race coefficients.

Counting "significant" equations is incorrect: Had all race coefficients been of the same sign, for example, the finding would have been "significant" regardless of individual equation diagnostics. Finding 13 negative race coefficients "significant" at the 5% level from 94 independent equations is itself improbable by chance at the 0.001 level. Thus the expert failed the court in both method and interpretation.

Although discussing expert obligations and proposing expert procedures is always worthwhile, in light of what actually happens in court (as in *EEOC v. IBM*) I am compelled to stress the paramount importance of having a correct understanding of the role of statistical evidence in law. The court is ill-suited to resolve technical differences between experts who otherwise agree on data, method, and concept. My goal as an expert is, where possible, to reduce apparent technical disagreements to issues of law and concept, areas the court can and properly should judge (see *Robinson,* 1983).

Harris Bank

The critical aspect of the *Harris Bank* analysis is not in the areas Conway and Roberts discuss, but in the areas they do not discuss. Why is regression being used? Why are the observations limited by hire date and organized into cohorts? What are the criteria for salaries? What does regression, direct or reverse, imply about management decisions?

My analysis as plaintiff's expert in *Harris* (1981) was formulated to correspond to a legal theory that recognized no time bar to events, and imposed an affirmative action obligation on the employer. I thought that my client's legal theory was incorrect. My job, however, was to formulate statistical procedures testing the institution within the context of the legal theory provided. For that, regression was appropriate.

No longer associated with that case, I am free to express my long-standing disappointment that the bank's expert chose to attack the *form* of my regression rather than the *fact* of my regression. The key to plaintiff's success was consistency between legal theory and statistical method. The key to the bank's failure was the lack of an articulate opposing theory, as indicated by its situation-based statistical analysis.

That the reverse regression debate has led, if anywhere, to a reaffirmation of the usefulness of direct regression (Ash, 1984; Goldberger, 1984) does not mean the effort expended in the debate was wasted. Good questions demand good answers. Unfortunately, the reverse regression discussion, regardless of its technical merits, has added little to the subject of statistical proof of discrimination. After all these years, allegations of discrimination at Harris Bank remain unanalyzed.

Case Citations

AFSCME v. State of Washington 578 F.Supp. 846 (W.D. Wa. 1983).

Capaci v. Katz & Besthoff 711 F.2d 647 (5th Cir. 1983).

Craik v. Minnesota State Univ. Bd. 731 F.2d 465 (8th Cir. 1984).

Eastland v. TVA 704 F.2d 613 (11th Cir. 1983).

EEOC v. IBM 583 F.Supp. 875 (D. Md. 1984).

EEOC v. D. Justin McCarthy et al. (Framingham State College) 578 F.Supp. 45 (D. Ma. 1983).

Furnco Construction Co. v. Waters 438 U.S. 567 (1978).

Harrison et al. v. Lewis (D. D.C. 1982) unpublished opinion; aff'd __F.2d__ March 1, 1984.

Kouba v. Allstate Insurance Company (E.D. Ca. 1984) Civil Action #S-77-99 LKK, unreported settlement.

Lilly v. Harris-Teeter Supermarket 720 F.2d 326 (4th Cir. 1983).

McDowell v. Safeway Stores, Inc. 575 F.Supp. 1007 (D.C. Ark. 1983).

Pouncy v. Prudential Insurance Company of America 499 F.Supp. 427 (D. Tx. 1980); 668 F.2d 795 (5th Cir. 1982).

Robinson et al. v. Polaroid Corp. 567 F.Supp. 192 (D. Ma. 1983), aff'd 732 F.2d 1010 (1st Cir. 1984).

Segar v. Civiletti 508 F.Supp. 690 (D. D.C. 1978): aff'd as *Segar v. Smith* 581 F.2d 815 (D.C. Cir. 1984), cert. denied sub nom Meese v. Segar __U.S.__ (1985).

Sobel v. Yeshiva University 566 F.Supp. 166 (S.D. N.Y. 1983).

Ste. Marie v. Eastern Railway Association 650 F.2d 395 (2nd Cir. 1981).

Texas Department of Community Affairs v. Burdine 450 U.S. 248 (1981).

Trout v. Hidalgo 517 F.Supp. 873 (D. D.C. 1981); aff'd. 702 F.2d 1094 (D.C. Cir. 1983), as *Trout v. Lehman;* vac, rem in part, (1984) __U.S.__.

U.S. Department of the Treasury v. Harris Trust and Savings Bank (1981) US Dept. of Labor Administrative Proceeding, Case No. 78-OFCCP-2, unpublished ALJ finding for plaintiff; (1983) remanded by Secretary of Labor, 31 FEP Cases 1223.

Valentino v. United States Postal Service 674 F.2d 56 (D.C. Cir. 1982).

Vuyanich v. Republic National Bank of Dallas 24 FEP Cases 128 (D. Tx. 1980); 723 F.2d 1195 (5th Cir. 1984).

Additional References

Abowd, J. (1984). Economic and statistical analysis of discrimination in job assignment." In B. D. Dennis (ed.), *Proceedings of the Thirty-Sixth Annual Meeting of the Industrial Relations Research Association,* 1983, pp. 34–47.

Ash, A. (1984). "The Perverse Logic of Reverse Regression," paper delivered at American Statistical Association meetings, summer 1984.

Baldus, D. C., and Cole, J. W. (1984). *Statistical Proof of Discrimination,* New York: McGraw-Hill.

Blattenberger, G., and Michelson, S. (1983). "Reverse Regression Analysis of Wage Discrimination," *Western Sociological Review* **14,** 95–110.

Fisher, F. M. (1980). "Multiple Regression in Legal Proceedings," *Columbia Law Review* **80,** 702–736.

Goldberger, A. S. (1984). "Reverse Regression and Salary Discrimination," *Journal of Human Resources* **19,** 293–318.

Oaxaca, Ronald L. (1976). Male-female wage differentials in the telephone industry. In Phyllis A. Wallace (ed.), *Equal Employment Opportunity and the AT&T Case* Cambridge, Mass.: MIT Press, pp. 17–47.

Special Committee on Empirical Data in Legal Decision Making (1984). "Recommendations on Pretrial Proceedings in Cases With Voluminous Data," *The Record of the Association of the Bar of the City of New York,* **39,** January/February, 49–78.

Comment

Arthur S. Goldberger
Department of Economics
University of Wisconsin—Madison

The paradox of the two regression lines has bedazzled generations of students in introductory statistics courses. I wonder whether it will be equally effective in the courtroom, where there are experts on both sides.

Called on by the defense in a discrimination suit, the statistical expert may want to use reverse regression, a device that almost invariably produces a smaller adjusted mean difference in salary by gender than does the more familiar direct regression. If so, the defendant's expert should be prepared to justify the choice of regression by reference to a coherent statistical model.

What stochastic assumptions will suffice to support reverse regression as a method for assessing salary discrimination? I confine attention to the empirically relevant situation where several measures of employee qualifications are available.

It will not do to suppose that those measures are, one by one, fallible indicators of corresponding "true qualification" variables, with the latter being the proper determinants of productivity and hence salary. For in that model, the Conway and Roberts reverse regression does not produce an unbiased estimator of discrimination.

On the other hand, it will do to suppose that all of the measured qualifications are fallible indicators of a single latent productivity variable, with errors that are independent of gender. In that model, the Conway and Roberts procedure will give an unbiased estimator of discrimination. Now this latent-variable model has an obvious implication: in the multivariate system of regression equations (in which each measured qualification variable is regressed on gender and salary) the coefficients must stand in a fixed ratio. This is a testable null hypothesis. If it is rejected, the scientific basis for adopting and presenting the Conway and Roberts procedure vanishes.

Documentation of the claims just made can be found in two recent articles of mine: "Redirecting reverse regression," *Journal of Business and Economic Statistics,* **2,** 114–116 (1984), and "Reverse regression and salary discrimination," *Journal of Human Resources,* **19,** 293–318 (1984).

Rejoinder

Delores A. Conway / Harry V. Roberts

We are grateful to both discussants for their comments. Goldberger's comment cites his two earlier papers, where he delineated two possible underlying models of employer behavior. These papers, and those of Dempster, highlight the importance of underlying stochastic assumptions for causal inferences about possible discrimination in analysis of observational data by regression models. Michelson's comment focuses on the tensions that exist between statistical analysis and legal considerations. Before addressing specific comments by each discussant, we should like to highlight the role of homogeneous job groups in causal inferences about possible discrimination.

1. Type 1 and Type 2 Employer Behavior

Fundamental to the development of a causal model of the employment process is the interplay between statistical analysis and model specification as components of an evolutionary process (e.g., see Box, 1980).

Models of the employment process are still evolving. No single current model describes the complex ways in which wages, employee qualifications, job market forces, unions, and other variables interact in the employment process. The lack of sufficiently detailed models poses a serious limitation in focussing discrimination studies.

Goldberger (1984a,b), Dempster (1984), and others have emphasized that underlying models of employer behavior, explicit or implicit, have direct bearing on statistical assessments of possible discrimination. Of particular importance are the assumptions concerning the stochastic elements of the employment decisions that relate income levels and job qualifications.

In an earlier reply to Goldberger (1984a), we suggested a different perspective of employer behavior that complements econometric formulations of stochastic assumptions (see Conway and Roberts, 1984). Two different viewpoints of employer behavior are suggested by direct and reverse regression. We originally stated these viewpoints for the case when the income measurement refers to the hiring decision of the employer. However, the framework is general and applies to other income variables, such as a salary or promotion. To write the behavioral models, we let Y represent a univariate or multivariate set of income measurements for the employee and X, the employee's set of qualifications that are related to job requirements and performance. The behavioral models given below are suggested by the two conditional distributions of X and Y. They are highly simplified descriptions of employer practices, and aspects of both are likely to be found in actual employment practices.

1. **Type 1 Employer Behavior.** An employer considers the qualifications of an individual candidate and chooses an income level that most closely matches his or her qualifications. The candidate's qualifications X are regarded as fixed, and the income level Y could vary across all income levels in the firm, although the employer may consider a narrower set of possible incomes.
2. **Type 2 Employer Behavior.** An employer considers a group of candidates for a particular level of income and selects the candidate who appears to have the best qualifications for the given income. In this case, the income level Y is regarded as fixed and the qualifications X vary across the candidates considered.

The difference between the two models turns on whether the employer starts with the qualifications of an individual candidate or with a level of income to be awarded. Random errors in each model are identified with repetition of employment decisions for different candidates (Type 1) and for different incomes (Type 2). Direct regression is the natural starting point for the study of employment practices under Type 1 employer behavior; reverse regression is the starting point for the study of Type 2 employer behavior.

As noted above, causal inferences about possible discrimination from statistical assessments depend on the stochastic assumptions specifying the relationship between potentially observable income levels Y and job qualifications X. If employment data were experimental rather than observational, the stochastic assumptions would be dictated by the randomization.

Model misspecification is a crucial problem for observational studies. Causal inferences turn critically on different assumptions about the random mechanism generating the data, and these assumptions cannot be completely verified or refuted from the data alone. Ideally, causal inferences should be relatively robust to different assumptions about the random mechanism underlying the data. This aspect of causal inference is difficult to treat and remains a major challenge for statistical analysis.

The Type 1 and Type 2 models of employer behavior offer a framework for formulating different possible random mechanisms. Specifically, Type 1 assumes that individual employment decisions are repeated randomly at a fixed level of qualifications, whereas Type 2 assumes that these decisions are repeated randomly at a fixed level of income. Direct and reverse regressions permit us to evaluate the conclusions about possible discrimination from each model. By looking at the data from both perspectives, we enlarge our viewpoint of the employment process and recognize potential biases in conclusions that arise from the use of observational data.

Conflicting assessments in statistical studies of discrimination can highlight limitations of the models or data, suggest additional refinements that may bring the results in closer accord, or point to external relevant information that may have been omitted. We believe that direct and reverse regression, when used in tandem, help to illuminate the data more fully and to focus attention on potential biases. In our own empirical work we find that checking out each perspective often exposes features of the data that might otherwise have gone unnoticed.

2. Homogeneous Job Groups

In studies of alleged employment discrimination and in commentary on those studies, inadequate attention has been paid to the role of the job itself. Models and analyses have often made no distinction among the different specialized tasks and levels of responsibility that are characteristic of workforces within large organizations. In some discrimination studies, job has been viewed as an income variable. This is one pertinent aspect of studies investigating possible discrimination in hiring, job placement, and job advancement. In some studies of salary discrimination, however, job has been viewed as a proxy for qualifications. For example, job may reflect omitted market factors that explain large occupational differences in observed salaries, such as those between the officers and clerical employees of a corporation.

The perspectives of Type 1 and 2 employer behavior suggest still another possible role for the job variable. The job itself may serve to delineate the range of candidate incomes in Type 1 employer behavior and the members of the candidate pool in Type 2 employer behavior. Consequently, job may serve as a catalyst in the employment process. That is, each particular job influences the outcome of the employment process without being changed by it. The job may provide the concrete circumstances within which the process operates.

We have come to believe that the existence of heterogeneous jobs in a workforce is central to the understanding of discrimination studies, including the concepts of Type 1 and Type 2 employer behavior. In the presence of substantial job heterogeneity, assessment of possible discrimination—whether by direct or reverse regression—could result in estimated sex coefficients that are biased. Differences between jobs might be confounded with differences between sexes, unless the distributions of qualifications across jobs are the same for both sexes.

Moreover, the assumption that an entire workforce constitutes a single homogeneous job group can be tested with the data. In our paper, we separated the clerical employees from the professional employees. We noted that apparent discrimination within each group—whether viewed from direct or reverse regression—differed substantially from the overall assessment for the pooled group. For example, the results in Tables 3 and 5 point out that the unexplained variance of the regression models decreased when the models were fit separately to the clerical group. This suggests that the within-group regressions have captured additional aspects of the process underlying the data that do not appear in the overall regressions. It is also important to note that the direct and reverse regression sex coefficients are closer in magnitude for the within-group regressions than the overall regressions. This effect is in part due to the greater homogeneity in income and job qualifications among employees within the same job.

Sample size considerations become more important when the data are disaggregated into homogeneous job groups. There will always be an implicit trade-off between the advantage of reducing potential bias in comparisons of heterogeneous job groups versus the disadvantage of losing precision in estimating possible sex effects within a homogeneous job group. In our paper, the sample sizes within each group are still adequate to detect potential sex effects. In fact, the larger sample size within the clerical group is sufficient to justify further disaggregation into more refined job groups.

Moreover, if there is a need to assess only the overall picture, one can aggregate the results of the subgroup regressions. For example, one can obtain an overall index by weighting the sex coefficients by the numbers of females on which they are based. The relevant sampling error is then the standard error of this index, not of the individual sex coefficients in the subgroups.

The most important implication of these arguments is that statistical analysis should be focused on relatively homogeneous job groups, consistent with the requirement of reasonable size of sample within each group. It may be possible to achieve the same effect in regression models by the use of indicator variables for job groups or titles or by quantitative indices such as salary grades or Hay points. Both direct and reverse regression can be implemented within each job group. If there is greater homogeneity in the distribution of income and job qualifications among males and females, we would expect a closer accord in the conclusions about possible sex discrimination from the two regressions.

Concern about possible discrimination in job entry, often referred to as "shunting," has often deterred the use of disaggregation into homogeneous jobs. The remedy for this concern is not to restrict attention to the entire workforce, but rather to study shunting on its merits. If data on rejected or ignored job candidates are available, the framework of direct and reverse regression can be used to audit both salary discrimination and shunting. Furthermore, this applies at all stages of the employment process: initial search and hiring, initial placement and salaries, promotion and job advancement, termination, and even forced retirement. We are currently fleshing out the details of this approach (see, e.g., Conway and Roberts, 1986).

One merit of a formulation by homogeneous job groups is that it forces the statistical analyst to focus on the realities of the job structure being studied. Nonsample information may be essential in deciding about job homogeneity or heterogeneity. The study of homogeneous job groups further enlarges our perspective of the employment process and highlights important aspects of the data that might otherwise go unnoticed.

3. Goldberger's Comments

In two previous papers, Goldberger (1984a,b) focused on two causal models of the employment process that may underlie statistical assessments of potential discrimination by direct and reverse regression. A primary contribution of these papers is to direct attention to underlying models of employer behavior. We find much common ground with Goldberger's development and agree that the implications of different causal models should be investigated to help determine an appropriate statistical model for assessing employer behavior.

The two models proposed by Goldberger assume that employers set salaries to equal assessed productivity plus a possible discrimination effect due to sex. The differences between the models appear in the stochastic assumptions that describe the imperfect relationship between assessed productivity and observed qualifications. Model A assumes that assessed productivity is determined by measured qualifications plus a random error that is uncorrelated with sex. For this model, the direct regression sex coefficient provides an unbiased estimate of potential discrimination. In Model B, observed qualifications are fallible indicators of assessed productivity. Specifically, each qualification is linearly related to assessed productivity plus a random error that is uncorrelated with sex. Goldberger points out that the sex coefficient in the reverse regression of the qualification index on salary and sex gives an unbiased estimate of possible salary discrimination.

Conway and Roberts (1984), Dempster (1984), Kamalich and Polachek (1982), and others question the realism of the stochastic assumptions specified by Model A. Specifically, the model assumes that there are no additional sex effects from omitted qualifications, not available to the statistician, but that may enter the employer's assessment of productivity. This is a strong assumption and may not be satisfied by the limited data typically available in discrimination studies. Seldom will the observed qualifications include all legitimate productivity factors that differ by sex. Finkelstein (1980) notes that this assumption has been challenged in almost every employment discrimination case that uses regression analyses for statistical evidence.

In Model B, the only sex differences in observed qualifications are due to a possible sex difference in assessed productivity. In the language of econometrics, there are no exogenous sex effects in individual observed qualifications. This leads to the restriction that the ratios of the sex coefficient to the salary coefficient in the individual reverse regressions of each qualification on salary and sex are the same. The test of proportional coefficients in the individual reverse regressions—now widely known as the Goldberger test—is used to test the validity of Model B as an underlying model of the data.

We agree that the Goldberger test can be used to test the validity of Model B with the data. However, we question his assertion that Model B is the only underlying causal model that establishes a scientific basis for reverse regression. There are several reasons for our skepticism about this conclusion.

First, we have always found it difficult to believe that the difference between a univariate and a multivariate set of qualifications can be so profound. (Stein's paradox is the only parallel that readily comes to mind; and the paradox can be understood in a more general formulation.) When there is only a single measured job qualification, Goldberger's test does not apply and has nothing to say about the scientific basis for reverse regression. If the Goldberger test reflects a basic principle for choosing between direct and reverse regression, we find it hard to understand how a basic principle could shed no light on the time-honored question of which way to run a simple regression.

Second, the Goldberger test could just as easily rule out direct regression as reverse regression. Income is inherently multivariate, as are job qualifications, and might include salary and measures delineating level of responsibility and job security. An extension of Model A would specify that the sex coefficients in the direct regressions of individual income measures satisfy similar proportionality restrictions.

Third, the proposed relationships of Model B between income, job qualifications, and sex are specified as multivariate relationships. The multivariate relationships naturally induce an implied univariate relationship of the qualification index to income and sex. However, it is not necessarily true that the univariate relationship of the qualification index to income and sex can arise only from the multivariate relationships specified by Model B. This is tantamount to a claim that the relationship between one variable and a univariate index summarizing a set of variables can arise only from a single multivariate model.

In a recent paper, Dempster (1984) questions the test of Model B as a general test of reverse regression and proposes a different causal model that justifies the use of direct and reverse regression. His model begins with a complete set of job qualifications, including both observed qualifications in the data and unobserved qualifications used by the employer to assess productivity. Under this model, reverse regression yields an unbiased estimate of possible discrimination whenever the sex differences in the indices summarizing observed and unobserved qualifications have the same sign and magnitude proportional to a measure of unexplained variance. The assumptions in Dempster's model do not imply any proportionality restrictions on the individual reverse regression sex coefficients. Dempster points out that there is no information in the data that can help to distinguish between the appropriateness of direct or reverse regression for his causal model.

In pursuing the issues at the level of the last few paragraphs, we risk giving the impression that the methodological dispute between direct and reverse regression can be resolved by the data to be analyzed, in the context of one compact and authoritative econometric model. We see here a parallel in the tendency, deplored by many statisticians, to regard a simple test of significance as the ultimate aim of a statistical study. One of the reasons that tests of significance are so much overused and abused is that a test gives the appearance of resolving the key questions of a study in one simple analysis. As a result, the really important questions often escape serious consideration.

In the present context, we have argued that the job structure of an organization is essential to the understanding of whether discrimination is or is not occurring. We have argued that salary discrimination is only one aspect of discrimination, and that shunting is another, equally important, aspect. To try to assess discrimination without addressing the role of jobs and of shunting could result in a dangerous simplification and risk a distorted impression.

Finally, there is a legal test of causal models. There are two statutory guides to national anti-discrimination policy. The one usually discussed is the Civil Rights Act of 1964. But there is also the Equal Pay Act of 1963, where studies of job structure are essential to a comprehensive examination of possible discrimination. Our view is that the difficult problems in understanding discrimination center on a clear understanding of job structures in organizations and on the application of statistical methods to these job structures. Once this central fact is understood, we believe that direct and reverse regression will evolve as complementary, not exclusionary, routes to statistical analysis of discrimination.

4. Michelson's Comments

Michelson questions the use of cohorts and proposes statistical analyses of "events" as opposed to "situations" for the study of employment discrimination. He also raises a number of questions regarding the use of regression analyses in legal cases, and the 1979 *Harris Bank* case in particular.

Michelson gives the impression that only one cohort was used in the study from which the data set of our paper was drawn. In fact, that study included analysis for the substantial majority of employees on the payroll as of 31 March 1977. These employees were disaggregated by date of hire into four specific cohorts, identified as those hired in 1965–1968, 1969–1971, 1972–1973, and 1973–1975. For legal reasons, employees hired before 1965 were excluded. The Harris Bank gave permission for public use of the data for the 1969–1971 cohort, and these data have been widely studied by students and other researchers. This explains our use of the 1969–1971 cohort for the illustration of methodology in our paper. The results for all four cohorts appear in Roberts (1979). This report also explains how Roberts *did* "aggregate results from different cohorts to a summary picture to be presented to a judge."

A major reason for disaggregating the data into cohorts was the difficulty of adjusting for widely different levels of seniority among employees. For example, over an extended period of time, there are often major shifts in market factors affecting incomes of employees in different occupations. As noted by Cochran and Rubin (1974), there are three principal ways to remove or reduce potential biases from major confounding variables in conclusions from observational studies: (a) stratification into more comparable groups, (b) regression adjustment, or (c) a combination of stratification and regression adjustment. Stratification of the data into cohorts, arranged by date of hire, was felt to adjust more satisfactorily for the effects of confounding variables than regression adjustments based on seniority alone. However, aggregation of results from the separate cohorts addresses the same objective as a large regression on the combined cohorts that specifies main effects for seniority and interactions between seniority and other job qualification variables.

Michelson prefers to analyze the seniority effects in the context of such a large model, and he did so in his own analyses at the 1979 hearing of the *Harris Bank* case. Based on calculations we made at that time, it is our belief that the divergences of conclusions between the plaintiff and defendant were little affected by this difference of strategy in statistical modeling. We therefore have trouble in seeing why Michelson places so much emphasis on "cohorts" in his comment on our present paper. We join with Michelson in opposing the dishonest statistical game of disaggregation to the point that there is "very low power to detect sex-based disparities." Even here, however, many insignificant differences running in the same direction can be tantamount to significance in the aggregate.

Throughout his comment, Michelson repeatedly "controls" for the presence of confounding factors that may affect conclusions about possible discrimination. It is more accurate to describe his "controls" as statistical adjustments. In our paper, we emphasized the difficulties in adjusting for potential confounding factors, when dealing with observational data. Statistical adjustments are essential for observational studies, but they seldom provide a completely satisfactory allowance for the influence of confounding factors.

Michelson's second methodological point concerns a distinction between analysis of "events" and of "situations." The distinction between the two types of analyses is not completely clear from his remarks, but it appears to parallel the standard technical distinctions between "dynamic and static" or "time series and cross-sectional." It is certainly *not* the distinction between "discrimination" and "unfairness," and we cannot see how it relates to the distinction between "structural model" and "reduced form".

Our understanding of Michelson's definition is that "event" refers to any change in a job situation. Hence, "event"-based analyses might refer to the statistical study of hiring, placement, promotion, or termination. If so, we have long recognized the relevance of hiring, placement, promotion, and termination as components of an overall examination of discrimination. The Harris Bank study, for example, examined placement and salary advancement as well as current salaries (Roberts, 1979). We also tried to deal with possible shunting.

We question Michelson's assertion that regression models have relevance only for a "situation-based" analysis of employment discrimination. After all, "regression" simply refers to conditional expectations, which provide a general approach for conducting discrimination studies. The income variable, target populations, and other factors may change as specific types of possible discrimination are the subject of inquiry; but regression involves conditioning, whether implemented by ordinary least-squares, nonlinear least-squares, logistic regression, or tabular analysis.

An employer's decision to hire, promote, or terminate an employee results in an outcome that can be related to a set of job qualifications and possible discriminatory effects due to sex. In this context, the income variable is a binary variable and logistic regression models are appropriate. Type 1 and Type 2 employer behavior and direct and reverse regression are relevant to these studies just as they are relevant to salary studies.

For example, in investigations of possible discrimination in promotion, the direct (logistic) regression conditions on a set of related job qualifications to compare the probabilities that male and female employees receive a promotion. The qualification index from the logistic regression can then be used for the reverse regression comparisons. Specifically, mean male and female qualifications can be compared to detect differences in qualification levels among those employees promoted and those not promoted. Causal inferences about promotion discrimination from logistic regression models will again turn on the potential effects of omitted variables and on assumptions about the disturbance in the model. In doing these studies, one must carefully delineate groups of employees who are candidates for promotion to specific jobs. (Again, we see the need for careful study of jobs in understanding the employment process that is under study.)

In our chapter, we focused on salary discrimination, since the intent was to highlight important aspects of the statistical methodology underlying regression studies. Michelson seems to imply that our analysis of salary discrimination was presented as a comprehensive investigation of employment discrimination. This is a misconception: we have always recognized the need for hiring, promotion, salary advancement, and termination studies in a comprehensive treatment.

In specific references to the *Harris Bank* case, Michelson makes statements, besides those already mentioned, that require correction or at least clarification. In more than one instance, he appears to link the Equal Pay Act with the *Harris Bank* case. The case was tried under the Executive Order, not the Equal Pay Act.

He further seems to cast doubt on our attribution of "legal considerations" in defining the target population for the statistical studies by the defense. Definitions of the target population involve considerations of subject matter in any statistical study. In legal cases, the definitions will take into account the specific allegations in the case and the interpretation of existing legislation, regulations, and court opinions. Michelson notes that he was required to "formulate statistical procedures testing the institution within the context of the legal theory provided," even though he "thought that my client's legal theory was incorrect." We understood and accepted the legal theory of the defense that delineated the target population and our four cohorts.

In another instance, Michelson says, "The key to the bank's failure was the lack of an articulate opposing theory, as indicated by its situation-based statistical analysis." The use of the word "failure" is clearly inappropriate, since the case was subsequently remanded for rehearing. Furthermore, he fails to give the correct reason for the remand; namely, that for reasons unrelated to statistics, the defense was not permitted to present its own statistical case at the 1979 hearing, but only to present rebuttal testimony.

Michelson states, "The critical aspect of the *Harris Bank* analysis is not in the areas Conway and Roberts discuss, but in the areas they do not discuss." In the case itself, in the 1979 paper referenced above, and in our subsequent work, we have addressed the specific questions that Michelson raises. Our article in this volume was not intended to provide a comprehensive account of all aspects of the *Harris Bank* case. Instead, the intent was to highlight important questions surrounding regression analyses, questions that transcend the particular circumstances of the *Harris Bank* case.

Michelson uses the term "productivity" for what might be more accurately described as "performance measure." We agree that performance measures, when available, fit into the general framework as an additional job qualification. However, performance measures do pose the possibility of random measurement error that could lead to underadjustment bias in direct regression.

We think that Michelson's discussion of omitted variables may be useful in understanding what may be left out or assumed by any model formulation. However, he provides no basis for his assertion that "without such a measure, reverse regression cannot be justified." As noted previously, there are underlying causal models that justify the use of reverse regression in studies of employment discrimination. Such models have been proposed by Goldberger and Dempster. Further, it is hard to see how shunting can be studied in the absence of some form of reverse regression.

We find an apparent contradiction in Michelson's discussion of this point in the case of *Kouba v. Allstate*. He starts with an underlying model justifying the use of direct regression (specifically, Model A proposed by Goldberger) and views reverse regression as a correction for underadjustment bias. Although the actual direct and reverse regression results are not presented, Michelson states, "I concluded that the reverse regression gave a better picture of the relationship between sex and salary than did direct regression."

The regression results presented are for a model that regresses "productivity" on sex and two components of salary. (It is difficult to judge the basis for his statement that "Allstate management has predicted productivity very well, using both objective and subjective criteria," since there are no results concerning the accuracy of the predictions.) Presumably, this equation provides a true assessment of actual discrimination since the dependent variable is "real productivity, not an indicator thereof." Let us ignore the potential biases in the dependent variable as a measure of true productivity and view the equation as an approximation for the underlying relationship between productivity, salary and sex. Michelson states that the reverse, rather than the direct regression assessment, based on indicators of productivity, came closer to the true assessment of discrimination from the productivity equation.

If Michelson's arguments are correct, they validate the reverse regression assessments that use only indicators of productivity. Equally, they invalidate the direct regression assessments with the same indicators. Rather than making a case against reverse regression, his arguments actually do the opposite. The empirical evidence and discussion provided by Michelson suggest that reverse regression gave a closer approximation to the true assessment of discrimination than did direct regression in this particular case. But if we are to accept his proposal that reverse regression can be used only in studies where true productivity measures are available for validation, then the same burden rests on direct regression.

Finally, in Michelson's comments on exchange of data bases, we do not think that such exchange, however desirable, is "usually accomplished in the process of discovery." In fact, we have encountered strong resistance to such proposals. We do endorse Michelson's suggestion for exchange of written reports.

Additional References

Box, G. E. P. (1980). "Sampling and Bayes' Inference in Scientific Modelling and Robustness," *Journal of the Royal Statistical Society, A,* **143,** 383–430.

Conway, D. A., and Roberts, H. V. (1986). "Homogeneous Job Groups, Candidate Pools and Study of Employment Discrimation," to appear in the *Proceedings of the Business and Economics Section, American Statistical Association.*

Dempster, A. P. (1984). "Notes on Reverse Regression and Salary Discrimination," Manuscript, Department of Statistics, Harvard University.

Goldberger, A. S. (1984a). "Redirecting Reverse Regression," *Journal of Business and Economic Statistics,* **2,** 114–116.

Goldberger, A. S. (1984b). "Reverse Regression and Salary Discrimination," *Journal of Human Resources,* **19,** 293–318.

Roberts, H. V. (1979), "Harris Trust and Savings Bank: an Analysis of Employee Compensation," Report 7946 of the Center for Mathematical Studies in Business and Economics, Department of Economics and Graduate School of Business, University of Chicago.

Reynolds v. C.S.N.
Evaluating Equipment Damage

G. A. Whitmore
Faculty of Management
McGill University
Montreal, Canada

ABSTRACT

In 1967, at the present Canadian Reynolds Metals Company aluminum-reduction plant in Baie Comeau, Quebec, electric power to the potrooms was cut during a labor dispute, resulting in an uncontrolled shutdown of several hundred aluminum-reduction cells. The shutdown eventually led to a legal action by the company against the workers' union and its confederation, Confédération des Syndicats Nationaux (C.S.N.), and others to recover losses and costs associated with lost production and damage to equipment. The case was heard in Quebec Superior Court and was complex and lengthy. A court decision finally came in 1979 after the case had been before it for more than a decade. This article describes (a) the role that statistical methods and experts played in the segment of the case concerned with estimating the extent of damage done to the aluminum-reduction cells, (b) the implications of the case for the use of statistical arguments and expertise in court cases of this type, and (c) the judge's decision and the ultimate resolution of the case.

On 16 May 1967, at the present Canadian Reynolds Metals Company aluminum-reduction plant in Baie Comeau, Quebec, workers walked off the job in a labor dispute. In the midst of this strike action, electric power to the plant's potlines was cut, resulting in an uncontrolled shutdown of several hundred aluminum-reduction cells (commonly called pots or furnaces). Even after the workers returned to work, the operating environment in the plant was unsettled for many months. The shutdown led to a $6 million damage suit by the company against the workers' union, and its confederation, Confédération des Syndicats Nationaux, and others to recover losses and costs associated with lost production and damage to the cells both at the time of the shutdown and during the unsettled operating period which followed. The case was heard by Judge Vincent Masson of the Quebec Superior Court without a jury. It was a complex and protracted case on which a decision was not rendered until 6 February 1979 (Société Canadienne de Métaux Reynolds c. Confédération des Syndicats Nationaux et al., 1980).

This chapter describes the role that statistical methods and experts played in the segment of the case concerned with estimating the extent of damage done to the aluminum-reduction cells. It also describes the principal court exhibits, the expert testimony, and the judge's written decision in the case. No reference is made to the names of experts involved in the case, however, as the aim of the article is not to judge them but rather to examine the substantive aspects of the case as they relate to the role of statistics in legal proceedings. The prefixes P and D are used to identify the plaintiff and defense experts and exhibits, respectively.

Some of the data from this court case appear in Whitmore and Gentleman (1982), in a case study in data analysis published by the *Canadian Journal of Statistics*. The reader is referred to the case study for the results of two analyses of the data by research statisticians who were invited by the journal to make their own independent investigations.

Section 1 gives the reader background information on the case, including a brief review of the characteristics of a potline and an aluminum-reduction cell. The section also introduces the data used in the equipment-damage segment of the case. Section 2 introduces the experts and the highlights of their testimony. Section 3 discusses the implications of the case for the use of statistical arguments and expertise in court cases of this type. It also describes the tests and analyses on which the testimony of the statistical experts was based. Section 4 describes the judge's decision and the ultimate resolution of the case.

1. Background

Pots and Potlines

An aluminum-reduction cell is a container in which aluminum is produced by an electrolytic process that operates at a very high temperature. Figure 1 shows a cross section of a cell. The top layer of contents consists of alumina (aluminum oxide) dissolved in a molten bath of cryolite. The cryolite dissolves the alumina and permits its reduction by electrolysis to aluminum. The molten aluminum accumulates in the bottom of the container. The container is carbon lined and acts as the cell's cathode. The anode is also made of carbon and is suspended in the cryolite bath just

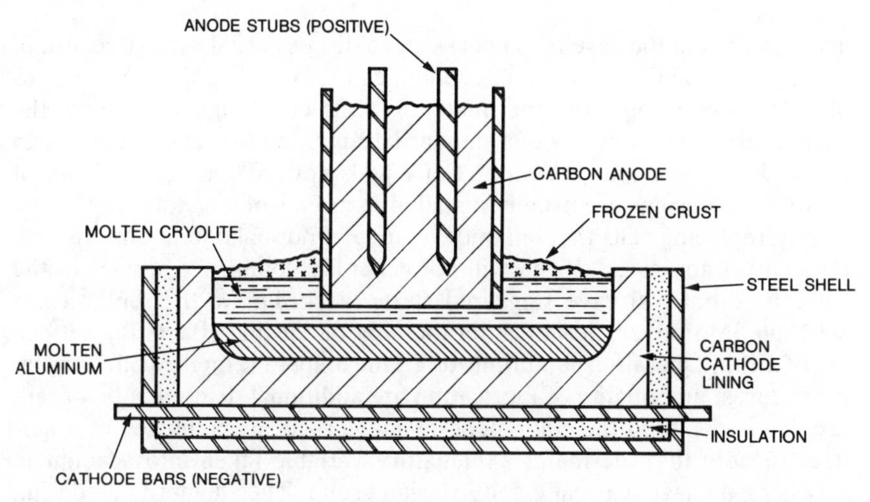

Figure 1. Cross section of an aluminum-reduction cell.

above the molten metal. Steel bars and stubs embedded in both the cathode and anode act as conductors for the cell.

A potline in a aluminum-reduction plant contains as many as 200 cells connected in a series circuit. At the time of the strike, the Baie Comeau plant had two potlines in operation with a total of 349 cells in service.

Cells generally have lives measured in years. Failure occurs eventually either because of serious distortion of the cathode or because of cracking of its carbon lining, which leads in some cases to iron contamination of the molten aluminum or to the rupture of the cell and the spilling of its contents.

The cutting of electric power to the cells on the day of the strike led to cooling and then to freezing (solidification) of their molten contents. The extreme thermal forces to which this cooling and freezing subjected them and the difficulties of restarting the cells after the strike were believed by the company to have shortened the effective operating lives of cells in service at the time of the shutdown. This loss of operating life is referred to subsequently as the *primary damage effect*. The dislocation of normal operations for many months after the strike ended was also believed by the company to have shortened the operating lives of cells installed during that period. This loss of operating life is referred to subsequently as the *secondary damage effect*. At the time of the case, the cost of rebuilding a cell was \$10,000. The primary and secondary damage claims together amounted to \$850,000, the equivalent of 85 cells.

To understand the manner in which cell-failure data were introduced and analyzed in the case, it is necessary to discuss cell design. Because of the desire to achieve operating economies by extending the service lives of cells, there is constant experimentation with cell design. Sometimes the designs are minor variants of a standard design. In other cases, the design is relatively new and untested. Cells with new designs or variants of established designs are usually installed in groups of a handful to several dozen, replacing cells that fail in the course of normal operations. Hence, the cells of any given design will not generally commence service at the same time but will have staggered starts, according to the replacement schedule. Moreover, design is not the only factor that affects the service life of a cell. Operating conditions (e.g., the amperage on the cell), startup procedures, and quality of fabrication are additional determinants of failure. To understand the equipment-damage proceedings, however, it suffices to note that the major explanatory variable taken into account in assessing damage was each cell's design group. The standard cell design that was in use at the time of the shutdown is referred to simply as *standard*. Other designs are referred to as *experimental*.

Data

Table 1 summarizes the categories of design groups whose failure data played a role in different parts of the testimony. Failure data from design groups in Category I played a role in the goodness-of-fit arguments of the experts. The 10 design groups in this category span a range of operating periods, from an early period (Groups 1, 2, 2A) to a late period (Groups 51D, 51E, 51F). The table shows the number of cells in each group.

Category II contains every design group that had one or more cells in service at the time of the strike. These data were central to the company's primary damage claim (i.e., damage to cells in service caused directly by the uncontrolled shutdown). The design groups in Category II are divided into those that were variants of a standard design and those that were

TABLE 1. Cell Design Groups Whose Failure Data Featured in the Court Case

(A) Category I—Design Groups Used to Establish the Typical Failure Pattern

Group	Number of Cells	Group	Number of Cells	Group	Number of Cells
1	22	24A	20	51D	30
2	20	30	28	51E	38
2A	42	35	23	51F	34
		37	17		

TABLE 1. (Continued).

(B) Category II—Design Groups Used to Establish the Primary Damage Claim

Standard Design Groups

Group	Number of Cells	Cells in Service at Strike	Group	Number of Cells	Cells in Service at Strike
18	20	3	39C	11	11
18A	21	16	39D	14	13
38	41	22	39E	12	12
38A	39	22	39F	8	8
38B,C	35	22	39G	11	11
38D	31	25	39H	14	14
38F	11	8	39K	23	23
38E,EW	31	23	39L	14	14
39	10	8	39M	3	3
39A,B	33	26	51	13	13
			Total	395	297

Experimental Design Group

Group	Number of Cells	Cells in Service at Strike	Group	Number of Cells	Cells in Service at Strike
24A	20	1	41,42,43	20	14
25–28	16	11	44A,B	4	3
32,33,46	9	4	45,45A	10	6
34	4	1	49,49A–C	9	6
38EX	10	4	50	2	2
			Total	104	52

(C) Category III—Design Groups Used to Establish the Secondary Damage Claim

Standard				Experimental	
Group	Number of Cells	Group	Number of Cells	Group	Number of Cells
39M	9	51G	11	52	5
39N	13	55	6	53	5
51	4	55A	5	54	8
51A	9	55B	8	Total	18
51C	13	56	24		
51D	30	56A	6		
51E	38	56B	2		
51F	34	Total	212		

SOURCE: Exhibits P-108, P-112, P-113.

experimental. For each design group, the table shows the total number of cells in the group and the number of these that were in service at the time of the strike. Cells of a group that were not in service at the time of the strike had failed before the strike. For example, Group 18 contained a total of 20 cells, of which 17 failed before the strike and 3 failed after the strike. For some groups, such as Group 39C, the entire set of cells experienced the strike. In total, Category II has 20 groups of standard design containing a total of 395 cells (297 of which were in service at the time of the strike) and 10 groups of experimental design containing a total of 104 cells (52 of which were in service at the time of the strike).

Category III contains groups of cells that were installed and operated during the unsettled period following the strike. Failure data for these cells were central to the company's secondary damage claim.

Table 2 shows failure data for an illustrative design group drawn from each category in Table 1. Space limitations make it unfeasible to present the data sets for all groups. The cells in each illustrative group have been ordered by failure age. For Group 18 in Category II, two age figures are given for each cell. The first figure is the failure age of the cell. The second figure is the age of the cell at the time of the strike or, in cases where the cell had failed before the strike, the age that the cell would have been if it had not failed by that date. For example, among the 20 cells in Group 18, 17 of the cells failed before the strike while three (18, 19, and 20) failed on or after the day of the strike. Cell 1, for example, failed at age 468 days and would have been 2236 days old if it had survived until the day of the strike. In contrast, cell 20 failed at age 2541 days, 254 days after the strike. On the day of the strike, its age was 2287 days.

2. Expert Testimony

In total, the plaintiff used three statistical experts in the equipment-damage segment of the case (P1, P2, P3). The defense employed two (D1, D2). A brief sketch of the role and testimony of each of these experts follows. The order of their introduction corresponds to that of the actual case.

Expert P1

The first expert called by the company (P1) was a senior company engineer who had extensive experience in the aluminum industry and, especially, at the Baie Comeau plant. He was responsible for the compilation of the failure and related data on cell operations that were used in the case. Indeed, much of the raw data on which subsequent statistical testi-

TABLE 2. Examples of Failure Data for Design Groups

(A) Category I—A Design Group Used to Establish the Typical Failure Pattern

Group 35

Cell	Failure Age (in days)	Cell	Failure Age (in days)	Cell	Failure Age (in days)
1	104	9	420	17	676
2	159	10	476	18	741
3	310	11	586	19	745
4	315	12	613	20	829
5	328	13	631	21	866
6	395	14	649	22	997
7	407	15	655	23	1053
8	411	16	665		

(B) Category II—A Design Group Used to Establish the Primary Damage Claim

Group 18

Cell	Failure Age (in days)	Age at Strike (in days)	Cell	Failure Age (in days)	Age at Strike (in days)
1	468	2236	11	1554	2192
2	725	2223	12	1658	2195
3	838	2197	13	1764	2195
4	853	2204	14	1776	2211
5	965	2226	15	1990	2268
6	1139	2209	16	2010	2253
7	1142	2196	17	2224	2281
8	1304	2199	18	2280	2280
9	1317	2275	19	2371	2245
10	1427	2245	20	2541	2287

(C)Category III—A Design Group Used to Establish the Secondary Damage Claim

Group 51F

Cell	Failure Age (in days)	Cell	Failure Age (in days)	Cell	Failure Age (in days)
1	229	13	1025	25	1373
2	319	14	1047	26	1377
3	437	15	1051	27	1407
4	635	16	1062	28	1429
5	740	17	1116	29	1438
6	808	18	1132	30	1453
7	814	19	1138	31	1472
8	849	20	1162	32	1561
9	935	21	1212	33	1624
10	960	22	1311	34	1691
11	1017	23	1328		
12	1019	24	1372		

SOURCE: Exhibit D-9.

mony was based was introduced by him. In addition to testimony on engineering aspects of pots and potline operations, his testimony also included explanations of the data records and record-keeping process, interpretations of the data and the findings of his own analysis of them. The report containing his findings became Exhibit P-9; a key reference document for subsequent expert testimony.

Complete records on cell design, installation date, failure date, and the like were kept routinely by the company. Fortunately for the company, a year and one half before the strike, P1 had prepared a four-year forecast of cell failures for all cells in service at the time (1 January 1966) as well as for all cells that would be installed in the course of routine replacement up to the end of 1969. The failure forecasts had been derived from plots on normal probability paper. The assumption that failures followed a normal probability law was not new in the industry. Browne (1968), for example, described analytical methods that were based on the assumption of normally distributed cell lives that had been in use for some time at the Aluminum Company of Canada. His article, introduced as Exhibit P-79, was important in the subsequent testimony of expert P2.

For each design group having cells in service at the beginning of 1966 and also having some failures by that date, P1 had used a normal probability plot to estimate the failure ages of the cells in the group that were still in service at the time of the strike. In essence, his graphical procedure is a method for estimating right censored observations in a normal sample. Figure 2 illustrates the basic steps of the procedure for a hypothetical group of 9 cells, of which 6 are presumed to have already failed. The failure ages of the remaining 3 cells are to be estimated. As the figure shows, the 6 known failures are plotted on normal probability paper at, say, percentage points 10, 20, . . . , 60. A straight line (representing the associated normal distribution) is fitted to the 6 points and extrapolated to the percentage points 70, 80, and 90. The failure ages corresponding to these 3 points are read from the graph and form the required estimates for the 3 cells still in service.

For each design group which was created after 1 January 1966 (for the purpose of providing replacement cells), the parameters of its straight line on normal probability paper were chosen by P1 on the basis of previous experience and judgment. The estimated failure age for each individual cell was then read directly from the plotted line as illustrated in Figure 2.

P1's four-year forecast, based on this procedure, showed the predicted total cell failure count for the plant on a monthly basis for the years 1966 to 1969 inclusive. In Exhibit P-9, he compared the actual failure counts with the predicted failure counts and noted that, in months following the strike, actual failure counts were sharply higher than the predicted

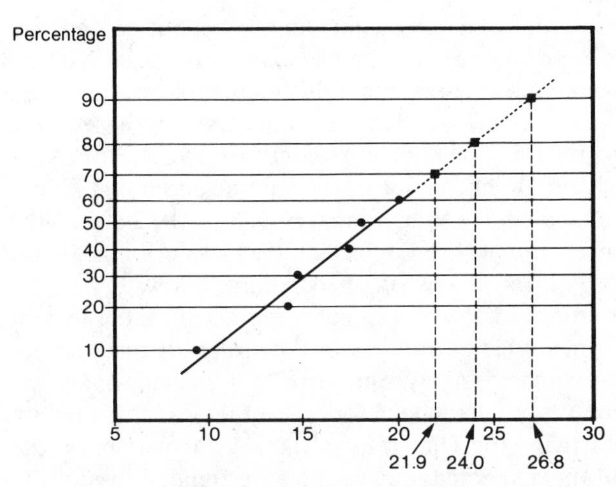

Figure 2. Illustrative use of a normal probability plot to estimate failure ages of cells still in service for a hypothetical group.

counts. Table 3 summarizes P1's figures based on this counting process. Note that in the 17 months from January 1966 to May 1967 his predicted failure count (104) almost matched the actual count (100). By the end of December 1969, however, the actual failure count had pulled ahead of the predicted number by 108 cells. P1 argued that the accuracy of his forecasts prior to the strike indicated that his forecasting system was reliable and, therefore, he concluded that the accelerated failure rate after the strike was caused by the strike.

TABLE 3. Expert P1's Four-Year Forecast of Cell Failures for 1966–1969[a]

	Number of Cell Failures				
	Predicted		Actual		Cumulative
Period	**Period**	**Cumulative**	**Period**	**Cumulative**	**Difference**
Jan.–Dec. 1966	72	72	67	67	−5
Jan.–May 1967	32	104	33	100	−4
Jun.–Dec. 1967	49	153	67	167	14
Jan.–Dec. 1968	93	246	157	324	78
Jan.–Dec. 1969	78	324	108	432	108

[a] Forecast prepared 1 January 1966.

SOURCE: Exhibit P-9.

P1's comparison of actual and forecasted failure counts preceded a more detailed assessment of the primary and secondary damage. Normal probability plots remained the cornerstone of his methodology, however, in this more detailed assessment. His damage estimates were based on a reconstruction of the probability plots from which the original four-year forecast had been derived. He chose not to employ the additional failure experience that had accumulated from January 1966 to the date of the strike in obtaining these estimates; an approach that was criticized later by the defense expert. A brief review of P1's findings follows.

First, consider the primary damage estimate, representing the loss of operating life for cells in service at the time of the strike. In cases where there were some failures in a design group prior to 1 January 1966, the normal probability plot used by P1 gave a forecast of the expected failure age of each of the cells that experienced the strike. A comparison of the actual failure age with this expected value gave an estimated loss of life for the cell that could be attributed to the shutdown. Where no failures or few failures had occurred in a design group prior to 1 January 1966, P1 proceeded in one of two ways, depending on whether the cell was one of standard or experimental design. P1 claimed that variants of the standard design at that time had an average service life of over 1500 days. To quote him (Exhibit P-9, p. 2):

> Initially at Baie Comeau, furnace lives were low (300 days or more). As a result of a continued experimental programme, a considerable improvement was effected. In particular, type 18 (20 cells) started in 1961 gave an average life of over 1,500 days and type 18A (21 cells) started in 1962 an average life of over 1,800 days. From early in 1963, the standard type of construction (type 38 and seq.) was changed to the above type 18 and 18A design and an average life for all standard type cells of over 1,500 days was indicated.

In his analysis, P1 used 1500 days as the assumed average life of a cell of standard design, presumably because this figure would yield a conservative estimate of damage. Thus, he took the arithmetic difference between 1500 and the actual mean life of a standard design group as an estimate of the average loss of life for cells in the group. For experimental cells, P1 argued that it was reasonable to assume that the loss of life was proportional to that experienced by standard cells, based on a direct ratio of the number of cells of each type that were in service at the time of the strike. The final results of his assessment of primary damage are displayed in the first line of Table 4. He concluded that 97,916 days of service had been lost from the 349 cells in service at the time of the strike. Based on the assumed mean life of 1500 days per cell, this damage quantity represented the equivalent of 65 cells (i.e., 97,916/1500 = 65).

TABLE 4. Summary of Experts' Estimates of Losses of Operating Life of Cells, Attributable to the Strike

Expert	Methodology		Primary Damage			Secondary Damage		
			Standard	Experimental	Total	Standard	Experimental	Total
P1	Normal censored-sample method-1	Cells	60.5[a]	4.8[a]	65	—	—	22
		Days	90,751[a]	7,165[a]	97,916	—	—	33,330
SOURCE: Exhibit P-1								
D1	Life-table method-1	Cells	24.6; 27.5	7.4; 12.7	32.0; 40.2	none		
		Days	38,651	11,597	50,248	none		
	Life-table method-2	Cells	—	—	20.2; 25.5	none		
		Days	—	—	30,619	none		
SOURCE: Exhibit D-37								
P2	Life-table method-3	Cells	64.8	12.7; 20.7	77.5; 85.5	not calculated		
		Days	104,707	20,521	125,228	not calculated		
	Normal censored-sample method-2	Cells	50.4	17.0	67.4	58.7	5.2	63.9
		Days	78,063	26,393	104,456	91,048	8,211	99,259
SOURCE: Exhibits P-111, P-112, P-113								

[a] Subtotals for standard and experimental cell groups calculated by the author from summary data in Exhibit P-9, Schedule E.

For the secondary damage effect, P1 compared the predicted failure counts in his four-year forecast with the actual failure counts, for cells installed in the time interval following the strike until the end of 1969. After a few adjustments of the counts (which will not be described further because the author does not fully understand them), he concluded that the probable loss of life was 33,330 days or, based on the mean of 1500 days assumed earlier, the equivalent of 22 cells. These results are also presented in Table 4.

Thus, on the basis of $10,000 per cell, P1's testimony established the primary and secondary damage amounts as $650,000 and $220,000, respectively. The total of these two amounts was rounded to $850,000 in the company's statement of claim.

Expert D1

To counter P1's testimony, the defense then introduced their first expert witness, D1. D1 was a university professor whose area of specialization was operations research applications in the field of mining and metallurgy. D1's critique of P1's report and the results of his own analysis constituted Exhibit D-37. In brief, D1's line of testimony proceeded as follows.

First, he claimed that the assumption of a normal probability law for cell failure age was unsupported by the data. To make this argument, he showed the results of some tests that he performed. His test methodology is discussed subsequently.

Second, once the assumption of the normal law was rejected, he proceeded to argue that actuarial or life-table methods should have been used. D1 then computed an estimate of the primary equipment damage based on life-table methods for censored data. The calculation of the estimate had two components. First, D1 estimated the survival distribution for cells of the standard design by pooling failure experience from all the standard design groups that experienced the strike. Next, he estimated the survival distribution for experimental cells by pooling failure experience from all of the experimental design groups which experienced the strike (see Category II of Table 1). The differences between the total days of service actually obtained from the cells in service at the time of the strike and the total days of service that would be expected on the basis of the estimated survival distributions gave D1 his primary damage estimate for standard and experimental cells, respectively. The estimates were 38,651 days and 11,597 days, respectively. D1 then converted these estimates into cell equivalents in two different ways: (a) based on the mean life of cells in service at the time of the strike and (b) based on the mean life of a new cell as estimated from the survival distribution. For standard cells, he calculated the mean values to be 1573 and 1404 days, respectively, and, for experimental cells, 1565 and 911 days, respectively. Therefore, D1 stated that the primary damage for standard cells amounted to either 24.6 or 27.5 cells and, for experimental cells, amounted to either 7.4 or 12.7 cells, depending on which mean value was used in the conversion from units of days to units of cells. Finally, toward the end of his report, D1 argued that the distinction drawn between standard and experimental designs was arbitrary and concluded that all design groups should be pooled in estimating the survival distribution. Following the same calculational steps as before, he arrived at a combined primary damage estimate of 30,619 days. This quantity was converted into two cell-equivalents: 20.2 cells, based on the mean life of cells in service at the time of the strike (1515 days), and 25.5 cells, based on the mean life of a new cell (1200 days). Table 4 summarizes D1's damage estimates.

Finally, with regard to the secondary damage estimate, D1 rejected P1's claim of damage amounting to 22 cells with a short statement that said simply that the claim was based on forecasts for cells not in service at the time of the strike and for which no data was available.

Expert P2

The lawyers for the company now decided to bring in an expert to bolster the testimony of the company engineer (P1) and to challenge the conclusions of the defense expert (D1). Expert P2 was also a university professor whose field of specialization was industrial applications of statistics.

P2's testimony had several parts. First, he reconsidered D1's challenge of the assumption of normally distributed failure times and offered evidence that the assumption was either adequately valid for the purposes of the case or, possibly, would lead to a conservative estimate of damage. Second, he claimed that although the life-table approach was sound in principle, D1's execution of the method was incorrect and, therefore, D1's estimates were in error. When P2 used what he claimed to be the correct procedure, he arrived at results close to those obtained by P1 using normal probability plots. Third, P2 went on to recompute the damage estimates using the assumption of censored normal samples but based on an analytical rather than a graphical procedure.

P2's estimates based on the life-table method are given in Table 4. Note that he did not use the life-table method to estimate secondary damage. P2 employed the estimated mean life for standard cells (based on his own actuarial calculations), namely, 1616 days, to obtain the estimate of 64.8 cells for primary damage to standard cells. The estimate of primary damage for experimental cells converted to 12.7 cells when he used the assumption that failing experimental cells would be replaced by standard cells (having a mean of 1616 days). He added in a technical note that the estimate converted to 20.7 cells when it was assumed that the experimental cells (having a mean of 992 days) were replaced in kind. He observed that the latter conversion had been used by the defense expert D1.

P2's estimates of damage based on the normal censored-sample approach are also shown in Table 4. He concluded on the basis of this method of analysis that the mean life for a standard cell was 1550 days (recall that P1 had concluded that the mean was over 1500). The conversion to cell equivalents was based on this mean value. His estimate of primary damage, computed by this method. was 67.4 cells. To estimate secondary damage, P2 considered all cells installed after the strike up to the end of 1968. His estimate of secondary damage for standard cells was based on the observed departures of group mean lives from the mean of 1550 days. Like P1, he calculated the secondary damage to experimental cells using the argument that damage to these was proportional to that experienced by the standard cells.

Company lawyers, anticipating that they could exploit the error that P2 testified D1 had made in his life-table calculations and discredit D1's estimates, had asked D1 in cross-examination for a published reference for his computational procedure. D1 cited the source as Dethoor and Groboillot (1968), indicating that he followed the numerical example on pp. 12–14 of the text. Subsequent testimony revealed that the company lawyers and expert P2 then contacted the authors of the text to determine if the text example had been discovered to be in error. Their correspondence with the authors was introduced subsequently by the defense as Exhibits D-81, 82, and 84. Unfortunately for the company lawyers and experts, they received a telegram in reply from Groboillot that (translated) said, "there is no error in the book" (Exhibit D-83) and a subsequent letter signed by both authors that (translated) said, "the estimate of $V(t)$ which you contest (page 14) is perfectly correct" (Exhibit D-85). $V(t)$ here refers to the survival distribution function. With this avenue of challenge blocked, the company lawyers set out to find a third expert witness.

Expert P3

P3 was a practicing actuary who held a graduate degree in statistics. P3's task was to examine the life-table procedures used by D1 and P2 and to decide which was correct. P3 concluded that P2 was correct and that the text example and, hence, D1's estimates based on that example were flawed. He testified to this effect, presenting his findings in Exhibit P-116. P3 was not asked to calculate his own damage estimates, simply to explain why the text example and D1's procedure were in error. It is worth adding that the company lawyers made a special point of establishing under oath that experts P2 and P3 did not know one another prior to the case and did not meet until after P3 had reached a decision about the correctness of the actuarial procedures used by D1 and P2.

Expert D2

The union lawyers now required support for D1 and introduced their second expert. Expert D2 was also a university professor whose field of specialization was mathematics and, in particular, actuarial mathematics. His testimony was a critique of the report of P3 (the company's actuarial expert) and a defense of the methodology of the Dethoor and Groboillot text. His written submission became Exhibit D-89.

Table 5 contains the main elements of the numerical example on pp. 12–14 of the Dethoor and Groboillot book. The disputed calculation is that of the survival function $V(t)$ in Table 5C. The reader is invited to study the example and perhaps will agree with this author's conclusion that Dethoor and Groboillot's numerical calculations of $V(t)$ are erroneous. Successive applications of the failure rates in the last line of Table 5B yield the estimates $V(1) = 1 - 0.011 = 0.989$, $V(2) = 0.989(1 - 0.068) = 0.922$, $V(3) = 0.922(1 - 0.135) = 0.797$, $V(4) = 0.797(1 - 0.300) = 0.558$, and $V(5) = 0.558(1 - 0.57) = 0.240$. These estimates of $V(t)$ differ sharply from those of Dethoor and Groboillot, which appear in the last line of Table 5C.

3. Discussion

Having now introduced the experts and briefly described their testimony, the discussion turns to (a) the constraints that bear on the expert in giving testimony, (b) the lessons in this case for the professional statistician who might be called on to act as an expert witness in a legal proceeding, and (c) a few comments on the tests and analyses performed by the experts.

Constraints on the Expert

To understand why the statistical testimony took the form it did, one must consider the contextual constraints of a legal proceeding for an expert witness. Some of the points made here are also noted by Barton (1983) in describing her personal experience as an expert witness in a different type of case. In this same connection, see the following sequence of articles read before the Royal Statistical Society, with discussion—Downton (1982), Newell (1982), Fienberg and Straf (1982), Napley (1982).

TABLE 5. **Main Elements of the Numerical Example on Pages 12 to 14 of the Dethoor and Groboillot Book**

(A) Survival and Failure Frequency Data for a Sample of Items

Age Interval	Surviving Items: In Age Interval	Surviving Items: Cumulative	Failed Items: In Age Interval	Failed Items: Cumulative
0–1	2	84	2	98
1–2	18	82	12	96
2–3	25	64	20	84
3–4	23	39	31	64
4–5	15	16	28	33
5–6	1	1	5	5

(B) Estimation of Failure Rate by Age Interval

Age Interval $(i, i + 1)$	0 to 1	1 to 2	2 to 3	3 to 4	4 to 5	5 to 6
Number of items failing in age interval $(i, i + 1)$	2	12	20	31	28	5
Number of items having attained age i	84 + 98	82 + 96	64 + 84	39 + 64	16 + 33	1 + 5
Mean failure rate between ages i and $i + 1$	0.011	0.068	0.135	0.300	0.57	0.84

(C) Estimation of Survival Function $V(t)$

Age i	0	1	2	3	4	5
Number of items having passed age i	84 + 98	82 + 96	64 + 84	39 + 64	16 + 33	1 + 5
Number of failed items + number of surviving items having passed age i	84 + 98	82 + 98	64 + 98	39 + 98	16 + 98	1 + 98
$V(i)$	1	0.99	0.91	0.75	0.43	0.06

1. In this court case, neither the presiding judge nor the lawyers had formal training in statistics. Explanations directed to the court therefore had to be presented in layman's terms using familiar analogies and illustrations. For example, one expert interpreted the P-value of a test in terms of the chance of obtaining certain patterns of outcomes in tossing a fair coin. As a second example, the relationship of a survival distribution to the corresponding failure rates in a standard life-table calculation was compared to the depreciation of a sum of $100 by spending successive percentages of the residual amount in each time interval.

Technical notes and references that were contained in the written and oral testimony of the experts were directed primarily at the other experts, although no direct communication occurred between opposing experts off the witness stand. The judge and lawyers infrequently asked for detailed comments on these notes. Indeed, in this case, the mathematical bases of the tests and estimation methods were rarely the subject of examination or cross-examination in court. For example, the mathematical correspondence of the graphical and analytical methods for normal censored samples was not examined in the proceedings. Also, the theoretical bases of the various tests for normality were not explained to the court. No direct mathematical arguments were used even in the attempt by the expert witnesses to explain the relationship of the life-table calculations to the mathematical formulas of reliability theory in the debate about the correctness of the Dethoor and Groboillot example.

2. Testimony in this trial was given in both French and English, depending on the mother tongue of the witness. Hence, the court had to contend with the presentation of statistical terms, concepts, and methods in two languages.

3. The experts were drawn into the case sequentially so each inherited the context and posture established by earlier testimony. Some of the subsequent comments on methodology will illustrate the sequential dependence of the testimony. The implication of this dependence is that the expert must be prepared to build on previous testimony and to consider methods, tests, and other demonstrations which dovetail with that testimony. In other words, the expert will not have a free hand in deciding how his or her expertise will be applied. Of course, the expert faces a predicament in this regard because he or she must strike a balance between a theoretically and conceptually best argument and one that, although not the best, maintains the continuity of argument and will be understood and accepted by the court. The need for a strong personal code of ethics and sense of professional responsibility is especially felt under this constraint.

4. The expert's testimony is presented in a question-and-answer format, which is standard in judicial and quasi-judicial proceedings. In direct testimony, this requires that the lawyer and expert work as a team. The basic pattern of the question-and-answer sequence is worked out in advance, but each must be prepared to prompt the other if the course of the proceedings dictates a departure from the script. Lawyers are generally afraid, however, to let an expert wander too far from the agreed format in case the opposing lawyers in cross-examination might drive a wedge into an otherwise tight argument. An inadvertent slip, an ambiguous phrase, or an incomplete explanation can lead to confusion, misunderstanding, or weakening of a case from which the expert may not be able to fully recover because of procedural limitations.

In summary, then, experts are not appointed by the court but instead are retained by the opposing sides in the case. They are not free, therefore, to reach a consensus among themselves on the best methodology, to apply it, and to report their common findings. As rational as that approach may seem, the adversarial justice system does not accommodate it. Moreover, the expert must be prepared to work within that adversarial system with "one arm tied behind his back" as far as his technical expertise is concerned. The task of the statistical expert is not for the faint of heart.

Lessons for the Expert

Several important general lessons for the prospective statistical expert arise from the case. The first lesson is that arguments or demonstrations that draw on established and accepted practices, are consistent with conventional wisdom, or are based on relevant practical experience are more readily accepted than those that fly in the face of tradition, custom, or experience. In essence, these accepted items constitute background or prior evidence that adds extra weight to the expert's testimony. For example, as the judge's written decision makes clear, the industry experience of expert P1 added a great deal of credence to his testimony. Particularly risky, it seems, is testimony that is viewed as reflecting an ivory-tower mentality, untried, naive, or theoretically speculative.

The second lesson is that published reference material also carries great weight in court. In a certain respect, carefully chosen references are equivalent to bringing supporting experts into court, namely, the authors of the works being cited. This material can act for you if it is introduced on your side or against you if an opposing expert introduces it. In this court case, several cited publications played a very important role. The Browne article and the Dethoor and Groboillot text are but two of many that might be mentioned. In his testimony, expert P2 pointed to errors in other published material submitted in evidence in order to support his contention that published works occasionally contain errors and, specifically, that the Dethoor and Groboillot text contained an important error. Interestingly, the other error pointed out by P2 was in a formula and associated calculations for a chi-squared test in an excerpt from a statistics text submitted in evidence by the opposing expert.

A third lesson is that any expert testimony that is given should be capable of being corroborated *independently* by another expert—and the stress here on the word independently is intentional. In this case, it was important that experts P2 and P3 did not know one another previously and did not draw their expertise from a common base. Moreover, no expert should be lulled into the belief that his or her demonstrations and arguments are beyond challenge by an opposing expert who is equally qualified in the eyes of the court. With finite evidence, it seems that both sides of any statistical question can be defended effectively and the expert who is careless enough to think otherwise will present a weaker argument in consequence.

A final important lesson that this case demonstrates clearly is the need for the expert to be extremely meticulous in his or her work and thorough in record keeping. Even the smallest error or anomaly in a written submission can cause serious trouble. To give several illustrations from the case, expert P1, in plotting n observations on normal probability paper recorded the ith-order statistic at percentage point $100(i/n)$. You can imagine the difficulty of explaining where the largest observation is plotted. Expert D1, according to the testimony of P2, had inadvertently omitted dividing by the sample size in computing coefficients of skewness and kurtosis that were submitted as evidence that the failure data were not normally distributed. Expert P2, through an oversight, submitted an exhibit that had marked on it, in pencil, "not proofread." Expert D1, in interpreting a P-value of 20% in a chi-squared goodness-of-fit test for normality, concluded that the result meant there was an 80% chance of the hypothesis being wrong.

Notes on Methodology

The case had a number of interesting methodological aspects to which the discussion now turns. The objective here is not to reargue the case or to examine the evidence or methods in detail. Rather it is to focus on a few of the more interesting substantive statistical issues that are raised by the case.

The assumption of normally distributed failure ages played a key role in the testimony of the experts. P1 had assumed it in using normal probability plots and, in fact, his testimony and that of other experts made clear that the assumption had been made by others in the industry. D1 used two types of tests to challenge its validity—a chi-squared goodness-of-fit test and a test of coefficients of skewness and kurtosis. The former he applied to the failure data of a single group, Group 30, and obtained a test statistic of 3.6 with 2 degrees of freedom. The latter he applied to the failure data in Groups 24A, 30, 35, and 37 (see Category I of Table 1). D1's selection of these test procedures led P2 to consider an expanded application of the same tests. Moreover, the use of normal probability plots by P1 had prompted the defense to question him about the acceptable extent of variation of sample points about a fitted straight line. The need to strengthen this aspect of the plaintiff's case led P2 to use the Lilliefors test (Conover, 1971) as well as the chi-squared and coefficient tests mentioned above. Briefly, P2's analysis began with the correction of D1's calculations of the skewness and kurtosis coefficients for Groups 24A, 30, 35, and 37. He found that the corrected coefficients did not deviate significantly from the normal theoretical values, and furthermore, he found that the larger deviations that did occur were produced by a single large failure age in the sample. The latter observation, he argued, made damage estimates based on the assumption of normality conservative. For both the chi-squared test and Lilliefors test of fit, he considered all 10 design groups in Category I of Table 1. He found no test statistics that exceeded the 5% significance level with either test procedure. He also added the test statistics for all 10 chi-squared tests to obtain an overall test statistic of 38.38 with 34 degrees of freedom. This concluded the tests that P2 performed to support the assumption of normality.

The life-table methods used by D1 and P2 involved nothing unconventional or noteworthy except (a) the dispute described earlier about whether or not the Dethoor and Groboillot text example was in error and (b) the arguments for and against pooling of groups of cells in the application of the method. Basically, the plaintiff argued that pooling was not sound, whereas the defense held the opposite view. To a large extent, the pooling question hinged less on statistical considerations and more on the validity of the company engineer's testimony that differences among experimental design groups and between experimental and standard groups were material.

The application of the normal censored-sample methodology is worth a few remarks. First, in terms of the probability plots used by P1, there was the difficulty noted earlier of selecting the appropriate percentage spacings for plotting the observations. There was also the pointed questioning by the defense on the reasons the normal model was used when, in fact, the plotted points for observed failures did not always appear nearly linear, especially at the ends of the data range where the plotted points tended to show more scatter. P2 testified that one reason he used the Lilliefors test was that it could be executed on the same normal probability paper, showed the potential for points to scatter at the end of the data range, and had the appeal of being based directly on the maximum deviation between the actual and theoretical survival distributions. Second, the normal censored-sample method used by P2 was that of Saw (1959). This analytical method yields estimates that are structurally like maximum likelihood estimates but are unbiased for all sample sizes. The property of unbiasedness was not taken up in the expert testimony. One might imagine, however, what impression would be left with a lay audience when an estimator is acknowledged to be biased—even if statistical theory might show that the bias is immaterial or that the estimator has other desirable properties. Here is an example of where the connotation of a technical term might prove to be crucial in expert testimony.

Several other interesting statistical issues that surface from the expert testimony and exhibits are the following.

1. There was a variety of methods used to convert damage quantities measured in days to cell equivalents. The conversion problem is not totally statistical as it requires judgments about how to measure the economic worth of lost service and what might be the cell replacement policy followed by the company after the strike. Still, in this case, it was the statistical expert who was required to make the conversion, and there was no consensus among the experts on the appropriate conversion method.

2. In estimating the loss of operating life for cells experiencing the strike, P1 estimated the failure ages of the cells in service at the time of the strike and deducted the actual failure ages to obtain the loss figure. In the life-table method and in P2's application of the normal censored-sample method the loss of life was computed as the difference between the estimated mean for the group based on the censored sample and the actual group mean. With any specific set of failure data, the two methods do not yield identical results. Neither the defense's nor plaintiff's experts discussed this potential discrepancy.

3. In standard design groups where few or no cells had failed at the time of the strike, P1 and P2 chose to use the mean value (1500 or 1550, respectively) drawn from experience with other standard groups in calculating the loss of operating life. Neither expert used the information content of the censored samples to obtain an estimate of residual mean life for each of the groups, even though this approach would have increased the corresponding damage estimates. Not surprisingly, therefore, the defense experts did not challenge the plaintiff's expert testimony on these grounds. For example, Group 39G (see Category II in Table 1) consisted of 11 cells that were all installed about a year before the strike, and none had failed by the strike date. Yet this information was not used in estimating the mean life for this group. Instead, as noted above, P1 and P2 took the means to be 1500 and 1550, respectively.

One might speculate that many statistical refinements such as these and others were ignored on purpose by experts because they were viewed as immaterial in their effect, would clutter an already complicated and technical presentation, or involved situations (such as censored-sample estimation) where intuitive statistical judgment tends to be faulty. The implication of the last point is that there are some statistical concepts that cannot be made meaningful to a lay audience.

4. Outcome

As for the ultimate resolution of the case, the company won its suit. A front page story in *The Gazette* of Montreal on 6 February 1979 summed it up as follows. "The 200,000 member union, one of its locals representing aluminum workers in Baie Comeau and 22 of the workers, were ordered by Quebec Superior Court to pay $5,981,424 plus interest for damages incurred in [the] illegal strike." The union announced that it intended to appeal the court ruling but subsequently agreed instead to an out-of-court settlement with the company amounting to $2.45 million.

As noted in the introduction, this case was complex and lengthy. The part of it that was concerned with the loss of operating life represented only $850,000 of the total claim of nearly $6 million. The judge's written decision makes informative reading. Here are a few of the highlights from his judgment on the equipment-damage segment of the case (translated from the French text).

1. The judge concluded that the strike was the direct and immediate cause of the damages. He branded the uncontrolled shutdown as an act of vandalism by the participating workers, an act for which they would have to bear full and total responsibility.

2. The judge found the plaintiff's demonstration of damages "not only impressive but most adequate," and was "totally satisfied with the evidence" presented. He also stated that the defense could not complain since they "had at their disposal all the documents and necessary explanations in order to present a contestation." The judge assented to the plaintiff's damage claims and, in particular, granted the claim of $850,000 for the loss of operating life for the aluminum-reduction cells.

3. Subsequently, the judge commented on the testimony of the defense expert(s) and made the following statement, one that every statistical expert should bear in mind: "An expert is not engaged in order to plead the cause but rather in order to enlighten the Court on the claims being made by his client." The judge then proceeded to state that the defense expert (presumably D1) "faced a team of men considered by the Court to be both very honest and very competent on the matter." He went on to note that this team had the advantage of having more practical experience in plant operations and greater familiarity with the incident on which the case was based. The judge's reference to a team here seems to include the company engineer (P1) and the other company witnesses but not the statistical experts. The judge added further, "The practical experience of these men must not make one forget their theoretical knowledge and the converse is equally true for the defendants' expert." These notes from the judgment seem to suggest that the judge gave a great deal of weight to the testimony of the experts who had first-hand experience. In contrast, he appears to have sought from the testimony of the statistical experts only a simple confirmation or refutation of the damage estimates calculated beforehand by the company engineer (expert P1).

Acknowledgments

My sincere thanks are extended to The Honourable Mr. Justice William S. Tyndale of the Quebec Court of Appeal, Professor J. D. Kalbfleisch of the University of Waterloo, and the editors for providing helpful comments on an earlier draft of this paper. Justice Tyndale, together with Mr. Casper Bloom, represented The Company in this case.

References

Barton, A. (1983). "My Experience as an Expert Witness," *The American Statistician,* **37,** 374–376.

Browne, P. (1967). "Methods of Making Early Estimates of Potlining Life." Paper presented to 96th Annual Meeting of the American Institute of Mining, Metallurgical and Petroleum Engineers, Los Angeles.

Browne, P. (1968). "Methods of Making Early Estimates of Potlining Life," *Journal of Metals,* **20** (December), 36–40.

Conover, W. J. (1971). *Practical Nonparametric Statistics,* New York: John Wiley & Sons.

Dethoor, J. M., and Groboillot, J. L. (1968). *La Vie des Equipements: Investissement, Renouvellement, Maintenance,* Paris: Dunod.

Downton, F. (1982). "Legal Probability and Statistics," *Journal of the Royal Statistical Society A,* **145,** 395–402.

Fienberg, S. E., and Straf, M. L. (1982). "Statistical Assessments as Evidence," *Journal of the Royal Statistical Society A,* **145,** 410–421.

Napley, Sir David (1982). "Lawyers and Statisticians," *Journal of the Royal Statistical Society A,* **145,** 422–438.

Newell, D. (1982). "The Role of the Statistician as an Expert Witness," *Journal of the Royal Statistical Society A,* **145,** 403–409.

Saw, J. G. (1959). "Estimation of the Normal Population Parameters Given a Singly Censored Sample," *Biometrika,* **46,** 150–159.

Société Canadienne de Métaux Reynolds c. Confédération des Syndicats Nationaux et al. (1980). *La Revue Légale,* 253–436.

Whitmore, G. A., and Gentleman, J. F. (1982). "Measuring the Impact of an Intervention on Equipment Lives. Case Studies in Data Analysis," *Canadian Journal of Statistics,* **10,** 237–259 (with contributed analyses by J. D. Kalbfleisch and C. A. Struthers and by D. C. Thomas).

A Question of Theft

William B. Fairley
Analysis and Inference, Inc.
Boston, Massachusetts

Jeffrey E. Glen
New York City Law Department
New York, New York

ABSTRACT

Several employees of Brink's, Inc., were convicted of theft from collections of the New York City's parking meters. The city sued Brink's for negligence in supervision of its employees and asked for compensation in the amount of the stolen coins. This chapter first discusses the legal background and criteria for proof in court that shaped the statistical question asked, and the precision required in its answer. It then lays out the statistical analysis presented on behalf of the city and a conflicting analysis presented on behalf of Brink's. The basis for the award given by a jury to the city in district court and its successful defense in the court of appeals is discussed.

1. Introduction

Brink's v. City of New York is a lawsuit in which statistical analysis was crucial to obtaining a verdict in a civil damages action involving frequently litigated questions of negligence and breach of contract (546 F.Supp. 403 (1982)). It is unusual for statistical analysis to be used by attorneys in "ordinary" cases like *Brink's;* the discipline of statistics has most frequently been called on in such specialized areas as antitrust law, where questions of market share have proved susceptible to statistical evaluation, and civil rights law, where proof of disparate racial or sexual impact often requires statistical analysis (Finkelstein, 1978; Baldus and Cole, 1980; Sugrue and Fairley, 1983; Fairley, 1983).

Lawyers rarely have enlisted statistical experts to prove damages in more familiar situations; our experience in *Brink's* indicates both that attorneys should consider such an approach to contract and tort cases and that statisticians should begin to ponder recasting their methods of analysis and presentation so as to make their findings usable, and persuasive, in a broad range of legal matters.

The facts in the *Brink's* case were summarized in the decision on appeal by federal Judges Oakes, Cardamone, and Winter of the United States Court of Appeals for the Second Circuit (717 F.2d 700 (1983), 702–718):

> The City of New York owns and operates approximately 70,000 parking meters, mostly on-street, but some in metered parking lots. Daily collections average nearly $50,000. After public bidding in March of 1978, Brink's was awarded the contract to collect coins from these meters and deliver them to the New York City Department of Finance Depository. Brink's was reimbursed at the rate of 33 cents per day for each meter it collected. Operations under the contract commenced in May of 1978.
>
> The process of collection involved collectors working in teams, with one individual going to each parking meter on a city-prescribed schedule. To collect a meter, the collector would insert a meter key, open the bottom portion of the meter head, and remove a sealed coin box. This coin box would then be placed upside down onto a gooseneck protruding upward from a large metal canister on wheels, resembling a dolly, that the collector rolled along from meter to meter. By twisting the coin box, an unlocking device called a "cannister key," attached to the canister by City personnel and theoretically not reachable by the collector, would allow the coins from the coin box to drop into the canister without the collector's having access to the coins. After the coin box was emptied the collector would replace it and lock the meter. Upon completion of a specific route the canister would be placed in a collection van. At the end of a given team's work on a particular day, the van was driven back to the PMD [Parking Meter Division] and the canisters were turned over to City personnel.
>
> As can be seen from the above, collectors were not to have contact with the coins in the meters. City personnel were to check each collection canister daily to ensure that it was in good working order; upon receipt of a canister the Brink's people signed a receipt attesting that it was in good working order. Under the contract, at the end of the day any canister malfunction or broken or uncollectible meters were to be reported to the City. Brink's was to provide ten three-person collection crews daily, the crews to be rotated in the discretion of the City Department of Finance as a security measure. The City directed that the rotation should be accomplished by a lottery system to prevent the formation of permanent teams, but the City's evidence was that the rotation system was ignored. Daily assignments were frequently made by the collectors themselves and the management of Brink's was aware of this but did nothing to correct it. As the trial judge stated, the jury "could readily have found that

the planned rotation system was honored more in the breach than in the observance.'' 546 F.Supp. at 409. The contract also provided that Brink's would provide supervisory personnel to oversee the proper performance of its obligations. Although the contract called for two supervisors and one field inspector, the full complement was never assigned.

In response to an anonymous tip, the City's Department of Investigation, in conjunction with the Inspector General's Office of the Department of Finance, began an investigation of parking meter collections. Surveillance of Brink's collectors revealed suspicious activity violative of both the City's and Brink's rules and procedures. The investigators then ''salted'' parking meters by treating coins with a fluorescent substance and inserting them into specific meters. ''Salted'' meters were checked after the coin boxes had been emptied by Brink's employees to make sure that all of the treated coins were collected; collections from the meters were then scanned to see if any treated coins were missing. The ''salting'' process indicated that a substantial percentage of coins collected by Brink's personnel were not being returned to the City. Surveillance at the 42 Franklin Street depository revealed that at the end of a day Brink's employees would often arrive in personal vehicles following their assigned collection vans, indicating some kind of ''drop-off.'' Brink's employees were also seen entering a parking lot in Manhattan in Brink's van, placing them in private automobiles and then returning to the vans to continue to the City depository. The City presented video tapes of these transfers at trial; Brink's collectors were shown straining to lift heavy bags into their cars. Brink's employees were also followed to a private residence and again observed carrying heavy bags from their vehicles into the building and emerging empty handed.

James Gargiulo, Trevor Fairweather, Richard Florio, Michael Solomon and John Adams were arrested and charged with grand larceny and criminal possession of stolen property when on April 9 they had in their possession over $4,500 in coins stolen *that* day from parking meter collections. Anthony DeNardo was arrested and charged with petit larceny and criminal possession of stolen property. A charge against Jorge Olivari was dismissed on motion by the district attorney, but the remaining six defendants were either convicted after trial (Florio, Solomon and Adams) or pleaded guilty before trial (Fairweather, Gargiulo, DeNardo). They were sentenced to varying jail terms and fines ranging from $1,000 to $5,000.

This chapter in outline proceeds as follows. Section 2 discusses legal criteria that governed proof in court of the amount of damages suffered by the city and incurred by Brink's. Section 3 defines the choice of a statistical comparison to yield legally admissible evidence of the amount of theft. Section 4 discusses aspects of the statistical analysis: issues of causal attribution; treatment of trend, seasonality and choice of comparison periods; and a linear model for revenues per meter-day. Section 5 gives the Brink's view of the facts in the case as presented by their statistical expert. Finally, Sections 6 and 7 tell the outcome of the case in the trial and appeals courts.

2. Law of Proving Damages in *Brink's*

The city faced three hurdles in its quest for a substantial verdict in this case. First, it had to persuade the judge that numbers of Brink's employees had stolen coins on numbers of occasions, and that the thefts could be found by a jury to be part of a pattern. If the judge were not persuaded of this, under the applicable law, damages could only have been awarded for thefts as to which there was individualized evidence, such as videotapes of bags being transferred or short-counts from the "saltings."

Fortunately, the judge allowed the city attorneys to demonstrate the pattern of theft as a legal matter through the introduction of evidence, such as videotapes of six days of observations and the results of the "saltings," before the jury. As the jury had heard and evaluated the evidence of each occasion of alleged theft, it is not surprising that the judge allowed the jury to conclude, if it so chose, that the thefts were repeated and in all likelihood occurred when unobserved by the cameras. To allow the jury to decide whether to infer continued misconduct—here theft—from the evidence is well supported in law. See *McFarland v. Gregory,* 425 F.2d 443 (2d Cir. 1979), and *Conner v. Union Pacific,* 219 F.2d 799 (9th Cir. 1955). But such an inference would not, under the case law, permit the jury to decide the amount of the damage caused by the misconduct; additional proof would be necessary.

Second, the city had to persuade the judge that the Brink's company could legally be held liable for the thefts of its employees. In general, an employer is responsible for the negligence or other dereliction of its employees, under a legal doctrine called *respondeat superior,* a Latin phrase that translates loosely as "let the boss be held responsible." This doctrine, under the law of New York that governed this case, permits a court to hold an employer liable even for the intentional "torts" (civil wrongs) of his employees. For example, an employer can be compelled to compensate a visitor to a factory who is punched for no reason by a plant security guard. But this doctrine does not, the trial judge in *Brink's* indicated, go so far as to hold an employer liable for thefts by his employees; as lawyers put it, theft is "outside the scope of employment" of an employee.

There are, however, two other principles that the city invoked in *Brink's*. The city argued that since Brink's had agreed in its contract to collect all the money in the meters and deliver it to the city, Brink's should be considered an insurer. Had the judge agreed with this theory, the city would have been entitled to collect the difference between the money put in the meters and the money that arrived at the counting facility, whether that difference was due to theft, carelessness, or any other reason. The statistical inquiry in that case would have been simplified, but the judge held, before the trial, that this theory would not be used in this case. The court permitted a contract claim to go to the jury, but required that the city prove willful or negligent failure to meet the supervision requirements written in the contract in order to prevail. The jury in fact found for the city on this claim, which had the result of starting interest on the judgment running from the date the contract was breached; under the common-law negligence claim, discussed immediately below, interest begins running only from the date the judgment is entered.

The city's remaining theory was that Brink's had been negligent in supervising its employees and in failing to fire them after it learned—or should have learned—that they were stealing. This breach of the historic common-law duty of an employer to the public is the theory on which the case was finally tried. Because it was relegated to this theory, however, the city was forced to call numerous witnesses from Brink's middle and upper management to establish the company's negligent supervision and retention. This had the effect of diverting the jury's attention from the problem of ascertaining how much was actually stolen.

As it turned out, the negligence of Brink's management, particularly in ignoring reports of its own security personnel that drivers and collectors were stealing, was deemed so gross by the jury that it awarded the city $5,000,000 in punitive damages. Such damages under the law are not to be calculated with reference to actual loss, and are meant to deter defendants and others in similar positions from such egregious behavior in the future. By definition, statistical evidence about the extent of the loss suffered is not relevant to the calculation of such damages.

Finally, the city had to prove the total amount of revenue it had lost from theft. In the absence of counters on the actual parking meters, the precise amount of money deposited could not be ascertained. The city knew that on the day of the arrests over $4000 had been recovered, but it felt sure that other thieves had gotten away with even more money that day. Without an actual count of the loot on any other day, the city lawyers believed that even if the jury were to find that there had been repeated and massive theft, it would have no way to estimate the total taken. Perhaps it would simply multiply the $4000 actually recovered by some number of days it would come up with on which it thought theft had occurred. From analyzing the reported appellate decisions, it seemed likely that such a result would be set aside by the court as mere guesswork, and thus not a legal basis on which to assess damages.

The city's attorneys decided, therefore, to seek the assistance of a statistical expert to attempt to prove how much money Brink's should have collected during its contract, and then, by subtracting actual revenues Brink's had turned in, to calculate the shortfall attributable to theft. How this was done is the subject of the body of this article. But from a legal point of view, it turned out to be crucial that an expert made that calculation. For both the trial judge and the appellate court in *Brink's* stated that the city had to first show (717 F.2d 700 (1983), at 712):

> evidence independent of the City's experts' testimony establishing that systematic theft had occurred over a long period of time,

and then must show a total revenue loss through competent expert testimony.

Where, as here, the expert's calculations are based on comparing revenues delivered in two different time periods by two different collection companies, that expert must devise and employ a methodology that, with a reasonable degree of certainty and in conformity with established and accepted statistical practices, enables him to state that (717 F.2d 700 (1983), at 711):

> differences in various nonculpable factors had been adequately accounted for and that conditions in the comparison periods were substantially the same.

Of course, the methodology must be shaped by the comparison to be drawn; if, for example, conditions had changed during the comparison period but, taking those changes into account, theft was still statistically demonstrable, the expert's testimony would be admissible and if believed by the jury, would support an award of damages. Revenue comparison over different time periods as a general approach was approved in the United States Supreme Court case of *Bigelow v. RKO Radio Pictures, Inc.*, 327 U.S. 251 (1946). Failure to use appropriate data, or to take varying conditions appropriately into account, yields not only bad statistics but bad legal results; an award of damages is likely in such a case to be reversed (see *Herman Schwabe, Inc. v. United Shoe Machinery Corp.*, 297 F.2d 906 (2nd Cir. 1962, cert. den. 369 U.S. 865)). Both the *Bigelow* and *Schwabe* opinions are well worth reading by any statistician contemplating using a comparison method in analyzing the effect of variables on revenue production.

The statistical evidence presented in the *Brink's* case met the basic test that had been established over 50 years ago for cases where the acts of the malefactor—such as theft of untallied receipts—prevent precise calculation of the extent of monetary loss:

> In such case . . . , while the damages may not be determined by mere speculation or guess, it will be enough if the evidence show the extent of the damages as a matter of just and reasonable inference, although the result be only approximate. The wrongdoer is not entitled to complain that they cannot be measured with the exactness and precision that would be possible if the case, which he alone is responsible for making, were otherwise. (*Story Parchment Co. v. Paterson Parchment Paper Co.*, 282 U.S. 555 (1931))

3. Framing of the Statistical Question: The Lawyering Consideration

In looking for a way to quantify the damages, the city's lawyers knew that there was no way to count the money actually put into the meters by drivers, and that the first actual count occurred after Brink's had delivered the coins. They also knew that the same system that had clearly failed to keep Brink's honest had been in use by the prior collection company, when revenues delivered had been substantially below those delivered by Brink's; in retrospect, city officials had their suspicions over the honesty of the predecessor collectors. Immediately after Brink's had been taken off the contract, however, on the day following the arrests, a new collector, the CDC Company, had been hired, and within six weeks had taken over all collection work. In its first 10 months of operation, CDC had brought in nearly $1,000,000 more than Brinks had in any 10-month period.

This company worked under strict surveillance by city investigators, and under a procedure of both routine and surprise polygraph examinations of its drivers and collectors. The city's faith that the new collection company was honest was strengthened when, about four months into the contract, a collector failed a polygraph and confessed that he had in fact stolen a coin box from a meter and kept some $20 in coins. It thus seemed that the new security system was effective in preventing any systematic theft.

The city lawyers thus went to the statistical expert with the idea that a comparison of Brink's suspect collections with CDC's honest collections would be likely to yield legally admissible conclusions. After all, the city knew it could prove at least the last day of theft during the Brink's period, and that it could introduce persuasive evidence of theft for a number of other days. The statistician asked what factors other than theft might account for an increase in collections after Brink's was terminated, since it was at least arguable that the videotapes and ''saltings'' had isolated a small ring of occasional thiefs, and the million dollar difference was otherwise explicable. The city's attorneys went to the heads of the Parking

Meter Division of the New York City Department of Finance, and learned that at least one high-ranking official had made the off-the-cuff comment—duly reported in the *New York Times*—that he thought most of the increase was due to better meter repair. Other supervisory-level personnel pointed to the conversion of some meters from a dime to a quarter and to possible increased car use by commuters because of the end of the 1979 gas shortage as possible causes for increased revenues.

The statistician, William Fairley, played a crucial role at this point, because he questioned whether the suggested factors would account for the dramatic climb in revenues between the last month of Brink's collections and the first month of CDC's. Brief investigation indicated that meter repairs and rate changes in any given month were insignificant, while calculations that had been done in the course of the preparation for trial of the arrested Brink's employees seemed to show extraordinary revenue differentials between Brink's and CDC's collection in certain collection areas frequented by the arrested men. The statistician recommended an in-depth study of comparative collections and a detailed analysis of the changes in revenue potential of the meter plant over as long a period as possible. He also urged the city to hire an expert on parking meter theft, because he could only evaluate the impact of factors that might affect revenue over time; he could not identify these factors.

In what for a budget-conscious municipality was, we believe, a startling leap of faith, the City of New York through its elected Board of Estimate hired the statistician, realizing that his extensive analysis might show that theft was one, but not the only, explanation for Brink's level of revenue production. The city also hired a nationally known expert in the analysis and design of security systems for parking meter plants and parking garages, Laurence Donoghue. When he identified over 40 possible causes of change in parking meter revenues, the analysis seemed doomed. When he was able to conclude that many of these variables either were not present in the New York area during the comparison period, such as the opening or closing of a unique shopping site, were equally present during the Brink's and CDC periods, such as the incidence of legal holidays, or were incapable of making a significant difference, such as the fact that snow emergency streets had parking bans for one more day in the Brink's period than in the CDC period, the picture brightened considerably.

As it turned out, and as the remainder of this chapter demonstrates, the analysis chosen enabled the statistician to conclude that theft accounted for the intercompany revenue difference, and that but for the theft Brink's collections would have exceeded those of CDC for a 10-month comparison period by some $1,400,000. And while one cannot probe the minds of jurors, the jury in fact returned a verdict for the city of $1,000,000 in compensatory damages, and this verdict was affirmed by both the trial judge and the court of appeals.

4. The City's Factual Case

Goals of Statistical Analysis

Theft of unknown dimensions had occurred over an indefinite period of months or years prior to the arrest 6 April 1980 of seven Brink's collectors. The factual question to be answered in determining the amount of damages suffered by the city was how much had been taken.

The law does not here, or in other contexts, spell out precisely the criteria it seeks for estimates in terms of the theories of statistical inference. However, the fact-finding goals of a trial and the roles assigned to the judge and jury suggest the following requirements for the precision and bias of estimators. As discussed above, in determining damages the law does not require any specified level of precision in the estimate, but it does require the use of available data and appropriate methods of estimation as opposed to pure conjecture or speculation as to the amount.

The law looks for an unbiased estimator in the subjective sense that the maker of the estimate have no more reason to believe the estimate too large than too small (see *Sargent v. Mass. Accident Co.* 307 Mass. 246 (1940)). Translated into the language of Bayesian inference, the maker of the estimate could if he chose report the median of a posterior distribution for the quantity. The goals of the statistical analysis could be summarized nontechnically as determining a reasonable estimate of the amount of theft, given available data and practical limitations on the depth of the investigation determined by time and money budgets.

Causal Attribution

The approach chosen to estimate the amount of theft was to compare revenues delivered by Brink's prior to the arrests with revenues delivered by the subsequent contractor, CDC, after the arrests. In making such a before-and-after comparison to determine the size of effect of a known causal intervention, it is useful to distinguish two types of threats to a causal interpretation. These threats to causally interpreting the observed difference as most reasonably attributable to the intervention—as opposed to other factors—are (a) a difference before and after could be due to *general time effects* of trend or seasonality, namely, a general trend in revenues for whatever reasons, or seasonal effects in the periods before and after; or (b) a difference could be due to *specific causes* that could account for the change before and after.

A two-step approach was therefore taken to investigate the validity of interpreting a before-and-after difference as a theft amount estimate. The first step was to investigate the existence of the general time factors of trend and seasonality. The second step was to consider whether other causes were reasonable candidates to explain the kind of difference observed.

Seasonality and the Choice of Periods of Comparison

The effect of seasonal differences on a difference between the two periods was controlled both by choice of periods and by seasonal adjustments.

The two comparison periods of 10 months each were chosen to be the same 10 calendar months, namely, June 1979 through March 1980 for Brink's and June 1980 through March 1981 for CDC. Since the same months were compared, seasonal differences in the two periods were eliminated when comparing the full two periods. Seasonal adjustment was still of interest for the purpose of comparing revenues in the months immediately before and after the transition. The two-month gap in April and May of 1980 between the two comparison periods was necessary because CDC was not fully collecting the same routes as Brink's until June 1980.

A period of 10 months was chosen because in the spring of 1981, when data collection for the study was undertaken, only data for 10 months up through March 1981 was available for CDC, the successor contractor to Brink's. The choice of these 10 months for CDC then suggested matching these with the same 10 months for Brink's one year earlier. Data earlier than this for Brink's was not obtained because the quality of the data was uncertain and a more distant period would likely be subject to more change in other factors that would be difficult to study.

Several adjustments for the purpose of comparing months' revenues immediately preceeding and succeeding the transition were made by a model for the data, as discussed below.

Trend

The question of a general trend over the 22-month period June 1979–March 1981, consisting of 10 months of Brink's collections, a 2-month gap, and 10 months of CDC collections, was investigated graphically and analytically.

Graphs of average monthly revenues delivered per meter-day of operation for the entire city and for each of the five boroughs, either without seasonal adjustments or with monthly seasonal adjustments, did not indicate a trend over time but rather indicated level revenues per meter-day within each 10-month period. Seasonally adjusted monthly revenue per meter-day data were displayed by dividing each city (or borough) actual average for the month by the linear model-predicted monthly revenue (by model defined below) and multiplying by the period average for the city (or borough).

Piecewise-linear regressions were fitted to data on monthly revenues over the 22-month period, one allowing a change in level between the two periods but fitting the same slopes, and the other a change in level as well as a change in slopes between the two periods. Both regressions estimated a positive jump in level and neither estimated an upward trend over both periods.

Fitting a piecewise-linear regression has the following advantage over fitting a single regression line. If there is a trend over the entire 22-month interval, the piecewise-linear regression will estimate it, and if there is a change in level, it will also estimate that. A single trend line, however, would not estimate a change in level if it were present. Thus, a single trend line fit provides no way to measure the amount of theft.

Figure 1 plots the residuals from a single line fitted to all of the data. The single trend line tends to underestimate the early months and to overestimate the later months in both periods, with this effect being most pronounced in the second period. The pattern of the residuals strongly indicates misspecification in the fit by a single line.

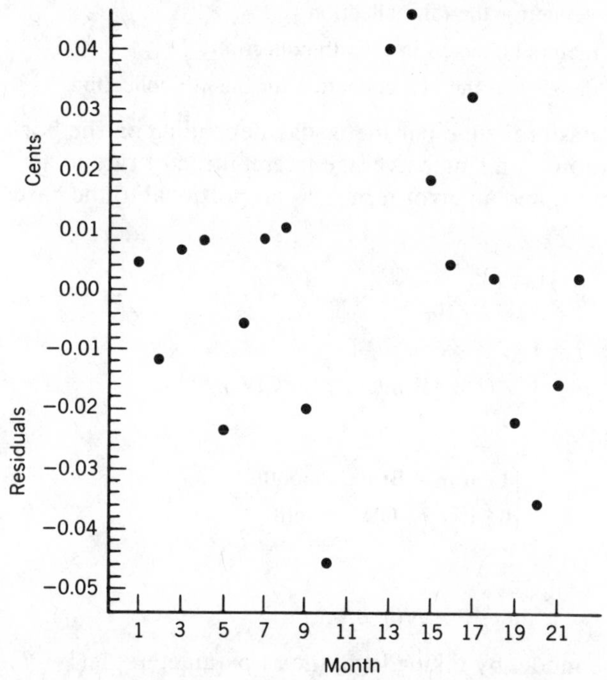

Figure 1. Residuals from trend line fitted to citywide seasonally adjusted revenues per meter-day.

The conclusion both from graphical inspection and from regression analysis is that there is no evidence of a rising general upward trend over the 22-month period that would account for the higher average revenues found in the second 10-month period versus the first 10-month period. Rather, there is evidence for a change in level at the transition.

Linear Model for Revenues Per Meter-Day

A model was specified to explain the dependent variable of average monthly revenues per meter-day for each contractor as a function of an intercontractor differential rate and of month-by-borough factors. The dependent variable monthly average revenues per meter-day by borough for the ith month, jth borough, and ath contractor is denoted by $w_{ij}{}^{(a)}$ and is defined as

$$w_{ij}{}^{(a)} = \frac{\sum_k c_{ijk}{}^{(a)}}{\sum_k (m_{ijk}{}^{(a)} d_{ijk}{}^{(a)})}$$

where

$c_{ijk}{}^{(a)}$ = reported revenue for the kth collection
$m_{ijk}{}^{(a)}$ = number of meters collected in the kth collection
$d_{ijk}{}^{(a)}$ = number of days since the last collection for the kth collection.

The model posits a base revenue per meter-day depending on the borough–month combination t_{ij}, a single average intercontractor rate in any borough and any month f, and an error term $e_{ij}{}^{(a)}$ proportional to the base rate:

$$\begin{aligned} w_{ij}{}^{(a)} &= (1-f)^{Z^{(a)}} t_{ij} e_{ij}{}^{(a)}, \\ i &= 1, \ldots, 10; \\ j &= 1, \ldots, 5; \\ a &= 1, 2 \text{ (1 for Brink's, 2 for CDC)} \end{aligned}$$

where

$$Z^{(a)} = \begin{cases} 1 & \text{if } a = \text{Brink's month} \\ 0 & \text{if } a = \text{CDC month} \end{cases}$$

and

$$\ln e_{ij}{}^{(a)} \sim N(0, \sigma^2).$$

After linearizing the model by taking logs, the 51 parameters, $\ln(1-f)$ and $\ln t_{ij}$, are estimated by least squares using the 100 observations for borough–month combinations.

TABLE 1. Residuals from Trend Line Fitted to Citywide Seasonally Adjusted Revenues per Meter-Day[a]

X	Y	YHAT	RESID
1	0.863	0.858351	0.004649
2	0.852	0.863208	−0.011208
3	0.875	0.868065	0.006935
4	0.881	0.872922	0.008078
5	0.854	0.877779	−0.023779
6	0.876	0.882636	−0.006636
7	0.895	0.887493	0.007507
8	0.903	0.892350	0.010650
9	0.878	0.897207	−0.019207
10	0.857	0.902064	−0.045064
13	0.956	0.916636	0.039364
14	0.968	0.921493	0.046507
15	0.944	0.926350	0.017650
16	0.936	0.931207	0.004793
17	0.968	0.936064	0.031936
18	0.943	0.940921	0.002079
19	0.923	0.945778	−0.022778
20	0.914	0.950635	−0.036635
21	0.939	0.955492	−0.016492
22	0.962	0.960349	0.001651

[a] Key: X = month; Y = revenues per meter-day, actual; YHAT = revenues per meter-day, predicted; RESID = residuals.

A constant intercontractor rate model is consistent with the observed indication of proportionality in the amounts by which CDC revenues per meter-day exceeded Brink's in every borough and in 47 of 50 borough–month combinations.

There are two principal advantages of this model over the piecewise-linear regressions for estimating an intercontractor rate. First, the model provides more degrees of freedom for estimating the parameters than do five separate regressions for each borough. Second, the model is parameter-rich, fitting all borough–month interactions, thereby making minimal assumptions about the way these two factors relate to revenues per meter-day.

The regression R^2 was 0.9937 and $s_E = 0.0438$, thus giving approximately a 4.4% prediction error for $w_{ij}^{(a)}$. Brink's predicted versus actual values shows good fit over the entire range of values of y (the dependent variable). The predicted values of average monthly revenues per meter-day for both Brink's and CDC closely match the actual values in each borough and in each month.

The borough–month parameters were almost all highly significant, and the intercontractor parameter $\hat{\beta} = \ln(1 - \hat{f})$ was highly significant:

$$\hat{\beta} = -0.0815$$
$$\hat{\sigma}_{\hat{\beta}} = 0.0088$$
$$t = -9.29.$$

The estimated intercontractor proportion $\hat{f}$ is

$$\hat{f} = 1 - e^{\hat{\beta}} = 0.0782$$

which, unless other factors indicate a different rate for diversion (see discussion below), is estimated diversion as a rate of expected revenue without diversion. The rate of expected revenue with diversion is

$$\hat{f}/(1 - \hat{f}) = 0.0849 \approx 8.5\%.$$

The total amount estimated diverted in the 10-month Brink's period can therefore be determined as

(Diversion rate from expected revenues)(Expected total revenues in 10 months)

$$= \hat{f}\left(\sum_{ij} \hat{w}_{ij}^{(2)} u_{ij}^{(1)}\right)$$
$$= 0.0782 \times \$17{,}674{,}608$$
$$\approx \$1.4 \text{ million}$$

where

$$u_{ij}^{(1)} = \text{number of meter-days collected by Brink's in month } i \text{ and borough } j$$
$$= \sum_k m_{ijk}^{(1)} d_{ijk}^{(1)}.$$

Equivalently, total amount estimated diverted in the 10-month Brink's period can be calculated as

(Estimated amount diverted per meter-day)(Number of meter-days)
= (Diversion rate from expected revenues)
× (Expected revenues per meter day)(Number of meter-days)
= 0.0782(0.9452)(18,698,946)
≈ (7.4¢ per meter day) × (18,698,946 meter-days)
≈ $1.4 million.

The model estimate of diversion rate is smaller than would be determined by first looking at the difference in revenues over 1, 2, 3 months.

For example, the jump in revenue in the gap between contractors between March 1980 and June 1980 was 13.9%. By adjusting for number of meter-days in each month and for seasonal effects in March and June the predicted revenue per meter-day for Brinks is 88.2¢ in March and for CDC is 98.4¢ in June, for an increase of only 11.6% instead of the observed 13.9%. The model therefore is useful for getting a better estimate of diversion by controlling for other factors.

Other Specific Causal Factors

The general time factors of long-term trend and seasonality do not explain a jump in level that occurred at the transition between Brink's and CDC. But perhaps there are specific causal factors other than theft that can explain the jump. Given the facts in the case, such factors should have the following four properties:

1. **Suddenness.** Be capable of causing an upward shift in level over a two-month gap.
2. **Sizableness.** Vary in magnitude or effect enough to be capable of causing a shift of about 8.5% over Brink's level over a two-month gap.
3. **Uniqueness.** Vary in a way that explained the shift in level at the two-month gap and the absence of a shift in level elsewhere.
4. **Uniformity.** Be capable of explaining a shift in level over the two-month gap in every borough.

Price changes of meters was a factor that could potentially have all four required properties. It was thought worthwhile therefore to investigate this factor. Six so-called Area Description tapes containing information on price changes for all 70,000 meters in the city were examined for changes that would increase or decrease the effective price. There were five types of such changes: size of coin accepted by meter; maximum allowable time limit; hourly rate; active days; and active hours. A count of changes expected to decrease and changes expected to increase revenues over the two-month transition period showed 73 increases and 158 decreases, so that the direction of effect on revenues in terms of the frequencies of such changes would be to decrease expected revenues in the month CDC took over below that expected from Brink's at the end of its tenure. The total of all changes in the two-month transition period was 231 or 0.33% of the 70,000 meter population. Over the entire 22-month period the number of changes expected to decrease revenues slightly exceeded the number expected to increase them, and the excess was greater in the CDC period. Therefore, again, the expected change in revenues was a decrease for CDC versus Brink's.

Thus, the net expected direction of effect of the factor of price changes was downward both at the transition and over the entire 22-month period. This factor in fact fails every one of the four tests of an explanatory factor.

A number of other factors were considered as potential explanations of the jump in level, including meter maintenance, meter back-up, city collections, installation and removal of meters, and the gas shortage. For some factors, pertinent data were available that threw light on whether changes in the factor satisfied the four properties of causal factors. The expert on parking and meter operations was asked for his judgment on the likely revenue effects of a list of factors that were not explicitly controlled for in the statistical model. He testified that he had no more reason to believe that the uncontrolled factor would cause an increase than a decrease, either over the transition or over the entire 22-month period.

In sum, no evidence was presented that demonstrated that specific causal factors other than theft would account for the jump in level between the two periods. Of course it is impossible to say that there were no such factors, for there could be. The jury, however, had to determine the most reasonable estimate it could of the amount of theft. Given the complexity of factors in the world that influences meter use, no complete causal model could ever be expected, certainly none within the time frame of the trial. A general statistical model, specified and estimated in such a way as to explicitly control for, or to estimate, the net effects of time-varying factors, is the only way that such an estimate could be made.

5. The Brink's Factual Case

Brink's had a different view of the facts. Their principal statistical witness was Bruce Levin, a statistician and professor at Columbia University. He testified after the city's experts, William Fairley and Laurence Donoghue.

Levin's conclusion was that other factors besides theft *could* explain the greater revenues delivered to the city by CDC in the 10-month period, June 1980 through March 1981, than delivered by Brink's in the same 10-month interval a year earlier, June 1979 through March 1980. He stated that on statistical grounds alone it was not possible to attribute either all or any part of the difference in revenues to theft.

These conclusions were reached not only by an analysis of the same monthly revenue data used by the city's experts, but also by using additional monthly revenue data going back to the beginning of the Brink's contract in May 1978. The analysis was done separately on data over the period of the Brink's contract for what is apparently the only area of parking meters in the city whose meters had been consistently collected by city personnel and not by either contractor. This area was coded in city records as Area 1A.

The analysis of revenue data was directed at the question of whether a "trend" or a "trendline" existed in the two areas consisting of (a) the city as a whole excluding Area 1A and (b) Area 1A by itself. Two approaches to this question were taken. The first was to calculate five-month moving averages of monthly revenues over the period of the Brink's contract and the CDC contract, May 1978 through March 1981, in the two areas.

The graph of moving averages for the city excluding Area 1A showed a generally rising series of points with some small jags downward and a fall in the last several months of the CDC contract. The graph for Area 1A showed some indication of rising level but with considerable up and down variability. The conclusion reached was, first, that there was an upward trend operating over the period of the Brink's contract, and second, that the increase in revenues in Area 1A in the CDC period in excess of the revenues in the Brink's period showed to a reasonable degree of certainty that other factors besides theft were accounting for the increase in revenues observed citywide.

The second approach to the data was to fit a regression line separately to monthly city revenue excluding Area 1A, and to monthly Area 1A revenue, in the period of the Brink's contract, May 1978 through March 1980. The line fitted to the monthly city revenue excluding Area 1A had a statistically significant positive slope of $17,960 per month revenue increase and a correlation coefficient of 0.69. When this line was extrapolated to the CDC period, the predicted excess of revenues in the 10-month CDC period over the corresponding 10-month Brink's period a year earlier was $2,155,000, compared to an observed excess of approximately $1,000,000. The significant positive slope of the fitted line established the existence of a trendline, and this trend could account for the existence of excess revenue in the CDC period over the corresponding Brink's period. The line fitted to monthly Area 1A revenue also had a positive slope. The conclusions from this regression analysis paralleled those from the moving average analysis. Trend accounted for by other factors than theft could explain the excess revenue delivered by CDC in its 10-month period.

Levin did not testify that he believed *specific* other causal factors explained the observed difference in revenues between the CDC and Brink's periods. He did testify that each of a list of several specific factors raised in questions by counsel to Brink's were potential biasing factors between the two periods, leading him to believe in a "strong possibility" that the two periods compared were "not comparable."

On cross-examination Levin agreed that there was no evidence of trend within either of the two comparison periods, the CDC 10-month period and the corresponding Brink's 10-month period.

Cross-examination also brought out the fact that there were some 45 meters in total in Area 1A, the total meter plant having some 70,000 meters. Since the analysis of monthly revenues was not on a per meter-day basis, this implied that a change of only one in the meter population of Area 1A due to disrepair, addition, or removal could be expected to cause about a 2% change in revenues.

6. Jury Verdict and Judges' Ruling and Opinion

The Federal District Court jury found Brink's liable for negligence with respect to supervising its employees and inspecting the process of collection. It awarded the city compensatory damages of $1,000,000 for loss of revenues due to theft and $5,000,000 punitive damages.

The trial judge, Judge Edward Weinfeld, ruled that the evidence sustained the jury's findings of negligence and the amount of damages, although he did reduce the amount of punitive damages awarded to $1,500,000. In his opinion he compared the methods of analysis presented by the two sides that produced conflicting evidence at trial and explained that such conflict was the jury's role to resolve, not by guess or speculation, but by finding a fair and reasonable estimate based on the evidence presented.

7. Appeal on the Factual Case

Brink's appeal of the case to the U.S. Court of Appeals on the factual aspect of the case contended that the trial judge had erroneously permitted the jury to hear expert testimony about the amount of theft, which was wrong because the CDC and Brink's periods were not comparable; that other factors the City's experts did not consider might well have accounted for the difference in revenues; and that the evidence of a trend in revenues could explain the difference. Brink's argued that the damage estimate was too speculative and the jury finding of a damage amount for the city was based not on a rational consideration, but on a delusive impression of exactitude generated by an array of figures.

In rejecting this basis for appeal, the court made several points. First, while Brink's advanced the *possibility* of other factors explaining the difference, it nowhere introduced evidence that these factors did explain the difference. On the other hand, the city's experts studied differences in various nontheft factors and Fairley concluded that these had been adequately accounted for, and that there was no apparent basis for thinking that factors left out biased the estimate made of difference attributable to theft. The jury believed the city's experts.

Second, the fact that the city's experts could not rule out the possibility that other factors explained some of the estimated theft amount and that Fairley could only offer what he thought was the most reasonable estimate, did not imply that the estimate presented required the jury to guess or engage in mere speculation to arrive at damages.

Third, the estimate was not used to establish the *existence* of damages due to theft. Systematic theft over a long period of time had been independently established, and an estimate was required of the amount of these damages.

Acknowledgments

The authors thank contributors to the project: Patricia Kruger, Wendy Love, Linda Lutz, Linda Scharff, JoAnn Tankard Smith, and Edward Tuozzo. Peter Kempthorne proposed the linear model discussed in Section 4.

References

Baldus, D. C., and Cole, J. W. L. (1980). *Statistical Proof of Discrimination,* New York: McGraw-Hill.

Fairley, W. B. (1983). Statistics in law. In S. Kotz and N. Johnson (eds), *Encyclopedia of Statistical Sciences,* New York: John Wiley & Sons, Vol. 4, pp. 568–577.

Finkelstein, M. O. (1978). *Quantitative Methods in Law,* New York: Free Press.

Sugrue, T. J., and Fairley, W. B. (1983). "A Case of Unexamined Assumptions: the Use and Misuse of the Statistical Analysis of *Castaneda/Hazelwood* in Discrimination Litigation," *Boston College Law Review,* **34,** 925–960.

Comment

Bruce Levin
Division of Biostatistics
Columbia University School of Public Health
New York, New York

Fairley and Glen (FG) have discussed a two-period comparison of gross revenues that raises two problems I wish to discuss in this comment. The first is their assumption that the data are best described by two horizontal lines without trend, separated by a discontinuity at the April–May 1980 juncture. The second is that the dollar size of the gap (or the intercontractor rate parameter in the FG model) is entirely attributable to theft. It cannot be too strongly emphasized that such assumptions are not to be granted without careful scrutiny in an observational study, for it is well known that potentially large biases can result from uncontrolled confounding factors. In my view the justifications advanced by FG are unconvincing on both methodological and substantive grounds, rendering their estimate of theft unpersuasive. I believe that, contrary to FG's conclusion, the data at hand do not support more than a speculative estimate of the amount of revenue differential attributable to theft.

Consider first the question of trend. FG address this issue via a hypothesis test of zero trend versus upward trend in a broken-line model (with and without a parallelism assumption) based on data limited to the two 10-month periods (with and without "seasonal adjustments"). Figure 1 presents the average daily revenue received per collection day for the *entire* period of Brinks' contract (May 1978–April 1980) together with the average daily revenues in the subsequent CDC period. The data exclude revenues received from Area 1A that was at all times collected by city employees and not exposed to theft by Brinks employees. (The revenue per meter-day unit used in FG is a figure representing the average revenue per meter per day. The unit of revenue per collection day used in this comment indicates monies returned by the collection crews on an average collection day. The latter unit seems more closely related to the measurement of theft. While it does not adjust for variations in the size of the meter plant or in the number of days between collections, it does not appear that either factor changed significantly between the two 10-month periods used by FG. Neither measure adjusts for other variables discussed in the sequel.) The trend line superimposed on the graph was produced with Cleveland's robust locally weighted regression algorithm for smoothing scatterplots (Cleveland, 1979). The smoothing fraction used was 0.3, the local weight function was bicubic, and one robust itera-

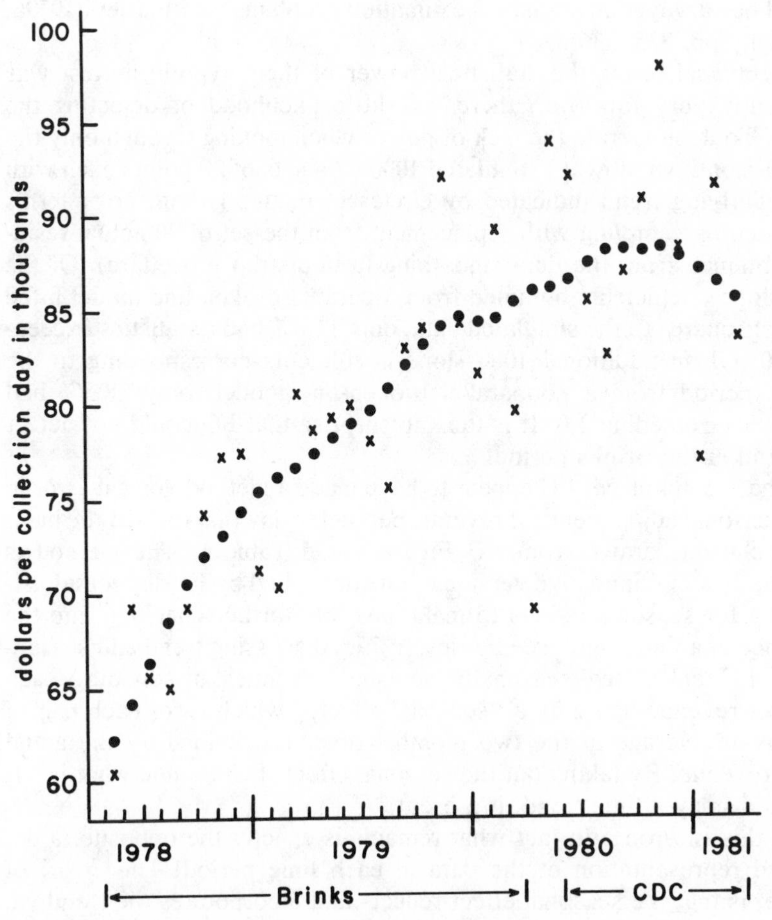

Figure 1. Average monthly revenue per collection day.

tion was used with a bisquare residual weight function (see the reference for definitions of these terms). The picture shows evidence of a general upward trend throughout the Brinks period tapering off into the CPC period. It is apparent that the total effect of all factors influencing revenues was *not* constant during the Brinks period.

How can FG have overlooked this trend? In the first place, a time base of only 10 months in the Brinks period is too narrow to establish the trend clearly. The reasons FG advance for failing to consider the previous year's worth of data are weak; one cannot ignore a major portion of relevant data simply because its "quality" was "uncertain" or affected by factors more "difficult to study." In fact Fairley has forcefully and correctly argued elsewhere that effects are not reliably estimated from a brief span of experience, and that the widest related base of experience should be surveyed in statistical estimation problems (see Fairley, 1979a, especially pp. 335–336).

In technical terms the statistical power of their hypothesis test was apparently very low (i.e., there was little likelihood of detecting the trend). To demonstrate the lack of power when looking through only the two 10-month windows, I simulated 1000 data sets of 20 points each with the underlying trend indicated by circles in Figure 1, with error terms generated by sampling with replacement from the set of 20 actual residuals obtained from the data smoothing (a bootstrap procedure). Of the 1000 slope coefficients obtained from a parallel broken-line model fitted by least squares to the simulated data, only 11.6% had t-statistics exceeding 2.0. Of an additional 1000 slope coefficients corresponding to the Brinks period from a nonparallel broken-line model, only 28.3% had t-statistics exceeding 2.0. It is thus no surprise that FG could not detect the trend in the Brinks period.

In the second place, FG appear to have made a methodological error in their seasonal adjustments of revenue per meter-day that formed the basis of conclusions drawn from FG Figure 1 and Table 1. The method is apparently a multiplicative version of one described by Fairley for adjusting data for seasonal effects to make possible further checking into the presence of a time trend (see Fairley, 1979b, discussing trend and seasonality). FG calculated seasonally adjusted revenues by dividing each month's revenue figure by a "seasonal effect," which is for each pair of months an average of the two months' revenues divided by the grand mean revenue. By taking out the seasonal effect, FG assume what is left should display a time trend, if present.

But this is wrong. In fact what remains is exactly the opposite: a detrended representation of the data in each time period! The error, of course, is that the seasonal effect reflects and incorporates the trend, so that when removed from the unadjusted data, the trend disappears. To see this clearly, suppose that the revenue figures W_{ij} for month $i = 1, \ldots, 10$ in period $j = 0, 1$ displayed a single log-linear time trend with arbitrary deterministic monthly effects denoted by d_i. (We ignore random error, which is here beside the point.) Thus

$$X_{ij} = \ln W_{ij} = a + b(i + 12j) + d_i.$$

Note there is no break assumed in the time trend between the two periods. The "seasonal effect" in Fairley's method is, for month i,

$$\begin{aligned} X_{i\cdot} - X_{\cdot\cdot} &= [a + b(i + 6) + d_i] - [a + b(5.5 + 6) + d_\cdot] \\ &= b(i - 5.5) + (d_i - d_\cdot), \end{aligned}$$

where we have used simple averages in the log scale. Then the seasonally adjusted figures become the antilogs of

$$X_{ij} - (X_{i\cdot} - X_{\cdot\cdot}) = (a + 5.5b + d_\cdot) + 12bj,$$

which takes the form of two perfectly flat lines separated by a gap between the time periods. The appearance of the data values in column 2 of FG's Table 1 may be a result of the defective deseasonalization. To properly deseasonalize a time series, the overall trend must be estimated and subtracted out first, and the seasonal effects calculated from the *residuals,* before subtracting from the unadjusted data. The result will then be a clearer picture of underlying trend.

Summarizing the first problem: to accept the null hypothesis of no trend in the Brinks period and no trend in the CDC period with a dollar gap in between, against the alternative model of positive trend (with or without a gap), may be an error of type II, due to low power, misleading adjustment, or both. On substantive grounds, to conclude that there is no trend in the Brinks period is to assert that the factors contributing to the trend line in our Figure 1 abruptly abated in May 1979. In my view this is not a reasonable inference to draw.

Turning to the second problem, the assumption that the difference between the two periods is primarily due to persistent theft must compete with another explanation that plausibly would account for a higher revenue level in the CDC period. There was an unusual gasoline shortage that prevailed in New York City from May to December 1979 that undoubtedly reduced automobile use and with it probably reduced metered parking. There was odd–even gasoline rationing during the four-month period June–September 1979, and automobile toll bridge and tunnel revenues showed marked drops during the shortage. In addition, there was higher revenue from New York City buses and rapid transit from June to December 1979 compared with CDC's corresponding period in 1980.

The gasoline shortage and rationing appear to have suppressed automobile usage in the Brinks period. There was an additional factor that may have increased automobile usage during the CDC period: an employee strike of the Port Authority Trans-Hudson subway line into New York City from June to August 1980, during which time automobile traffic between New York City and New Jersey increased by 10.7% over the previous year's period.

FG's model assumes a constant percentage increase in revenue from each month in the Brinks period to the corresponding month in the CDC period, and a constant percentage increase in all boroughs. The data do not support the first of these assumptions, and the second seems implausible. Over the two 10-month periods, the percentage increase in average revenue per collection day was 5.9%. However, in the first four-month subperiod (June–September 1979 and 1980, respectively) the percentage increase was 9.9%. This greater percentage increase appears plausibly related to the presence of odd–even gas rationing in the first period. By contrast, in the later six-month period (October 1979–March 1980 and similarly in 1980–1981) the percentage increase was only 3.5%. The change in percentage increase is thus fully consistent with a causal explanation based on the gasoline shortage and the transit strike, and is inconsistent with FG's model. As for the second assumption of their model, I think it unreasonable to assume that the same percentage of receipts were stolen in every borough in every month.

With regard to factors affecting the meter plant itself, FG show in their article that meter price changes were not responsible for period revenue differences. Additional factors unrelated to theft were introduced at trial that are not discussed by FG, however, such as a reduction in the number of parking meters with "revenue-affecting defects" in the CDC period, and the continuation of the program of relocating meters from low-revenue, high-vandalism areas to high-revenue, low-vandalism areas. We have only the testimony of the city's expert on parking meters that *he* could not determine the direction of these biasing effects. This seems surprising, since as a matter of common sense any effect of these changes would be in the direction of increasing revenue. The expert did state that his consulting firm had not made any in-depth studies of New York City's parking meter system in order to make such a determination.

There was another line of evidence concerning Area 1A. This metered parking area in the vicinity of the federal courthouse in lower Manhattan was always collected by city employees and therefore formed a natural "control" area for comparative purposes. The reasoning was that if one found percentage increases and trends in this area comparable to those in the areas collected by Brinks, then an attribution of theft would be less compelling. In fact, in the 10-month period comparison, the average revenue per collection day received by the city from Area 1A increased by 6.9%, an even larger percentage increase than the 5.9% increase in the city overall, excluding Area 1A. A simple linear trend line fitted to the entire period as in Figure 1 showed a highly significant slope coefficient ($P < 0.001$). While numerically smaller than the citywide slope coefficient, the effect is still present and clearly is not attributable to theft. As FG point out, one should be cautious about conclusions drawn from Area 1A since it is a nonrandom sample of the city's meter plant. It does raise serious concern about causal attribution, however, and certainly similar caution should be elicited over the estimate of theft.

On the general issue of comparability, each factor that has been cited above might reasonably be expected to cause a difference in meter revenue between the two comparison periods. Whether they *did* cause a difference in the presumed direction, and to what extent, is largely unknown, and I could not therefore testify that these factors did indeed have the presumed effect. The absence of such testimony was apparently interpreted as a showing of period comparability. This is an error of inference. Given the factors as they have been presented, and given the parking meter expert's failure to quantify the effect of all but the most trivial of these, there must be a strong presumption of noncomparability, because to assert the contrary would require a coincidental cancellation of effects that is difficult to justify. This position is consistent with our understanding that uncontrolled observational studies will, in the presence of a preponderance of biasing factors, produce biased results unless exceptional circumstances prevail.

FG note that a causal factor other than theft should be able to account for the occurrence of a sizable, sudden upward shift across a two-month period that is unique to that period and that occurs in every borough. We have already seen that precisely these effects may arise as an artifact of the seasonal adjustments FG used. Beyond this point, however, it appears that sizable, sudden upward shifts are common in these data. Referring to Figure 1, the inherent month-to-month variation in the data produces several instances of such shifts: see, for example, February–March 1979 during Brinks' tenure, or January–February 1981 in the CDC period. In fact, the latter jump is of greater magnitude than the jump from March to June 1980. It is apparent that the scatter of points in the CDC period is greater than the scatter of points in the Brinks period. Under these circumstances the appearance of a sizable jump across a two-month period by chance alone is not improbable. The larger point to be made here is that we are observing a highly noisy process in which the contribution of theft was one possibly small component. When data are highly regular, for example, in well-controlled comparative trials, the ascertainment of even small effects is relatively easy; when the signal-to-noise ratio is low, then estimation of specific effects becomes inaccurate. Because of systematic period noncomparability and relatively large inherent variability,

I believe the statistical data in this case can be used to provide little more than speculative estimates of theft.

What statistically valid techniques might have been employed to provide a reliable estimate of the extent of theft? One possibility would have been proper use of the "salting" technique. In this technique coins that are specially marked with an invisible fluorescent dye are deposited directly into the coin boxes and later counted under ultraviolet light in the returned revenue. This technique was in fact used in the case, but there was dispute over the evidentiary status of many of the saltings. Brinks' principal objection was that "checkbacks" had not been made to guarantee that uncounted marked coins were indeed missing. The method might have been used to advantage had sound statistical procedures been used, including randomized salting across temporal, geographic, and collecting-team strata; data quality control, including checkbacks on all salted meters; and periodic replications to gauge the rate of theft at various times. With these procedures the city might have obtained a quantitative estimate, free of confounding biases, and with the authority of a well-made scientific measurement.

I have criticized two aspects of Fairley and Glen's analysis. First, the methodology used to address the question of trend is defective because the time intervals they study are too narrow to accurately gauge the trend effect, and because their method of seasonal adjustment artifactually produces a pattern which is improperly interpreted as a meaningful signal in the data. Second, the attribution of period revenue differences entirely to theft remains largely speculative due to the presence of uncontrolled systematic biasing factors and substantial month-to-month variability. Reasonable alternative explanations for the revenue differences are presented that are unrelated to theft and that cannot be easily dismissed.

Additional References

Cleveland, W. S. (1979). "Robust Locally Weighted Regression and Smoothing Scatterplots," *Journal of the American Statistical Association*, **74**(368), 829–836.

Fairley, W. B. (1979a). Evaluating the "small" probability of a catastrophic accident from the marine transportation of liquefied natural gas. In W. B. Fairley and F. Mosteller (eds.), *Statistics and Public Policy*, Reading, Mass.: Addison-Wesley, pp. 331–353.

Fairley, W. B. (1979b). Accidents on Route 2: two-way structures for data. In W. B. Fairley and F. Mosteller (eds.), *Statistics and Public Policy*, Reading, Mass.: Addison-Wesley, pp. 24–33.

Rejoinder

William B. Fairley / Jeffrey E. Glen

Setting aside several technical issues, which we address below, we find that Levin's basic theme is that factors other than theft "might have" caused most of the observed increase in revenues, so that "the contribution of theft was one *possibly* small component" (italics ours).

This point, while true, is not responsive to the determination of the damage amount that the jury was asked to make. As Judge Edward Weinfeld said in his District Court opinion, concurred in by the U.S. Appeals Court, the question of damages before the court was not what a precise and accurate estimate of the damages was, but rather, what was the *most reasonable* estimate of damages that could be made *given* that it *had already been established* that "Brink's employees engaged over an extended period in concerted action in the pilferage of meter coin deposits" (546 F.Supp. 403 (1982), at 407). Indeed, reading Levin's comment one might not have guessed that substantial investigative evidence had been introduced in court to support an estimate of $1,400,000 theft over 10 months and even more. Judge Weinfeld's opinion noted that

> The evidence offered by the City included surveillances of suspected employees who were observed to have deviated from their assigned routes carrying heavy bags from the transport vehicles into an apartment house where one of the employees resided; the results of a salting test which showed substantial unaccounted for coins at various dates and involving different teams of collectors; videotaped surveillances of occasions prior to and on April 9, 1980, when a number of employees were arrested, which showed them acting in what appeared to be a concerted, clandestine, and secretive manner with respect to heavy bags allegedly containing meter collection proceeds; the arrest on April 9, 1980 of the group of employees who had in their possession almost $5,000 in coins; and the conviction of a number of the arrested employees either upon pleas of guilty or by a jury verdict for stealing meter coin deposits. (at 407)

Therefore, Judge Weinfeld observed

> To argue, as Brink's does . . . that there was a complete absence of evidence to support the verdict on this issue . . . is to ignore the force of the totality of evidence and to adopt an ostrich-like pose. (at 407–408)

In seeking completely comparable periods demonstrably "free of confounding biases," Levin erects a pristine standard for statistical control. While such a goal is our ideal, the real world must sometimes be studied by observational studies where control is less than perfect. In an earlier paper, one of the present authors (Fairley, 1978, pp. 794–795) said:

> The limitations of purely observational data and the parallel limitations on the capacity of statistical methods to control for important confounding effects have not been widely understood.
>
> . . . Yet we should avoid the untenable position that experiments alone can support cause-and-effect inference. Much scientific advance, not to mention practical knowledge, has been based on observational studies—and the "observational" sciences of astronomy and geology are not alone in this . . .

In the same vein Hoaglin *et al.* (1982, p. 74) note:

> Comparative observational studies may offer the only means of ever collecting any data at all about a treatment. Some events, such as earthquakes, are impossible to deliver as designed treatments, and certain treatments are unethical, as in many medical and social investigations. In these extremes, investigators must do their best with data from observational studies.

Levin offers no alternative estimate of theft in the fact of the inherent difficulties. All that he offers is a reference to a possible systematic use of "salting," which, *had* it been used, would have provided a quantitative estimate free of confounding biases. Wonderful! But it had not been done, and so the court had to deal in the real world with the data and evidence at hand.

The root problem is that Levin is importing into a court decision-making context a set of standards and conventions for "evidence" that are *not* the same as those that are required by the applicable law. Specifically, Levin is willing to remain in a *dubitante* position and not offer any alternative estimate that he believes is more reasonable. He clearly doesn't want an inference threatened by real-world possibilities of the error of confounding bias. None of us wants such threats, but the court plainly cannot remain in a *dubitante* position. It must decide. The ruling legal doctrine in a case like this, where the *existence* of theft was not in dispute, and where the jury *had* determined tortious conduct on the defendant's part, is that (546 F.Supp. 403 (1982), at 410, and quoting *Bigelow v. RKO Radio Pictures, Inc.* (327 U.S. 251 (1946)) "when a defendant's tortious conduct is of a nature that precludes precise ascertainment of damages, the jury may make a 'just and reasonable estimate of the damage based on relevant data.' "

As support for the *dubitante* position Levin claims that the estimate of theft that we advanced was "little more than speculative." In the relevant case law "speculation" is likened to a "guess" and is not viewed as legally sufficient evidence to support an estimate of damages. However, the statistical and subject matter study of parking meter theft done by the experts testifying for the city was substantial. It was clearly not a "guess" and therefore it was not mere "speculation." In fact, although no observational study could demonstrate that *no* other explanation than theft was possible, the study did rule out or diminish important alternative explanations. No *more* reasonable estimate of the amount stolen was ever proffered.

Were every situation in which a difficult estimate had to be made deemed a "guess" and no estimate were allowed, then defendants in these situations would go scot-free, despite their admitted wrongful conduct. Better an estimate based on a deliberate and careful attempt to provide the most reasonable figure permitted by the evidence than no estimate at all. At least that is what the law requires, as the District Court and the Appeals Court agreed. Levin's requirements for an estimate differ from the law's, and clearly it is the law's requirements that apply.

We turn now to an examination of Levin's specific criticisms of the statistical estimates of theft described in our paper.

Levin says in the first half of his comment that "trend" could account for the increase in revenues in the CDC period over the Brink's period and that we overlooked "this" trend.

We do not agree that there is a "trend." First, if we look, not at monthly revenues in the period from April 1978 through March 1981 selected by Levin and graphed in his Figure 1, but rather at the longer 48-month period May 1977 through March 1981 as provided in a Brink's exhibit, the picture presented is very different. Figure 1 in this rejoinder graphs the monthly revenues for the longer period. The points in the early period are heading down, not up. Thus the longer period displayed in our Figure 1 does not convey the same impression of "a general upward trend" as does Levin's Figure 1.

However, which figure to use is not the point. While we agree that looking at historical context can be useful, one *can* and should here discuss the question of whether one observes, statistically, a trend within *any* specified interval of time. In particular, we can ask if, within each of the two 10-month periods on either side of the arrest of Brink's employees, there is evidence of trend. The answer, graphically or by test, is clearly no. This answer, furthermore, from the point of view of inference, is not tainted by selection bias because the two 10-month periods were selected before looking at the time series, not after. (Our Figure 2 in this rejoinder gives city-wide revenues per meter-day over the two periods.

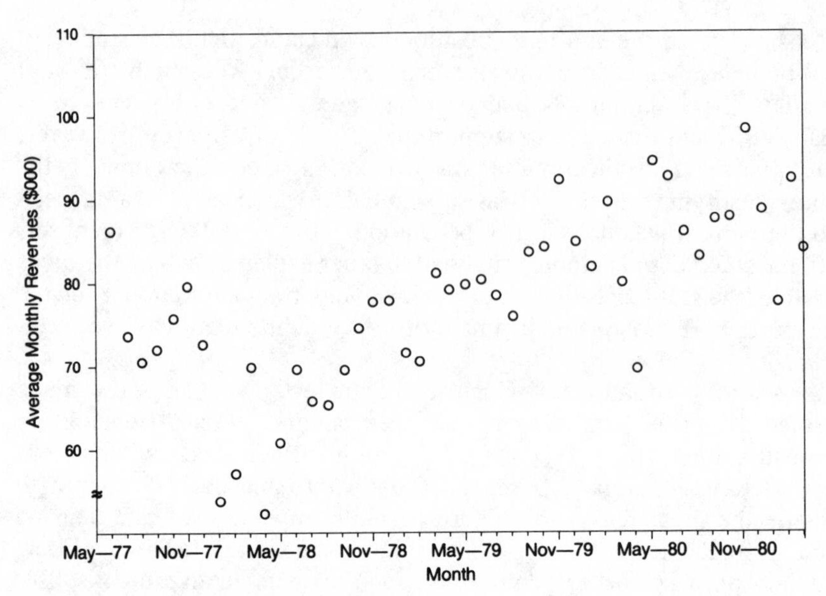

Figure 1. Revenues per collection day.

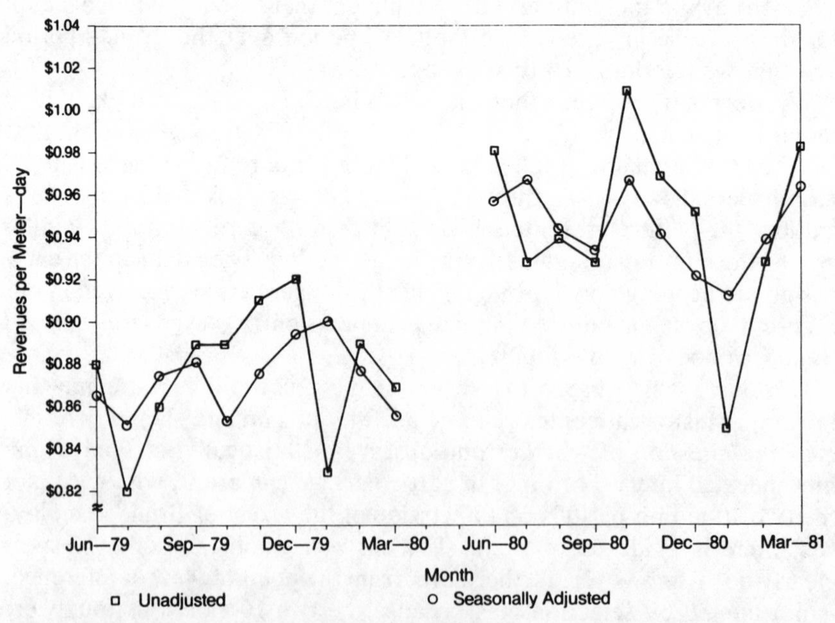

Figure 2. Revenues per meter-day, city-wide.

The absence of clear trend within the periods is also true for each of the five boroughs separately).

Thus, looking at the broadest historical context made available through data at the trial does *not* lead to a conclusion that there was a trend that *would* explain a difference in revenues before and after the arrests of Brink's employees. The possibility that factors other than theft, whether manifesting themselves in a secular trend or not, *could* explain the difference remains, but is not bolstered by any demonstration of a statistical trend purportedly offered by Levin's Figure 1.

Third, demonstration of a *statistically* observed trend, even assuming this were accomplished, is much less weighty as a counter-causal explanation of an observed difference than a demonstration that *specifically identified* causal factors are operating over the 22-month period June 1979–March 1981 to produce that difference. Levin did not establish any such specific factors. Thus, although there is no reason to believe that the total effect of all factors influencing revenues was constant or exactly constant during the 22-month period, the direction of effect of factors other than theft is unknown. That is, no known nontheft factors were shown to have caused the observed increase in revenues. While Levin advances four specific factors that he believes "might reasonably be expected to cause a difference in meter revenue between the two comparison revenues," he acknowledges that, "Whether they *did* cause a difference in the presumed direction, and to what extent, is largely unknown, and I could not, therefore, testify that these factors did indeed have the presumed effect." We are unconvinced that Levin's specific factors are even plausible explanations, as we discuss below.

Fourth, Levin says that the power of our broken-line model "test" for trend was low and therefore if there was a trend in the 22-month period, we would not have detected it. Here again, Levin has a tool for another job, but not for the job at hand. Levin's tool is a test of a single "trend" line fitted to the 36-month period April 1978–March 1981. This test produces a "highly significant slope coefficient ($p < 0.001$)" (Levin, this volume). As we observed in our chapter, fitting such a line can give no estimate of the effect of the *known* causal factor of theft *at* the transition between the Brink's and CDC periods. Such an estimate, however, is what the court required. Further, finding a significant slope coefficient for the line over 36 months, for reasons given above, does not establish that factors other than theft, whether identified or not, caused the observed change in revenues over the 22-month period.

By contrast, our tools, graphical exploratory analysis and a least-squares fit of a broken-line model to the monthly revenues (per meter-day) over the 22-month period, were suited to observe and to estimate a jump, if there was any, and also at the same time to observe and to estimate a statistical trend over the whole period, if there was any. The data do *not* show, as we explained in our chapter, evidence of trend over the entire 22-month period that *cannot be explained* and is not best explained by a jump in the middle. Thus, the claim that our "test" has low power is not a real criticism unless there is another test clearly suited to our job that has interestingly higher power. Levin has not exhibited such a test.

Fifth, Levin says that we "appear" to have made a methodological error in that our method of seasonal adjustment of the monthly revenue data appears not to be appropriate for the data and might have artifactually removed a trend actually present in the data. The method Levin believes we "appear" to have used we did not use. As we stated the seasonal adjustments "were made by a model for the data, as discussed below." Since the linear model described determined the seasonal adjustment, the error that Levin refers to was not committed, because all the linear model parameters, including the season–borough parameters and the intercontractor jump parameter, were determined simultaneously by least squares. We did then, and appropriately so, define an average model-predicted revenue per meter-day in each month for each contractor [defined for each month as the antilog of the sum over boroughs of the model-predicted monthly revenues per meter-day (in log scale) for each contractor and divided by the meter-days pooled over boroughs for the month]. These average model-predicted revenues per meter-day for each month were then used in the standard way to *display* seasonally adjusted revenues by month for each contractor by dividing them into the observed monthly revenues and multiplying by the 10-month period average. Note that this standard display procedure was adopted for a descriptive use long after the existence of a linear trend of the kind Levin postulates was explicitly rejected. ("Graphs of average monthly revenues delivered per meter-day of operation for the entire city and for each of the five boroughs, either without seasonal adjustments or with monthly seasonal adjustments, did not indicate a trend over time, but rather indicated level revenues per meter-day within each 10-month period." Fitted piecewise-linear regressions *also* did not indicate a trend such as Levin postulates).

Finally, Figure 1 in our chapter displays residuals from a trend line fitted to seasonally adjusted revenues per meter-day. The purpose of the fitting done for Figure 1 was not to estimate a trend nor to establish that no trend existed, but to *illustrate* for the reader through a residuals display how city-wide data are not well described by a single linear trend line. The *same* conclusion emerges from examining the residuals from a line fitted to unadjusted data, and the same conclusion emerges for every borough considered separately.

Even had it been true that our choice of seasonal adjustment was inappropriate, the criticism would have no weight because our results would be only trivially changed by using unadjusted data. Figure 2 gives the unadjusted monthly revenues per meter-day for the 22-month period for the entire city. No model is indicated for this data different from that for the seasonally adjusted data. This is also true for each separate borough. None of our conclusions changes. The purpose of the seasonal adjustment was to make a fair comparison between revenues per meter-day in the last month or months of Brink's, beginning March 1980, and the first month or months of CDC, beginning June 1980. For this purpose, the adjustment was correct and appropriate.

The second half of Levin's critique is a discussion of four factors that "might have" caused the observed increase in revenues. Three of these four factors can be immediately dismissed because, even were their existence and importance established, they do not explain the change in level that is observed on either side of the transition between Brink's and CDC. This change in level, measured by the estimated intercontractor parameter in the linear model, can also be described by the difference between one-month, two-month, three-month, etc., averages of revenues per meter-day on either side of the transition time between Brink's and CDC.

The first two of Levin's factors are (a) a presumed reduction in meters with revenue-affecting defects in the CDC period and (b) a presumed program of relocation of meters from low- to high-revenue-generation areas. Brink's did not present an expert who could establish, quantitatively or otherwise, that these presumed programs could be expected to produce a difference in revenues per meter-day of the size observed. Furthermore, these effects, if any, were presumably operating, if at all, over the entire 22-month period and would not then have a discontinuous effect at the transition.

The third factor cited is the gas shortage in the latter part of 1979, which involved odd–even rationing in the period June–September 1979. The effect of this factor on revenues per meter-day was debated at the trial, and in any case it would not explain the change in level at the transition, which occurred several months later.

The fourth factor cited was the strike on the Port Authority Trans-Hudson (PATH) trains between New Jersey and Manhattan in June–August 1980 and an increase in automobile traffic between New Jersey and New York City in that period, presumed to be associated with the strike. The parking meter expert, Laurence Donoghue, could not tell what the effect of these events would be on meter revenue. He noted that an increase in vehicles going into the often saturated metered parking areas of Manhattan could very well *decrease,* not increase, revenues if these vehicles parked for several hours in poorly enforced areas and did not comply as well as the customary parkers in these areas. Also, these vehicles might well have used meters more efficiently than in-and-out shoppers did, thereby decreasing revenues.

It is not possible in these pages to reproduce the discussion of specific causal factors that occupied considerable time at the trial, so we do not believe that any reader should be convinced one way or the other about the likely direction of effect on revenues of the four factors cited. However, at trial the city did present a considerable amount of detailed factual research on specific causal factors, whereas Brink's by and large presented hypothetical possibilities of effects, or rested, in Levin's words, on "common sense" as the arbiter of differences over the direction of effect of these factors on revenues. In sum, we were not, and the jury and judge evidently were not, moved to believe that causal factors other than theft were, on balance, any more likely to cause an increase than a decrease in revenues between the two comparison periods. Furthermore, the conclusion as just stated is the relevant one for determining a reasonable estimate of damages. It was *not* incumbent on the city to show, as Levin believes, that during the 22-month period there was "a coincidental cancellation of effects [of causal factors other than theft]." The legal burden, once tortious conduct had been established, was on Brink's to show that other factors probably did account for the jump that was observed at the transition when the contractors changed.

We close by observing that, while Levin is impressed by the difficulties of controlling for causal factors other than theft, his own analysis of revenues fails to control for two important factors that could be controlled, namely, the number of meters and the number of days they were in operation. Contrary to his assertion (Levin, this volume), we found that the analysis of revenues per meter-day instead of revenues uncontrolled for meter-days in operation was a useful device. To illustrate its use, we note that the jump in unadjusted monthly revenues from $77,500 in January 1981 to $92,000 in February 1981, which Levin cites as an example of a jump in revenues larger than that observed at the transition between Brink's and CDC, is reduced, in meter-day units, to a jump of $0.85 per meter-day in January 1981 to $0.93 per meter-day in February 1981, when meter-days are controlled for. The latter difference of $0.08 now is *smaller* than the jump in revenues per meter-day of $0.11 observed at the transition. This fact is not essential to the conclusions we reached, but illustrates that the conclusions reached in our study were based on a good deal more than "speculation."

Additional References

Fairley, W. B. (1978). Statistics and public policy. In W. H. Kruskal and J. M. Tanur (eds.), *International Encyclopedia of Statistics,* Vol. 2, New York: Free Press.

Hoaglin, D. C., Light, R. J., McPeek, B., Mosteller, F., and Stoto, M. A. (1982). *Data for Decisions: Information Strategies for Policymakers,* Lanham, Md.: Abt Associates.

How Much Is Enough?

Applying Regression to a School Finance Case

John Pincus **John E. Rolph**
The Rand Corporation
Santa Monica, California

ABSTRACT

This case study reports on a school finance case in which the authors testified for the defense. The purpose of the testimony was to address the question: what is the relationship in the State of Washington between the amount of money school districts spend and their students' scores on standardized tests? Our approach was first to carry out a critical review of educational production function literature and second to do our own empirical study addressing this relationship in the State of Washington. We concluded that there is no consistent relationship between school resources and student outcomes, at least for the range of resource use found in most U.S. public schools. This conclusion was cited by the defense (the State of Washington) in arguing against the plaintiffs' (several school districts') position for more school funding by the state.

1. Introduction

This chapter describes a regression study and a literature review of the relationship between school funding and student outcomes. The study was commissioned in 1982 by the State of Washington as part of its defense in a school finance case brought against the state by many of the state's school districts. The first author of this paper testified for the defense, on the basis of the data and analysis from the study. The second author testified briefly on appropriateness of the statistical methods used.

Besides describing our study, we attempt to point out how the particular legal setting of the case affected the statistical logic and reasoning we used to reach our conclusions.

Suit and History

The case in question, *Seattle School District v. State of Washington* (Thurston County No. 81-2-1713-1, filed in 1981, argued in 1982, and decided in 1983), was the second of two recent constitutional law cases relating to school finance brought by Washington's school districts against the state. The lawsuits were an attempt to shift virtually all of existing local burdens of public school financing from the school districts' local taxpayers to the state government.

Before describing the cases, we explain briefly the educational and fiscal background that led to the legal actions. From 1950 to the early 1970s, school-age population and spending on schools grew rapidly in Washington. In 1950 there were 401,000 public school children in the state, and the annual level of spending per student was $250. By 1971 public school enrollment had risen to 785,000, and annual spending to $985 per pupil, far outpacing the rate of inflation since 1950. Starting about 1960, more and more of the increased costs of schooling were paid by local tax levies, approved by voters residing in a school district, and funded by a local property tax. Until the early 1970s these levies were generally approved by substantial majorities, reflecting the interests of parents in the postwar baby-boom decades. Thus, between 1960 and 1975 local tax levy funding in Washington rose from $8 million annually to $299 million.

But by the mid-1970s the combination of inflation, existing tax levy burdens (which rose from 5% of statewide public school spending in 1960 to 25% in 1975), and slower population growth brought a change in voters' willingness to support school tax levies. The percentage of voters with school-age children declined in the 1970s (the public school population in Washington actually declined about 8% during the 1970s). At the same time, inflation was reducing many voters' purchasing power, leading them to oppose higher school taxes.

Consequently, during the 1970s, Washington voters began to vote down school tax levies. The school districts were therefore faced by unprecedented and apparently chronic budget crises. The crises were dominated, in terms of demand for funds, by teachers' union efforts to keep members' earnings up in face of inflation, and by sharply increased material costs, notably petroleum prices. The principal constraints on supply of funds were the unwillingness of state government to meet teacher and school district demand for funding, and the repeated refusal of voters to approve new levies as the 1970s wore on.

Faced with this dilemma, the school districts turned in 1976 to the courts for relief, relying on a provision of the Washington state constitution (Article IX, Section 1):

> It is the paramount duty of the state to make ample provision for the education of all children residing within its borders.

The Seattle school district, joined by parent and student plaintiffs, claimed that under this provision the state was required to fund a basic program of education for all public school children. The Superior Court, in a 1977 decision, later upheld by the Washington Supreme Court (90 Wn.2d 476, p. 2d71 (Sept. 1978)), agreed with the districts, and required the legislature to define and fund 100% of a basic program of education (*Seattle School District v. State of Washington,* Superior Court, Thurston County (March 17, 1977)). The definition of basic education was a subject of controversy, the tasks of definition and of appropriate funding levels being left to the legislature, which passed a Basic Education Act in 1977.

In 1981, Seattle, joined by many other districts, returned to court, to argue that the state was spending too little on basic education and that a number of additional items, including programs for the handicapped, bilingual education, special programs for slow learners and gifted students, pupil transportation, and extra costs of school desegregation and of operating large urban school districts, be included in the definition of basic education. The school districts specifically argued for increases in staff salaries, in teacher–pupil ratios, and in costs for equipment and supplies.

Despite these problems facing the districts, it should be noted that as a consequence of the 1977 decision their position improved markedly from 1977 to 1981, in terms of reduced dependence on local levies. In 1977 districts raised $343 million in special levies, which included about one-fourth of the costs of basic education. By 1980, special levies raised only $124 million, none of it for basic education. In other words state funds substituted substantially for local basic education funds, so that increases in state spending were not associated with comparable increases in total spending. It is hard to say what the net fiscal effect was. On the one hand, taxpayers in districts with high special levy rates benefitted the most. These were likely to be high income districts, and if state taxes are not progressive, those who benefit from reduced levies would not pay commensurately through state taxes for the higher levels of state spending. In other words, the state's de facto assumption of special levies starting in 1977 may have been somewhat regressive in fiscal effect.

Purpose of the Testimony

In the 1967–1977 trial, the first author had testified for the state on the relationship between levels of spending, types of resource use (e.g., class size, teacher educational background and experience) and student outcomes as measured by standardized test scores. He argued on the basis of the so-called educational production function literature (regression analysis of the relations between resource use and outcome) that there was no compelling evidence to support the view that increases in resources were associated with better student outcomes. The judge ignored this evidence in his 1977 decision, because the decision was made on constitutional grounds, to which efficiency questions were irrelevant.

In the 1981 case, such testimony was more relevant, because the legal issue was no longer solely that of 1977: whether or not the state had to provide each student with a basic education. It also revolved, among other things, around such questions as, Should the courts specify how much the state must spend for basic education? Should they prescribe appropriate teacher–pupil ratios? Should they establish minimum teacher salary levels? For such issues, evidence on resource-use effectiveness would be more germane in the new case than in the earlier one. However, there were additional questions in the 1981 case (e.g., whether bilingual education, remedial education, or the costs of school busing should be considered as part of basic education) to which our statistical evidence on educational "productivity" was not relevant. We therefore do not dwell on such issues here, restricting ourselves to those issues for which our testimony was relevant.

The Approach

Since the 1977 decision had established that the State of Washington was required to provide each student with a basic education, the central question in the 1981 case was whether the state was providing for such basic education and, if not, whether the court should require the state to do more. The plaintiffs, the districts, argued that the state should be required to provide more education resources in order to bring the level of education up to this basic education level. While the plaintiffs brought forth various witnesses who testified that increased levels of resources would be appropriate in their districts, the plaintiffs made no effort to define the output or product of the education system and thus what level of this output would satisfy the basic education requirement. Instead the plaintiffs focused on the inputs to the education system; more resources—teachers, equipment, administrators and higher pay for district staff—were needed. Implicitly, at least, they argued that these increased resources would lead to better student outcomes.

Our role in the case grew out of the defendant's argument that "basic education" is an output of the education system. Thus in order to justify increasing the inputs to the education system—teachers, salaries, facilities, and the other resources money can pay for—the defense argued that the connection between inputs and outputs must be established, and our testimony therefore focused on what could be said about the relationship between them. Could the plaintiffs demonstrate that the additional resources they were asking for would in fact increase the level of education (student outcomes)? Our study and the testimony based on it argued that the connection between the inputs and outputs could not be established at least under the circumstances that governed this case.

Thus our evidence questioned the plaintiffs' implicit proposition: If the education system is not producing a "basic education" for its students, can such a goal be achieved by moderately increased funding?

Our approach was to investigate whether added funds, or the added resources they buy, will improve educational outcomes. We began by defining the purpose of the study and standard of proof that applied in our work. The purpose of the study then was to offer evidence on how an increase in the inputs would affect the outputs (student outcomes) of the public education system in the State of Washington. That is, we aimed to describe the education production function for Washington school districts. We took two tacks: a literature review and critical assessment of what the research community could say about educational production functions; and our own study of the relationship between resource inputs and student outputs using Washington data, in order to confirm or refute whether the Washington education production function might differ in any important way from what had been found in other studies. We now describe the logic of how these two activities fit together.

As we shall see in Section 2, the education production literature is plentiful and does not produce an unambiguous picture of how student outcomes are affected by resource inputs. As experts offering evidence for the defense, however, the burden was not on us to precisely estimate an input–output relationship. In our investigation, we found no evidence of a consistent positive relation between the inputs of the educational system and its outputs. We were, in effect, investigating the plaintiffs' contention and unearthing no statistical evidence to support it. This finding was helpful to the defense since we concluded from the literature review that for the range of resource increases being considered in Washington, there was no compelling evidence that increased state funding would result in increased average student test scores. However, a court might question whether data and research on non-Washington schools is compelling and relevant evidence for the present case. To allay this concern, we decided to learn what we could about the educational production function for the State of Washington. To this end, we did a regression study using Washington data, to assess, within time and funding constraints, whether there was any reason why the conclusions of the literature review should *not* apply to Washington. That is, are the Washington data consistent with other studies?

It was necessary to review the production function literature, because our own Washington regression study was limited in scope, reflecting the demands of the trial schedule that allowed us only eight weeks to conduct the study. Therefore it was essential to verify whether our brief analysis of Washington data was consistent with the body of statistical findings on the subject. Our results could not stand alone, particularly if faced by a preponderance of other findings that differed from ours.

The regression study was designed to investigate whether the data refuted the hypothesis that increases in school spending, and consequent increases in resource use, were not associated with improved educational outcomes for Washington public school students in 1976 and 1980. The study used each of the 88 Washington school districts with over 2000 enrollment (accounting for 83% of Washington's total public school enrollment in these years) as the unit of analysis. The regressions were used to compare different school districts in the same year (where 1980 spending per pupil ranged from $2018 to $3706, as among the 88 districts), and thereby to observe the association between the districts' average standardized test scores at selected grade levels and *variations in resource use* (e.g., class size, spending per student, staff salary level), *socioeconomic characteristics* of the district (e.g., percentage of college graduates and percentage of poverty population in district), and *characteristics of students* in the district (e.g., percentage of students of Hispanic origin and district average test scores in prior years).

If we found in both the literature review and in our regression very little evidence of a positive association between resource use variables and outcomes as measured by test scores, then we would argue in our testimony that the plaintiffs' case for more funds, more staff, and higher salaries was not supported by the existing evidence.

This chapter presents our testimony and the study it is based on. We begin (Sec. 2) with the literature review, describing the evidence on the connection between resource use and educational outcomes. We then describe our Washington regression study and its results (Sec. 3). The following section describes both our testimony and the plaintiffs' rebuttal testimony (Sec. 4). We conclude by describing the court's ruling in the case, and by discussing the apparent effect of our testimony on the ruling (Sec. 5).

2. Literature Review

We reviewed some 35 production function studies of elementary and secondary education (see Bibliography). Most of them were regression or correlation analyses. A few used other multivariate statistical methods. Most of the analyses were based on a model that assumed that educational outcomes are a function of background variables (largely socioeconomic), peer group variables (average characteristics of fellow students at the classroom, school, or school district level) and school or school district variables (including resource use levels and distribution, and characteristics of individual students and staff).

Sources of Variation in Findings

Unit of Observation

The most common unit of observation found in the literature, for both dependent and predictor variables, is a single grade level at a given school. The single grade level, rather than a group of grades, is dictated by the nature of the dependent variable, which is virtually always a standardized test, administered to a particular grade, varying in content at different grade levels. The fact that half of the studies we surveyed used grade level data by particular schools suggests that the school level data is the least aggregated data that can be conveniently collected. The other half of the studies were about equally divided among district level aggregate data bases and those using individual student or classroom level data.

In general, disaggregation to the individual student and teacher level (e.g., Hanushek, 1972; Summers and Wolfe, 1977; Murnane, 1975) offers the most sensitive information on resource–outcome relations. When the school or school district is the unit of observation, the averaging procedure involved reduces the extent of variation in both dependent and predictor variables, with a resulting loss in the ability of the regression procedure to pick up actual relationships. As is well recognized in the literature, the relationship between averages is not the same as that between individual values. An elementary exposition of this "ecological correlation fallacy" is given in Freedman et al., 1978, Chapter 9).

This does not imply that analysis of aggregated data is not useful, nor that its results are likely to be inconsistent with those of disaggregated data analysis. Aggregated data, say at the district level as in our study, are likely to be useful when the policy questions of interest must be resolved at this level of aggregation. For example, if the issue of interest is school district behavior or policies, then district level data may represent the appropriate unit of observation.

Our literature review showed no necessary inconsistency between studies using aggregated or disaggregated data. One of many such examples is that studies at both the school or student level (Perl, 1973; Murnane, 1975) and the district level (Pincus and Rolph, 1983) support the hypothesis that teachers' postgraduate education is not associated with variations in student outcomes even after controlling for a variety of other relevant factors. A more complete and comprehensive description of both the literature review and of our regression study of Washington schools is available from the authors in Pincus and Rolph (1983).

Differences in Data Coverage

Another source of discrepancy among findings in the literature is difference in the geographic, racial, economic, or social coverage of the predictor variables or in the kinds of achievement tests used as the dependent variable. Yet as in the case of units of observation, wide divergence in characteristics of the data are compatible with consistent findings. For example, an increased teacher verbal ability is associated with a significant positive increase in outcome for black students (Bowles, 1970; Hanushek, 1972), for students from blue-collar families (Hanushek, 1972), and for white students of varying backgrounds (Hanushek, 1972; Levin, 1970; Michelson, 1970). Different dependent variable measures also can provide compatible results. When different standardized tests are taken by the same subjects, the results were usually highly correlated.

In summary, both the unit of observation and the type of population observed can affect the regression results, even when similar models are used. Nevertheless, the literature generally shows consistent results across level of aggregation and populations observed (Averch et al., 1974; Bridge et al., 1979).

Differences Associated with Explanatory Variables

For purposes of discussion, we divide our description of the literature into three groups of independent variable categories consistent with the approach used by us in our own regression analysis in Pincus and Rolph (1983): background variables, peer group variables, school resource variables. For discussion of the reasoning behind such variable groupings, see Coleman et al. (1966) and further remarks in Mosteller and Moynihan (1972). For detailed literature reviews, which discuss the different models used, see Averch et al. (1974) and Bridge et al. (1979).

The *background variables* most frequently used in the 30 studies that used background variables are (in descending order of frequency of appearance in the literature): parents' educational level, parents' occupational level, family size, family income, kinds of possessions in the home, and parents' educational expectations for their children. The association (partial correlation) between all but one of these variables and student outcome was statistically significant and positive in the great majority of studies. Out of 84 appearances in regressions, these variables were positive and statistically significant in 69 cases, positive and not statistically significant in 15, and were never negative and statistically significant. The one variable that did not show a positive association with outcome was family size; it consistently had a negative regression coefficient and was almost always statistically significant. Because each investigator tends to use somewhat different variable definitions, comparison of regression coefficients, or even of partial correlations, can be misleading. Therefore the review literature, like this chapter, is unfortunately restricted to using statistical significance as a measure of importance.

Peer group variables (e.g., percentage of minority students, social class composition of school, ability composition of one's classmates, turnover of classmates) usually give mixed results, with each variable being statistically significant in both a positive and negative direction in different studies, but more often the estimated coefficients were not statistically significant. The estimated coefficient for percentage of minority students is usually (in two-thirds of the regression studies) negative by a statistically significant amount. Coleman et al. (1966) found some other peer group effects to be significant in his study (e.g., proportion planning to go to college, average classmate hours of homework), but these variables have not been widely used since. In general, the evidence on peer group variables is scanty and inconclusive, except in the case of the percentage of minority students. While intuitively it seems likely that the characteristics of fellow students could be systematically related to one's school performance, the evidence is weak and mixed for the peer group measures that have been used in the literature.

The seven *school resource variables* most frequently used in production function regressions are class size, teacher postgraduate education, years of teaching experience of teacher, teacher salary levels, spending per pupil, teacher verbal skills, and quality of teacher education. Only two of these, teacher verbal ability and the quality of the colleges a teacher attended, usually have positive coefficients in the regressions. The other five variables that appear most frequently in the literature show a common pattern. Four of 32 analyses find the estimated coefficient of one variable, teacher postgraduate education, to be positive by a statistically significant amount; 17 other studies find a nonsignificant result for teacher postgraduate education; and three even report significant negative results. Similarly, for teacher experience, 19 studies report significant positive results, while 15 are not significant.

In general, the school resource findings are consonant with the view that, as compared to background factors, no school resources are consistently associated with student outcomes. "The particular resources that seem to be significant in one study do not prove to be significant in other studies that include the same resources in the analysis" (Averch et al., 1974).

It is probable that these inconclusive findings stem in part from deficiencies of the models and data with respect to measuring school resource use. For example, it may be that school resources are being used at varying degrees of effectiveness among schools and school districts. This would create systematic problems for the economic efficiency model underlying the analyses described here, because the averaging process of the regression would, in this event, make it impossible to estimate the true marginal productivity of resource use in different units of observation of varying efficiency. To put it differently, teachers' education, as an example, may contribute to student outcomes in a given school or district but not in another one. Regressions that average across schools or districts may well produce nonsignificant results when results are actually idiosyncratic. Strictly speaking, including appropriate interaction terms (or other created variables) in the regression should capture this phenomenon; however, studies in this literature rarely use interactions in their regressions.

The persistent inability of regression analyses to detect consistent positive or negative differences associated with school resources, as contrasted with the strong and consistent significant results for most background variables and a few peer group variables, may also imply the obvious: that the school resource variables are less consistent in their effects than the other variables.

Of course, acceptance of a null hypothesis of "no difference" does not prove that the alternative of some difference cannot hold. It might be that the data are too imprecise to distinguish between a meaningful difference and no difference—the test has little power. It may be that the regression equations are misspecified to a degree that they cannot capture differences that do exist even with large sample sizes. Finally, some workers in education research argue that the cross-sectional education production studies done to date have designs that may preclude reliable findings (Bowles and Levin, 1968; Gilbert and Mosteller, 1972). In particular using longitudinal data rather than cross-sectional data might reveal a genuine relationship between school resource variables and student achievement (Dyer, 1972; Luecke and McGinn, 1975). Such longitudinal data are difficult and expensive to collect and analyze. If in the future, a number of careful longitudinal studies are carried out, this conjecture can be tested. Meanwhile we must conclude that the case for such a relationship is "not proved."

In summary, at the present stage of the research, it seems useful to cite two conclusions from Averch et al. (1974, pp. 171–175):

1. Research has not identified a variant of the existing system that is consistently related to students' educational outcomes.

2. There is no evidence that increasing expenditures on traditional educational practices will improve educational outcomes substantially.

Their conclusion applies to expenditure ranges typically found in public education. Naturally zero expenditure would be associated with lower outcomes, and very high expenditures might be associated with higher outcomes (Glass and Smith, 1978). It would be premature to forecast whether analytical models and research designs of increased sophistication and scope would lead to changes in these conclusions.

3. The Washington Regression Study

This section describes the regression study we carried out to investigate the relationship between school resources and student achievement scores for Washington state school districts. Time constraints precluded primary data collection; thus we used only existing data: census data, Washington state school data, and state achievement test data. We used the traditional multiple regression approach of modeling of educational outcomes (student achievement on test scores) as a function of background variables, peer group variables, and school variables.

In this section, we first describe the data and present the conceptual model. We then describe the model-building process by using one of our 16 regression models as an example and present the results and interpretation. Finally we compare the results from this one model to the others that are not presented in detail. The reader interested in the complete results of this regression study should consult Pincus and Rolph (1983).

Data and Conceptual Model

The majority of the educational production function literature described in Section 2 is based on the following model: educational outcomes are a function of background variables, peer group variables, and school variables (including individual student and staff characteristics). These relationships can be expressed as a multiple linear regression equation. The actual variables chosen as candidates for inclusion in the regression equation depend primarily on two factors: (a) the data available to the researcher from existing sources or new data collection; and (b) the researcher's views about different independent variables, or stated more formally, his conceptual model of the education production function.

Since no new data collection was feasible, we first had to decide whether existing data would allow us to conduct a useful study of the state-wide resource–outcome relationship. There were several limitations.

Existing data sources were:

1. The 1970 and 1980 U.S. Census for community income, occupation, poverty level and education as background variables.[1]

2. Data from the State of Washington for peer group variables (Title I enrollment, minority group enrollment, prior test scores, total enrollment), for school variables (numbers of and salaries of administrators and teachers, degree levels for teachers, spending per student, pupil–teacher ratios, teacher experience), and for math and language scores for 1977 and 1980 on the fourth-grade California Achievement Test (CTBS test in 1977).

3. Data from Washington Pre-College Testing Program on 11th-grade quantitative and language test scores for 1976 and 1980. Test were taken largely by college-bound students, 38% of all 11th graders in 1980. The possible selection effects of which students took the test makes interpreting regressions using this outcome variable a risky endeavor.

From these sources we developed 255 variables, and winnowed them down to 39 independent variables and 8 dependent variables (these were test scores for two different subjects, two grade levels, and two years as discussed below). The winnowing consisted largely of four processes: (a) eliminating intermediate quantities of no direct relevance to the study; (b) a decision to use 4th-grade school resource variables only for 1976 and 1979, and to use 11th-grade school resource variables only for 1976 and 1979; (c) a decision, wherever possible, to use independent variables that related to the grade levels of interest and not to the district as a whole; (d) other judgmental decisions. Students took tests in October 1980. Our data included 1979–1980 resource use, because the 1980–1981 resources had hardly been used by the date at which the students were tested. The same procedure was followed for the earlier year. The above four decision categories eliminated approximately 150 of the variables. The rest were eliminated mainly because we did not believe a priori that they were conceptually or statistically appropriate. Thus we arrived at 39 candidate independent variables without any formal statistical analysis.

Various descriptive statistics and plots were made for most of the 255 variables that were created at various stages of the analysis. Our purpose was to check the quality of the data as carefully as possible and to detect instances when variables needed rescaling. A few data errors were dis-

[1] Census data by census tract was mapped onto school district boundaries by Mr. George Shepard.

covered and corrected, and some candidate predictor variables were transformed. On the whole, we believe that the quality of the data is good.

The Model Building Process

Our goal in model building was to construct 16 equations, each containing background, peer group, and school resource explanatory variables, as follows:

Two Years: 1976 and 1980

Two dependent variables sets: math and language test scores

Two grades: 4th and 11th

Two regression models: one emphasizing independent variables that attempted to measure school resource inputs and another measuring finance or money inputs.

We chose not to use both expenditure variables and resource variables (e.g., teacher–pupil ratio, years of teacher experience) in the same model, because the expenditures were used to purchase the resources. This would make interpreting the regression difficult and appear to be a form of double-counting.

The crossing of these four pairs gives the 16 equations. There was variation in choice of independent variables for several reasons, mainly related to data availability.

The regression model specification process used the conventional set of statistical tools: critical examination and significance testing of coefficient estimates, residual plots, and calculation of influence statistics on a case-by-case basis (Weisberg, 1980). This was an iterative process and resulted in trying various transformations and interactions of the candidate variables as independent variables in the various equations. Rather than attempt to chronicle this process in detail, we mention a few guiding considerations to give the reader a feel for the process.

First, in defining the predictor variables for the regression, we tried to avoid extreme dependence (collinearity) between variables—such dependence can create both interpretation and computational problems. We carefully examined correlation matrices, cross-tabulations, and other measures of dependence and tried to make sensible choices when several variables seemed to measure only one or two attributes. For example, median income, percentage below poverty level, percentage college graduates, and percentage professional are four variables that are highly correlated with one another in our data. After looking at several possibilities we decided to use only the two variables, median income and percentage professional, in most of our final equations.

Second, when entertaining a particular regression specification we used a "reasonableness" test in examining estimated coefficients. For example, if test scores appeared to be negatively related to parents' education, we would look critically for an explanation. Scientific insight and knowledge of the subject are the keys to developing sound models.

Third, we did extensive "diagnostics" on the fitted equations. These began with various plots of residuals looking for ways the model could be improved: Have important variables been omitted or has an independent variable been entered on the wrong scale? That is, if the error terms in the equation as reflected in the residuals have recognizable pattern, this pattern can be included into the independent variables to improve the model. See Weisberg (1980), Draper and Smith (1981), or Morris and Rolph (1981) for discussion. In addition, we computed Cook's distance as well as several other commonly used influence statistics for each case or school district for our first complete set of regression runs. See Weisberg (1980). The idea here is to assess (a) the "leverage" of each district in driving the fit of the regression, and (b) whether any district is an outlier—unduly far away from the fitted equation. None of the 88 districts was identified as being unduly influential. This was reassuring since a priori we were concerned that the larger districts, especially Seattle, might have very different patterns of independent variable values and hence be high leverage points and drive the regression results.

Despite some variation, there was a general consistency in the final variable structure of the equations. This consistency resulted from three considerations: (a) our desire to include in each equation appropriate and, where possible, consistent measures of background, peer group, and school resources (consistent with those used in other studies); (b) our desire to include variables that were relevant to policy issues raised in the trial (e.g., spending per student, class size); (c) the limitations on the number of usable variables occasioned by the fact that our ultimate list of "candidate" variables was relatively small, and that we were limited to district-level data.

One wrinkle particular to this study was our presumption that the error term did not have constant variance. A priori since the outcome variables are average scores for whole districts, one would expect the error term for a district with many students to have a smaller variance than one from a district with few students. That is, if individual student test scores have variance σ^2, the variance of the district average is σ^2/n, where n is the number of test takers. This observation prompted us to check the residual plots for nonconstant variance. Finding evidence of nonconstant variance in these residual plots, we then did a "variance residual regression" to estimate what weights to use in a weighted least-squares analysis. This is a regression of the absolute value of the residual against the weighting variable—one over enrollment in our case. If the relationship is linear and the true intercept of the line is zero, using 1/enrollment is appropriate. One can make statistical inferences about the intercept and linearity. We did this and concluded that a weighted least-squares regression with weights equal to 1/enrollment was consistent with the data as our a priori reasoning indicated.

Results and Interpretation of One Model

We shall not give a complete description here of all the fitted regressions; we present the results in detail of only one of the 16 regressions that we constructed in order to give some feel for the final fitted equations. While the 16 equations differed in detail, the same general pattern emerged in all regressions. We compare the 16 fitted equations in the next subsection.

Choosing an equation to discuss means choosing a year (1976 or 1980), a grade level (4th or 11th), an outcome variable (math or language), and an equation type (resource or finance variables). We choose the resource model for 1980 4th-grade language test scores. The year 1980 is the most recent, the 4th-grade exam has broader coverage than the 11th-grade one, the choice of language over mathematics is somewhat arbitrary, and the resource model is more similar to other education production function studies than the finance model.

As described above, the form of the final equation reported in Table 1 is the result of an extensive model building process. This regression equation estimated average test scores for districts on the 1980 4th-grade language exam. It is based on data from the 88 Washington state school districts that had at least a 2000 student enrollment and uses 13 explanatory variables to achieve a coefficient of determination (R^2) of 0.64. We turn now to discussion of the variables.

Throughout this paper, we use *statistical* significance as our primary measure of the size of estimated regression coefficients. We would prefer to use a measure of how large the coefficient estimate is compared to other educational production function studies, tempering our conclusions by the width of the confidence interval around the estimate. As noted earlier, investigators in other studies use different dependent variables (test scores) than ours, not to mention independent variables. Further, the different levels of aggregation (district, school, classroom, individual student) make comparisons across studies difficult. For these reasons, published reviews of this literature (Averch et al., 1974; Bridge et al., 1979) have restricted themselves to statistical significance as a measure of importance of estimated coefficients. We reluctantly do the same.

We decided on the final specification using a combination of data analysis and the literature review. Thus independent variables whose estimated

TABLE 1. **Regression for 1980 4th-Grade Language Scores**

	Estimated Coefficient	t-Values	Variable Mean	Variable Standard Deviation
Dependent Variables				
4th-grade language score	—	—	60.14	8.63
Background Variables				
Occupational index (%)	0.458	3.08	20.91	6.67
Median income ($)	−0.000141	−0.59	2069.6	3763.4
Peer Group Variables				
Title I enrollment (%)	−58.47	−4.02	0.09	0.06
Hispanic enrollment (1 if over 5%)	−3.668	−1.73	0.17	0.38
Logarithm of 4th-grade enrollment	−2.942	−2.34	5.67	0.53
Prior test score (district average, same grade, 1976)	0.200	3.60	53.40	12.31
School Variables				
Administrator–teacher ratio	17.93	0.42	.09	.01
Pupil–teacher ratio	0.689	0.42	28.35	3.63
Certificated staff–pupil ratio	−0.403	−0.27	0.45	0.50
Teacher Experience:				
Level A (%) (less than three years)	−0.0614	−1.30	69.03	16.61
Level B (%) (more than six years)	−0.0382	−0.58	14.64	12.71
Constant	70.67	6.08	—	—
Standard error of the estimate	5.57		—	—
R^2	0.64		—	—

coefficients were small and did not differ from zero by a statistically significant amount were included only if these variables were either of interest in the case (most of the school variables) or were traditionally used in education production function research (median income). About half of the 13 explanatory variables were so included.

The 4th-grade 1980 regression presented in Table 1 has a similar pattern to the 15 other regression models. Among background variables, percentage of managerial and professional workers in the community ("occupational index") has a statistically significant coefficient estimate, while median income does not. These findings are consistent with the literature, which shows a much more consistent relationship between occupation and education outcome than between income and education outcome.

All four peer group variables (Title I enrollment percentage; logarithm of 4th-grade enrollment; Hispanic enrollment; prior 1977 4th-grade test score) consistently have estimated coefficients that differ from zero by a statistically significant amount (significance probabilities range from about 0.10 to below 0.0001 based on a two-sided Student's *t*-test). These estimated coefficients are generally consistent with earlier studies, except that our estimated coefficients for enrollment variables are larger than in most other studies. The prior test score variable, which does not appear in most other studies, may indicate a strong school district effect in addition to the effects of the other variables in the equations. As noted in Section 2, this finding parallels other findings at the teacher and school levels (Murnane, 1975; Hanushek, 1972).

None of the five school resource variable estimated coefficients is either statistically or substantively significant (elementary school administrator–teacher ratio, 4th-grade pupil–teacher ratio, elementary school certificated staff–pupil ratio greater than 50 per 1000, 3rd-grade teacher experience 0–3 years and 6+ years). Because we used district level data, however, the range of these variables is limited (see Variable Standard Deviation column). Teacher postgraduate education, excluded from our final equations, came in weakly in earlier specifications. Interestingly, the teacher experience variables suggest that, all other things equal, teachers with an average of 3–6 years of experience were more closely associated with higher test scores than the other two teacher experience categories.

These school resource findings are consistent with the literature, which shows predominantly nonsignificant and mixed findings for school resources (Averch et al., 1974; Bridge et al., 1979).

Comparing the Different Models

While comparing regression models having different dependent variables has an apples and oranges quality about it, we justify doing it here by noting that all our outcome variables are efforts to measure school district output in terms of student achievement. In this section we briefly summarize the other equations and note the differences.

The sixteen regression equations used in the study are as follows:

Years: 1976 and 1980
Grade levels: 4th and 11th
Subject tested: language and math (quantitative in 11th)
Regression models: resource and finance

Resource model refers to equations in which the services that money buys, are used as school variables (e.g., numbers of district staff); finance model refers to equations in which per-capita spending itself is used as a proxy for school resource variables.

In general, the regression results as a whole were consistent with the body of the professional literature. Of the three background factors we used, two (adults' occupational status and percentage poverty population) were statistically significant in the same direction in all equations where they appeared. The third background factor, per-capita income, was statistically significant in only two of 16 equations.

Of the peer group variables, Title I enrollment and percentage Hispanic students always had negative estimated coefficients, but had much higher *t*-values for 4th grade than 11th grade, presumably because relatively few Title I and Hispanic students take the Washington precollege test. The percentage of total district enrollment that were black students was not statistically significant. The total enrollment variable was statistically significant (negative) in 1980 4th-grade equations, and generally not statistically significant elsewhere. The percent of 11th graders taking the precollege test was statistically significant (negative) in the majority of equations. The fact that 1976 district average test scores were highly significant in all 1980 equations suggests that district quality may be strongly associated with student outcomes after controlling for all other independent variables. This variable, prior test scores of earlier 4th- and 11th-grade students, does not appear in the literature, but may be of some value for policy research if its implications can be followed up.

An interesting difference of our peer group findings from some others in the literature is the lack of statistically significant differences associated with black enrollment; this is no doubt attributable to the small black enrollment in most Washington districts, and the high level of data aggregation (district level instead of school, classroom or student level) in our regressions. We attempted to isolate the possible effect of a relatively large percentage of black enrollment in a district by using a dummy variable for having over 5% black enrollment. Even this method did not produce a statistically significant coefficient estimate.

With respect to school resource and finance variables, our findings were generally consistent with the literature. No variable was consistently related to student outcomes. In two of the eight finance equations (1980 4th- and 11th-grade math), spending per student had a statistically significant positive estimated coefficient. In all four 1980 11th-grade equations, the ratio of administrators to teachers entered in a statistically significant negative way. The number of years of teacher experience *per se* was never statistically significant, but using dummy variables showed that teachers with 3 to 6 years experience were always more positively associated with high student outcomes than other teachers. Certificated staff per thousand students was significant (positive) in the 1976 11th-grade resource equations; classified staff per thousand students was not significant.

Finally all the regression equations achieved comparable coefficients of determination, ranging from 0.53 to 0.63. These relatively high values may well be due to our using aggregated district level data.

4. Direct and Rebuttal Testimony

Our testimony, which was given by the first author, followed the description presented in this paper, although it included also an introductory explanation of regression analysis as a basis for explaining the literature review and our Washington regression study. This material, covered in one day of testimony, was developed in lay language as much as possible, with extensive use of explanatory charts and graphs. This material was presented in lecture form rather than in traditional question-and-answer form as it did not lend itself to traditional courtroom treatment.

The second day of testimony began with a review of the literature as in Section 2 above, leading us to the conclusion that there was no consistent estimate that school resources affected student outcomes as measured by standardized test scores, at least at the range of resource use found in U.S. public schools.

During the second day we also summarized the results of our Washington regression study, which were consistent with the findings of the literature review.

We finished this part of the testimony by stating (a) that in our study high-quality districts (those that had high test scores consistently) were not correlated with levels of resource use; and (b) that if additional resources were spent in traditional ways (more teachers, more administrators, smaller classes), the data are consistent with the main effect on Washington public schools being a transfer of wealth from the community at large to educators, with negligible effects on educational outcomes.

The testimony concluded with cross-examination aimed at demonstrating that the literature review had omitted some relevant studies and that our regression analysis was deficient on several grounds (failure to include other measures of outcomes than test scores; no consideration of handicapped, bilingual, minorities, gifted, etc.; possible failure of our regressions to meet assumptions of the regression model).

Following the testimony, the judge asked the witness questions designed to make clear (a) that our analysis cast no light on handicapped students' benefits from higher spending; (b) that there was no apparent benefit to average test scores as a result of increases in school spending from 1976 to 1980; (c) that the source of the funds could not be shown to have an independent effect on outcomes. He also wanted to know how teacher experience, teacher educational level, and changes in teacher–pupil ratio would affect student outcomes.

Our testimony, based on the literature review and the regression study, offered the plaintiffs two alternatives in rebuttal: either to ignore the testimony on the grounds that it was not germane to the constitutional issues or to present rebuttal testimony that would impugn the quality or the relevance of our testimony. In effect, the plaintiffs tried both options sequentially, by first presenting a rebuttal witness and subsequently by ignoring our testimony and that of the rebuttal witness in their final submissions to the court (findings of fact and posttrial belief).

The decision to call a rebuttal witness was motivated, according to the plaintiff's account, by the concern that our testimony might persuade the judge to require less than full funding of the state's basic education formula, on the grounds that our evidence showed no need for full funding.

The rebuttal witness's main argument was that our study was not comprehensive enough. He defined a comprehensive study as one having independent data collection, soliciting the views of experts and parents about educational goals, undertaking more extensive work on the specification of variables and benefitting from systematic peer review. He argued that no conclusions could be drawn from an educational production function study unless it met these criteria. Thus our eight-week effort of data collection and analysis could not provide useful evidence for a trial.

The rebuttal witness responded to defense counsel's point that our findings were substantially in accord with those in the literature, only by stating that nonfindings (accepting the null hypotheses of no effect) are often of little significance in social science.

In any case, it is impossible to quarrel with the witness's plea for more comprehensive studies, but a basic question is whether such a comprehensive approach as he described in his testimony would produce substantially different results. The closest analogy to the rebuttal witness's criteria is presumably the Coleman report (Coleman et al., 1966) with associated criticisms by Bowles (1970), Bowles and Levin (1968), Michelson (1970), Cain and Watts (1970), and Smith (1972). But the results of the Coleman report do not differ systematically in their findings from less comprehensive research as reported by Averch et al. (1974) and Bridge et al. (1979) in their reviews of the literature. In other words, although the witness was clearly correct in advocating that one collect, as appropriate, one's own data, and allow ample time for professional criticism and suggestions, his rebuttal did not focus on the specific biases introduced into our results, as compared to those of more detailed studies, by our failure to follow this procedure.

In discussing the model, the rebuttal witness also stated without explanation that these so-called production function models, which relate resource use to outcomes, are, although widely used, of no value for policymaking.

He stressed several other lesser points, only one of which we describe here. He stated that our study is not consistent with the literature because it failed to show a significant negative effect for black enrollment, which all other studies have found. [In fact the most comprehensive review (Bridge et al., 1979) cites 18 studies that include this variable, of which only 8 had significant (negative) findings.] The probable reason for any discrepancy is that the median school district in our study had only 0.5% black students. The social conditions of black children in these districts are likely to be quite different from those where there are many more black students, and such issues as white flight and ghetto conditions are more prominent.

Under cross-examination, the witness stated that he was not familiar with educational data for the State of Washington. He and the defendants' counsel disputed at some length and failed to agree as to whether it was ever appropriate to base public policy recommendations on less than the complete and exhaustive forms of data collection and analysis proposed by the witness. In conclusion, the witness stated that our study, although an admirable job under the circumstances, was carried out under restrictions that were too great for the job to be done well enough to be useful.

Once the testimony was over the two sides prepared findings of fact and posttrial briefs for the court, summarizing their arguments. In these documents, the plaintiffs cited neither our testimony nor that of their rebuttal witness, implying either that our evidence and the rebuttal were not relevant or that citation of them would not help the plaintiffs' case. The defendants' brief cited several of our findings, stressing those that bore on the uncertain relation between increases in spending and student outcomes. Although the defendants' brief did not cite our findings relative to teacher salaries and pupil–teacher ratios, the defendants' findings of fact did refer to our testimony on these issues. In subsequent final oral agreement, defense counsel cited school resource effects from our testimony as well as the data on fiscal effects.

Although in the final arguments the plaintiffs ignored both the statistical testimony and the rebuttal, and the defendants tended to concentrate on the fiscal implications of our testimony, it was ultimately up to the superior court judge to weigh that evidence and rebuttal in reaching his decision.

5. Ruling and Effect of the Evidence

The Court's Opinion

The superior court judge in the case, Robert J. Doran, gave his decision on 29 April 1983. During and after the trial, attorneys for both sides had properly emphasized the constitutional issues. The plaintiffs sought to expand the definition of constitutionally mandated basic education to be paid by the state to include bilingual education, school bus costs, extracurricular activities, costs of desegregation, special costs of urban districts, teacher salary increases, larger staff–pupil ratios, remedial education, education of the gifted, food services, and several other school district expense items. What light could our testimony and the rebuttal cast on issues, from the judge's viewpoint?

In his decision the judge in effect divided the issues of the case into two types: first, whether a given claim by a plaintiff deserved recognition and support under the state constitution or existing legislation; second, whether a given claim merited a court order specifying that the state provide the district with specific levels of funding, specific pupil–teacher ratios, or specific staff wage levels. There were also specific issues of whether the legislature had underfunded basic education in the recession years of 1981–1983, and whether the legislature had an appropriate policy on supplies and equipment cost. These issues were not related to the testimony we prepared and are not discussed here.

As to the first set of issues, the judge, guided by the constitutional provision and by past legislative actions (which largely took place after the 1977 school finance decision), held, as in 1977, that the state was constitutionally responsible for funding basic education (for the so-called normal range ability students and vocational education programs). He also held that an obligation devolved on the state, either on constitutional grounds or because of existing legislation, for education of the handicapped, bilingual education, pupil transportation, and remedial education. He did not uphold the plaintiffs' contention that food services, education of the gifted, extra costs of the urban districts, costs of desegregation, and a variety of other claims also merited mandatory state funding. He also found that the legislature could not reduce its basic education spending once it had appropriated a specified amount at a legislative session.

On the whole these aspects of his opinion extending state constitutional responsibilities favored the plaintiffs, by assuring a regular and dependable source of funding to the school districts, one that would provide the overwhelming majority of their operating funds.

By the same token, this aspect of his opinion was unsatisfactory to the defendants, who had hoped to win on such issues as remedial education and bilingual education.

But in the second aspect of his decision, comprising those passages of his opinion in which appropriate levels of funding, staffing, and wages were discussed, the field was reversed. Here his judgment was that the court would not order the legislature to appropriate any specified sums for these activities but whatever program the legislature chose to approve should be fully funded. It was left up to the legislature to decide pupil–teacher ratios and staff salaries, although in the future the state would have to tell the districts how much money there would be for salaries in advance of districts' budget preparation time.

In other words the court enhanced school districts' constitutional rights to state funding of educational programs but gave the legislature very wide authority to determine the scope and costs of those programs. From the plaintiffs' view, the court could have appeared to donate with one hand and to take back with the other. The result was to give each party some of what it had wanted. From the defendants' viewpoint, the court's upholding of legislative authority to determine the content of educational programs and their levels of funding, subject to some provisos, would make it harder in the future for school districts to raid the state treasury, at least on constitutional grounds. It should be noted that plaintiffs state that it was never their intention to remove legislative discretion over funding levels for education, but only to establish the constitutional basis for a number of programs, where this had been in doubt.

In the event, neither party chose to appeal the 1983 decision to the state Supreme Court as they had in 1977, perhaps fearing the uncertainties of appeal more than the consequences of Judge Doran's decision. Furthermore, and of real importance, both parties won much of what they sought. The school districts no longer had to fear the prospect of sudden changes in funding levels, while the state retained legislation control over educational policy and funding.

Effect of Our Evidence

It is easy enough to appraise the effects of our evidence on those elements of the opinion that bore on whether certain programs should be considered as belonging to basic education. The court's main criterion for these judgments was whether the program in question (e.g., bilingual education) was found by the legislature to be necessary for all or some of the state's school children, or to be a requirement that the legislature had imposed on itself.[2]

Thus our testimony was not germane to this set of issues. It was clearly relevant, however, to such issues as whether the courts should determine levels of funding and staffing, or variations in program content. On virtually all such issues in the case the judge's opinion was that the legislature should use its discretion.

Judge Doran's oral decision of 29 April 1983, (*Seattle School District v. State of Washington,* Thurston County No. 81-2-1713-1) makes no reference to our testimony, or, indeed, to virtually any testimony; however, his decision is clearly consistent with our testimony. On the basis of our literature review and the regression study, we stated there was no clear evidence that the mandating of funding levels, staffing levels, teacher educational levels, pupil–teacher ratios, and other such variables would

[2] See memorandum dated 4 May 1983 from Murphy and Davenport, Attorneys at Law, Olympia, Washington, to Governor John R. Spellman et al., re *Seattle v. State of Washington*. M. R. Murphy was chief counsel for the state in the case.

affect student achievement, as measured by test scores. In fact, in view of other earlier testimony that current teacher salary levels in Washington were adequate to recruit and retain competent staff, we stated that the chief effect of mandated salary increases would be to transfer wealth from the community at large to the educational sector. We also stated that these traditional ways of spending more educational funds were not likely to affect student outcomes in the aggregate. Progress was more likely to stem from new pedagogical approaches than from business as usual.

In a letter to one of the authors, after the judge's decision, the chief counsel for the defense stated "Judge Doran could not have arrived at the conclusion he did with respect to the breadth of the definition of basic education, and the legislature's discretion in that regard, without having fundamentally accepted at least the basic premise of your testimony." Although this view may be speculative, it does seem clear that on all matters having to do with level and content of programs, the judge's decision was consistent with our agnostic findings about the merits of mandating more resources unless accompanied by more effective practices. It is arguable that the testimony also made some contribution to a more general thesis found in the decision: namely, that the adequacy of educational programs should not be a subject for judicial determination. The plaintiff's view of our testimony is that it bolstered the defendants' case, although they believe that it is difficult to know what the effect was, because they felt that the decision favored their cause.

Some Reflections

The authors, in this case and in others, have had occasion to develop evidence based on statistical analysis. In our opinion such testimony is most likely to be effective if it is demonstrably germane, if it is presented in lay language whenever possible, and if it is possible to explain the statistical techniques used clearly and simply to a lay audience. These are not easy assignments. In our testimony we spend an entire courtroom day explaining how multiple regression works. This laborious process is justified by the fact that otherwise the court is likely to downplay the importance of assertions whose provenance is, in its eyes, murky.

Once the analytical methods have been clarified, the same must be done for the results. We have found it useful, for both purposes, to make extensive use of charts, tables, and graphs, which can be referred to whenever clarification is needed. In testimony on technical elements, it is often a good idea to repeat testimony in the form of summary. Key points in an argument can even be written out and repeated in oral testimony.

If time permits before the trial or deposition stage, it is extremely helpful to explain your statistical approach in lay terms to your attorney, as he is then in a better position to conduct direct examination of his own expert witness. However, this educational task is not easy to accomplish. Attorneys are often too busy during the final days of a case to invest the time and intellectual effort. Nor do legal training and experience necessarily offer fertile seedbed for statistical understanding. The statistical expert witness must be prepared to be flexible and, if necessary, to handle his testimony without much assistance from his lawyer.

The requirement for being germane hardly needs explanation. However, it may be worth pointing out that what is germane to a statistician may be irrelevant to a lawyer or a judge. The common failing is overinvolvement with the details of one's model, rather than with the question, What does the judge (and/or jury) want to know? This is clearly one danger involved in explaining the statistical methods at any substantial length; hence the importance of lay language in explanation.

Moving from these general considerations to one more particular to statistical analysis per se, we believe that statistical analysis of data arising directly out of a case is likely to be much more persuasive to the judge and the attorneys than more global analyses. In the present case, this view led us to undertake our own regression study of Washington data even though the time constraints and data availability dictated that the study would necessarily be less thorough and less comprehensive than many studies that we cited in our literature review. We felt that to offer evidence about the proposition: "there was no compelling evidence of relationship between school resources and student outcomes" required us to go beyond data from other jurisdictions and make a good faith effort to find a relationship in Washington data. Our view about case specific information was further reinforced by the chief defense counsel's tactics in his cross examination of the rebuttal witness. Defense counsel questioned the rebuttal witness at length about the witnesses' lack of familiarity with Washington schools. This appeared to be an effective tactic. One possible explanation of judges' and attorneys' strong preference for "case-specific" statistical analysis is that legal training and legal reasoning emphasize case-by-case analysis of precedents as contrasted with the more inductive approach that statisticians use.

Another and different set of reflections arises from the first author's involvement with school finance cases in other states and both authors' experience in the Washington case. Essentially, government has, in terms of educational productivity, no sound criteria for judging how much to spend in schools or what programs to support. The production function models described in Sections 2 and 3 are clearly rough approximations of the educational "production" process. Therefore their value for guidance to courts and policymakers is limited.

To the extent that courts order financial changes on equity grounds (e.g., equalizing spending per pupil among school districts in a state) it is difficult to oppose the policy strongly, except on technical grounds. After all, there is no evidence that equalizing spending leaves any student worse off in educational outcomes. However, once a court case turns on whether to spend more money on the schools and hence, less on other things, there are singularly few criteria for deciding how much if any more is better. In some cases it might be shown that salaries are too low to attract qualified teachers, but this may not be self-evident, as the Washington case elicited. Ordinarily, very little can be demonstrated in terms of the relation between money, at current ranges of public school spending, and educational effectiveness, as measured by achievement test scores.

When it comes to the relative claims of different programs, ranging from desegregation and bilingualism to programs for the gifted, school lunches, running the bus system, and coaching slow learners, all is in the realm of political judgment, whether in courts, legislatures, or school boards. Consequently the most that statistical evidence can offer the courts in this area is the impression that there is no clear evidence that, within a wide range, any particular pattern or level of spending will improve schooling. Accepting this view, educational policy should therefore not be frozen by court order. Rather it should be left to the political arena where in the present state of our knowledge, these matters are best resolved. The Washington court espoused this view in the case described here.

Acknowledgments

We want to acknowledge the participation of many people for doing the work that allowed us to complete our research on a demanding two-month schedule in connection with Seattle School District v. State of Washington. *They include Thomas J. Blaschke, Jan Hanley, and Frank Berger, who provided intensive computer programming, and Nanette Brown and Marian Oshiro, who served as research assistants. In Washington, we acknowledge the collaboration of Superintendent of Public Instruction staff in providing data: Gordon Ensign, Alfred Rasp, Jean Antonio, and Michael Fuller. In the Office of the State Attorney-General, Timothy Malone was invaluable in coordinating the data gathering effort. The LEAP staff, under the supervision of Don Tierney, provided us substantial data from its computer file, which was programmed by Carolyn Lindsay. The Washington Pre-College Testing Program kindly supplied test scores and related data from its files, under the supervision of Remy Greenmun. Michael Hoge, attorney for Seattle School District, described the outcomes of the case to us from the plaintiff's viewpoint. Finally, Malachy Murphy, Office of the Attorney-General, suggested that we prepare the written report of this research. He also helped us understand how our research fit into his presentation of the case and the court's decision, and later reviewed the manuscript of this paper, offering valuable suggestions for improvement. The research and the preparation of the original report (Pincus and Rolph, 1983) was supported by the Office of the Attorney-General, State of Washington, in part through JurEcon, Inc. The Institute for Civil Justice of The Rand Corporation provided secretarial support for preparing this paper.*

References

Averch, H. A., Carroll, S. J., Donaldson, T., Kiesling, H., Pincus, J. (1974). *How Effective Is Schooling? A Critical Review of Research,* Englewood Cliffs, N.J.: Educational Technology Publications.

Bowles, Samuel S. (1970). Toward an educational production function. In W. L. Hansen (ed.), *Education, Income and Human Capital,* New York: Columbia University Press, pp. 11–61.

Bowles, S. S., and Levin, H. M. (1968). "More on Multicollinearity and the Effectiveness of Schools," *Journal of Human Resources,* **3**(3), 393–400.

Bridge, R. G., Judd, C. M., and Moock, P. R. (1979). *The Determinants of Educational Outcomes: The Impact of Families, Peers, Teachers and Schools,* Cambridge, Mass.: Ballinger.

Cain, G. G., and Watts, H. W. (1970). "Problems in Making Inferences from the Coleman Report," *American Sociological Review,* **35,** 228–241.

Coleman, J. S., Campbell, E. Q., Hodson, C. F., McPartland, J., Mood, A. M., Weinfeld, S. D., and York, R. L. (1966). *Equality of Educational Opportunity,* Washington, D.C.: U.S. Government Printing Office, pp. 384–422.

Draper, N. R., Smith, H. (1981). *Applied Regression Analysis,* Second Edition, New York: John Wiley & Sons.

Dyer, H. S. (1972). Some thoughts about future studies. In F. Mosteller and D. P. Moynihan (eds.), *On Equality of Educational Opportunity,* New York: Random House, pp. 384–422.

Freedman, D., Pisani, R. and Purves, R. (1978). *Statistics,* New York: Norton, New York.

Gilbert, J. P., and Mosteller, F. (1972). The urgent need for experimentation. In F. Mosteller and D. P. Moynihan (eds.), *On Equality of Educational Opportunity,* New York: Random House.

Glass, G. V., and Smith, M. L. (1978). *Meta-Analysis of Research on the Relationship of Class-Size and Achievement,* San Francisco: Far West Laboratory for Educational Research and Development.

Hanushek, E. A. (1972). *Education and Race: An Analysis of the Educational Production Process,* Lexington, Mass.: Heath.

Levin, H. M. (1970). A new model of school effectiveness. In *Do Teachers Make a Difference?,* Washington, D.C.: U.S. Department of Health, Education and Welfare, Office of Education.

Luecke, D. F., and McGinn, N. F. (1975). "Regression Analyses and Education Production Functions: Can They Be Trusted?," *Harvard Education Review,* **45,** 325–350.

Michelson, S. (1970). The association of teacher resourcefulness with children's characteristics. In *Do Teachers Make a Difference?,* Washington, D.C.: U.S. Department of Health, Education and Welfare, Office of Education.

Morris, C. N., and Rolph, J. E. (1981). *Introduction to Data Analysis and Statistical Inference,* Englewood Cliffs, N.J.: Prentice-Hall.

Mosteller, F., and Moynihan, D. P. (eds.) (1972). *On Equality of Educational Opportunity,* New York: Random House.

Murnane, R. J. (1975). *The Impact of School Resources on the Learning of Inner City Children,* Cambridge, Mass.: Ballinger.

Perl, L. J. (1973). "Family Background, Secondary School Expenditures, and Student Ability," *Journal of Human Resources,* **8,** 156–180.

Pincus, J., and Rolph, J. (1983). *Production Function Study of Washington School Districts.*

Smith, M. S. (1972). Equality of educational opportunity: The basic findings reconsidered. In F. Mosteller and D. P. Moynihan (eds.), *On Equality of Educational Opportunity,* New York: Random House, pp. 230–342.

Summers, A. A., and Wolfe, B. L. (1977). "Do Schools Make a Difference?," *American Economic Review,* **67,** 639–652.

Weisberg, S. (1980). *Applied Linear Regression,* New York: John Wiley & Sons.

Bibliography on Educational Input–Output Relations

Armor, David, et al. *Analysis of the School Preferred Reading Program in Selected Los Angeles Minority Schools,* The Rand Corporation, R-2007-LAUSD, Santa Monica, California, 1976.

Averch, Harvey A., et al., *How Effective Is Schooling? A Critical Review of Research,* Educational Technology Publications, Englewood Cliffs, New Jersey, 1974.

Berman, Paul, and Milbrey W. McLaughlin, *Federal Programs Supporting Education Change, Vol. IV: The Findings in Review,* The Rand Corporation, R-1589/4-HEW, Santa Monica, California, 1976.

Bidwell, Charles E., and John D. Kasarda, "School District Organization and Student Achievement," *American Sociological Review* **40,** 55–70, 1975.

Boardman, Anthony, Otto A. Davis, and Peggy R. Sanday, "A Simultaneous Equations Model of the Educational Process," Carnegie-Mellon University, School of Urban and Public Affairs, Pittsburgh, Pennsylvania, 1973.

Boardman, Anthony, and Richard J. Murnane, "The Use of Panel Data in Education Research," Fels Discussion Paper No. 110, Department of Economics, University of Pittsburgh, Pennsylvania, 1977.

Bowles, Samuel S., "Educational Production Function: Final Report," U.S. Department of Health, Education and Welfare, Office of Education, 1969.

Bowles, Samuel S., Towards an educational production function. In W. L. Hansen (ed.), *Education, Income and Human Capital,* Columbia University Press, New York, 1970.

Bowles, Samuel S., and Herbert Gintis, *Schooling in Capitalist America,* Basic Books, New York, 1976.

Bridge, R. Gary, Charles M. Judd, and Peter R. Moock, *The Determinants of Educational Outcomes: The Impact of Families, Peers, Teachers and Schools,* Ballinger, Cambridge, Mass., 1979.

Brown, Byron W., and Daniel H. Saks, "The Production and Distribution of Cognitive Skills Within Schools," *Journal of Political Economy* **83,** 571–593, April 1975.

Burkhead, Jesse, Thomas G. Fox, and John W. Holland, *Input and Output in Large City High Schools,* Syracuse University Press, Syracuse, New York, 1967.

Centra, J. A., and D. A. Potter, "School and Teacher Effects: An Interrelational Model," *Review of Educational Research* **50,** 273–291, 1980.

Central Advisory Council for Education, *Children and Their Primary Schools,* Her Majesty's Stationery Office, London, 1967.

Cohn, Elchanan, "Economies of Scale in Iowa High School Operations," *Journal of Human Resources* **3,** 422–434, 1968.

Cohn, Elchanan, and S. D. Millman, *Input-Output Analysis in Public Education,* Ballinger, Cambridge, Mass., 1975.

Coleman, James S., et al., *Equality of Educational Opportunity,* U.S. Government Printing Office, Washington, D.C., 1966.

Froomkin, Joseph T., et al., *Education as an Industry,* Ballinger, Cambridge, Mass., 1976.

Garner, William T., "The Identification of an Educational Production Function by Experimental Means," Ph.D. dissertation, University of Chicago, 1973.

Glasman, Naftaly S., and Israel Biniaminov, "Input-Output Analyses of Schools," *Review of Educational Research,* **51**(4), 509–540, 1981.

Glass, Gene V., and Mary Lee Smith, *Meta-Analysis of Research on the Relationship of Class-Size and Achievement,* Far West Laboratory for Educational Research and Development, San Francisco, California, 1978.

Hanushek, Eric A., "Conceptual and Empirical Issues in the Estimation of Educational Production Function," *Journal of Human Resources* **14**(3), 351–388, 1979.

Hanushek, Eric A., *Education and Race: An Analysis of the Educational Production Process,* D. C. Heath, Lexington, Mass., 1972.

Heim, John, and Lewis Perl, *The Educational Production Function,* Cornell University, Ithaca, New York, 1974.

Henderson, Vernon, Peter Mieszkowski, and Yvon Savageau, *Peer Group Effects and Educational Production Functions,* Economic Council of Canada, Ottawa, 1976.

Jamison, D. T., P. C. Suppes, and S. J. Wells, "The Effectiveness of Instructional Media: A Survey," *Review of Educational Research* **44,** 1–67, 1976.

Jencks, Christopher, and Marsha D. Brown, "Effects of High Schools on Their Students," *Harvard Educational Review* **45,** 273–324, August 1975.

Karweit, N., "A Reanalysis of the Effect of Quantity of Schooling on Achievement," *Sociology of Education* **49,** 236–246, 1976.

Katzman, Martin T., *The Political Economy of Urban Schools,* Harvard University Press, Cambridge, Mass., 1971.

Kiesling, Herbert J., *The Relationship of School Inputs to Public School Performance in New York State,* The Rand Corporation, Santa Monica, California, 1969.

Kiesling, Herbert J., *A Study of Cost and Quality of New York School Districts,* U.S. Office of Education, Washington, D.C., 1970.

Kirst, Michael, and K. Jung, "The Utility of a Longitudinal Approach in Assessing Implementation: A Thirteen-Year View of Title I ESEA," *Educational Evaluation and Policy Analysis* **2**(S), 17–24, 1980.

Klitgaard, Robert E., and George R. Hall, "Are There Unusually Effective Schools?" *Journal of Human Resources* **10,** 90–106, 1975.

Levin, Henry M., "A New Model of School Effectiveness," *Do Teachers Make a Difference?,* U.S. Office of Education, Washington, D.C., 1970.

Levin, Henry M., "Measuring Efficiency in Educational Production." In William H. Sewell, Robert M. Hauser, and David L. Featherman (eds.), *Schooling and Achievement in American Society,* Academic Press, New York, 1976, pp. 291–307.

Levin, Henry M., "Cost-Effectiveness Analysis of Teacher Selection," *Journal of Human Resources* **5,** 24–33, 1970.

Levy, Frank S., "The Racial Imbalance Act of Massachusetts," Ph.D. dissertation, Department of Economics, Yale University, 1967.

Link, Charles R., and Edward C. Ratledge, "Social Returns to Quality and Quantity of Education: A Further Statement," *Journal of Human Resources* **10,** 78–89, 1975.

Maynard, Rebecca, and David L. Crawford, School performance. In *Rural Income Maintenance Experiment: Final Report, Vol. VI,* Chapter 12, Institute for Research on Poverty, University of Wisconsin, Madison, Wisconsin, 1976.

Michelson, Stephan, The Association of Teacher Resourcefulness with Children's Characteristics. *Do Teachers Make a Difference?,* U.S. Office of Education, Washington, D.C., 1970.

Mosteller, Frederick, and Daniel P. Moynihan, *On Equality of Educational Opportunity,* Random House, New York, 1972.

Murnane, Richard J., *The Impact of School Resources on the Learning of Inner City Children,* Ballinger, Cambridge, Mass., 1975.

Perl, Lewis J., ''Family Background, Secondary School Expenditures, and Student Ability,'' *Journal of Human Resources* **8,** 156–180, 1973.

Pincus, John, and John Rolph, ''Production Function Study of Washington School Districts,'' prepared for the Office of the Attorney-General, State of Washington, Olympia, Washington, January 1983.

Ritzen, Josef M., and Donald R. Winkler, ''The Production of Human Capital Over Time,'' *Review of Economics and Statistics* **59,** 427–437, November 1977.

Silkman, Richard, *The Diffusion of Educational Innovation,* Ph.D. dissertation, Yale University, 1980.

Summers, Anita A., and Barbara L. Wolfe, ''Equality of Educational Opportunity Quantified: A Production Function Approach,'' paper presented at the winter meetings of the Econometric Society, December 1974.

Summers, Anita A., and Barbara L. Wolfe, ''Do Schools Make a Difference?'' *American Economic Review,* **67,** 639–652, 1977.

Tuckman, H. P., ''High School Inputs and Their Contribution to School Performance,'' *Journal of Human Resources,* **6,** 490–509, 1972.

Welch, Finis, ''Black-White Differences in Returns to Schooling,'' *American Economic Review,* **63,** 893–907, December 1973.

Welch, Finis, ''Measurement of the Quality of Schooling,'' *American Economic Review* **56,** May 1966.

Wiley, David E., Another hour, another day: Quantity of schools, a potent for policy.'' In William H. Sewell, Robert M. Hauser, and David L. Featherman (eds.) *Schooling and Achievement in American Society,* Academic Press, New York, 1976, pp. 291–307.

Winkler, Donald R., ''Educational Achievement and School Peer Group Composition,'' *Journal of Human Resources* **10,** 189–205, 1975.

Capital Market Analysis in Antitrust Litigation

Martin S. Geisel **Jean L. Masson**
Graduate School of Management
University of Rochester
Rochester, New York

ABSTRACT

This paper discusses the theory and statistical analysis of capital market tests in the context of the analysis of litigation. Economy-wide capital market analyses of antitrust policies toward mergers are reviewed, and the use of these methods in individual antitrust cases is illustrated. Capital market tests provide a relatively straightforward means of assessing the competitive impacts of antitrust violations.

1. Introduction

Antitrust litigation has been a major growth industry during the past three decades. This growth has been primarily due to large increases in the number of antitrust actions initiated by private parties. Posner (1970) estimates that 1092 private antitrust cases were initiated during the 60-year period from the passage of the Sherman Act through 1949. The *Annual Report of the Director of the Administrative Office of the U.S. Courts* (1983) indicates that during 1950–1959, 2146 cases were started. In the 1960s, 6490 new actions were brought, including 2233 electrical equipment cases. In the 1970s, 13,162 cases were initiated, and in the period 1980–1983 4978 more private cases were brought to court.

The growth in antitrust litigation has induced similar growth in the expert witness business. Today there are numerous economic consulting firms that earn a large fraction of their revenues by providing expert consultation in antitrust and other litigation. Similarly, many academics in business schools, economics departments, and statistics departments engage in litigation consulting on at least an occasional basis.

One might anticipate that this consulting activity would lead directly to a large volume of research publications in scholarly journals. Although there are notable exceptions such as Evans (1982) and Fisher et al. (1983), this does not appear to be the case on a broad scale. A variety of factors may contribute to this low publication rate. Much of the econometric work involved is routine application of standard techniques such as multiple regression, the scope of a typical study is limited to a specific part of a specific industry or market, and publication may be prohibited by confidentiality agreements. Publications may also appear in "disguised" fashion so that one cannot immediately ascertain that the work began in consulting activity.

While individual antitrust case studies do not permeate the scholarly literature in economics, it would be inappropriate to conclude that economists are generally uninterested in or have nothing to say about antitrust policy. On the contrary, there is a substantial literature analyzing the effectiveness of the antitrust laws in promoting competition. See Elzinga (1969), Green et al. (1972), and Pfunder et al. (1972) for typical analyses of antitrust law enforcement. Furthermore, during the past decade or so, financial economists have developed and applied a simple but powerful methodology that uses capital market data to effectively address these concerns. Most capital market tests of antitrust policy have concentrated on economy-wide analyses of mergers and other possibly anticompetitive activities. The capital market approach, however, is also appropriate to analyses of individual firms and markets and is increasingly being applied in such instances.

In Section 2 we provide a brief introduction to capital market test methodology. We then review two studies that analyze the behavior of security prices of firms involved in horizontal mergers to assess the impact and effect of antitrust litigation. Finally, we present the results of capital market tests in two cases involving a merger and a price-fixing conspiracy.

2. Capital Market Theory: A Brief Review

Quantitative assessment of the competitive impact of a collusive agreement or a merger by means of conventional econometric methods is in general a nontrivial task. A model capable of describing the industry's competitive price and output under a wide range of market and external conditions is needed. In the case of a collusive agreement, the model should be able to discriminate between price changes due to the collusion and those due to changes in market or external conditions, typically in the presence of uncertainty regarding the effective dates of the collusion. In the case of a horizontal merger, the model should be sufficiently detailed to predict the impact of a change in the number of sellers on market price. To our knowledge, few studies of antitrust cases approach this degree of econometric detail and precision. Whether for reasons of time, data limitations, or model inadequacy, simpler approaches such as time series comparisons of price behavior with other related firms or markets or ad hoc regressions are typically employed. While such approaches have the virtue of simplicity, they frequently leave considerable room for disagreement regarding the central object of the analysis. Capital market tests forgo much of the detail that a well-specified econometric model might provide, but by focusing on firm values rather than on prices and outputs, they can yield more accurate estimates of the bottom line competitive impact of an antitrust violation than other simple approaches.

If mergers, collusive agreements, or other alleged antitrust violations are to have an anticompetitive impact, they must grant market power to the participating firms. This market power will enable the firms to increase or stabilize product prices so as to increase profits. These increased profits will in turn be reflected in increased values of the participating firms. Thus, if good time series data on the market value of the participating firms are available, they may be analyzed to test for the presence or absence of competitive impact of an alleged antitrust violation. This is the approach used by capital market tests, and the econometric work involved is much simpler than that required to estimate competitive prices and outputs.

The capital market test approach to measuring competitive impact is most directly applicable in cases in which the firms involved have equity securities that are actively traded on organized exchanges. In such cases, there is voluminous evidence that supports the hypothesis that the firms' security prices reflect all available relevant information. This hypothesis regarding security price behavior is known as the efficient markets or rational expectations hypothesis. Fama (1970, 1976) presents a formal discussion of the efficient markets model and a review of the evidence regarding it. Schwert (1981) also provides a good review of these issues in the context of studying the effects of regulation. The efficient markets hypothesis implies that when investors perceive (anticipate) a change in the fortunes of a firm, they will make investment decisions that result in prompt adjustment of the firm's security price to an unbiased estimate of the new market value of the firm, thereby eliminating systematic profit opportunities. In the present context this means that security prices will promptly reflect the market valuation of the prospective increase in profits flowing from the alleged anticompetitive activity. The possibility thus exists that we may infer the presence or absence of the anticompetitive impact of an alleged antitrust violation by examining the time series of the firm's security prices for abnormal returns at relevant times during the evolution of the activity.

There are, of course, factors other than alleged antitrust violations that will affect the returns on a firm's securities. In order to estimate the anticompetitive impact of an alleged antitrust violation, it is thus necessary to have a model of "normal" security returns. The most commonly used model of normal returns is the market model that is given, for firm *i*, by

$$R_{it} = \alpha_i + \beta_i R_{mt} + \varepsilon_{it}$$

where R_{it} and R_{mt} are the daily continuously compounded rates of return of firm *i* and the value-weighted market portfolio in period *t*. The market model posits that the return on the market portfolio is a common factor in the equilibrium returns of all securities. The coefficient β_i is the measure of covariation between R_i and R_m. Conditional on the information set available prior to day *t* and the contemporaneous return on the market portfolio, the disturbance term ε_{it} is a zero-mean serially uncorrelated variable. An extensive discussion of the market model and of the empirical distribution of continuously compounded stock returns is given in Fama (1976). Fama finds that daily stock return have a somewhat more leptokurtic distribution than the normal and that market model parameters are probably not stationary over long time periods.

Abnormal returns for security *i* are defined as the forecast errors from the market model regression:

$$\text{PE}_{it} = R_{it} - (\hat{\alpha}_i + \hat{\beta}_i R_{mt})$$

where $\hat{\alpha}_i$ and $\hat{\beta}_i$ are least-squares estimates of α_i and β_i. The hypothesis that an alleged antitrust violation has no anticompetitive impact implies that the time series of abnormal returns should behave approximately as white noise. In practice, various statistical tests are employed to analyze the behavior of the PE_{it} series around the times of important events (e.g., merger announcements, filing of complaints, and meetings between officials of colluding firms) to ascertain whether there is evidence of significant departures from randomness. The most common of these, which involves cumulating the prediction errors over time, will be illustrated in Section 3.

The market model typically accounts for a substantial portion of the variance of the daily or monthly returns of an individual security or portfolio of securities. In studies involving a single firm or a subset of the firms within a single industry, however, an index of the security returns for that industry is frequently included as an additional independent variable in the market model. This permits more precise estimation of truly "abnormal" returns by eliminating factors that had a common impact on all firms in an industry. King (1966) finds that an average of about 20% of the variance of security returns is explained by factors common to an industry. When small numbers of firms are under consideration it is also common practice to search sources such as the *Wall Street Journal Index* for other extraneous events that might have influenced the return on the firms' securities.

There are alternative models of "normal" returns used to study the effects of mergers. The Sharpe–Lintner and Black versions of the capital asset pricing model (CAPM), described in detail in Jensen (1972), are used frequently. Essentially, these models impose some restrictions on the intercept term of the market model. The Sharpe–Lintner version of the CAPM restricts α_i to be equal to $(1 - \beta_i)R_f$, where R_f is the risk-free rate of interest. The Black CAPM restricts α_i to be equal to $(1 - \beta_i)E(R_0)$, where $E(R_0)$ is the expected return on a portfolio uncorrelated with the market portfolio. From this perspective, the market model is simply an unrestricted version of the CAPM. Some recent studies have used Ross's arbitrage pricing theory to estimate normal returns. In this approach normal returns are described by a multiple-factor linear model, but the independent variables or factors are determined empirically using factor analysis. Factors that approximate market and industry factors typically appear with this methodology. See Ross (1976) and Roll and Ross (1980) for detailed discussions of arbitage pricing theory.

Brown and Warner (1980), in a survey of various event study methodologies, conclude that the single-factor market model is probably superior to more complicated methodologies under most circumstances.

3. Capital Market Tests of Horizontal Mergers

Section 7 of the Clayton Act prohibits one firm from acquiring the stock or assets of another if "the effect of such acquisition may be substantially to lessen competition or tend to create a monopoly." When the staff of the Antitrust Division of the Department of Justice or of the Federal Trade Commission believes that an anticompetitive acquisition will be or has been undertaken, it seeks "relief" from the firms involved through negotiations or legal proceedings. "Relief" may take the form of merger cancellation, divestiture, and/or a variety of regulatory restrictions. The sole intended purpose of relief is to restore robust competition in the affected market.

Some lawyers and economists have reservations about the effectiveness of antitrust activity in promoting competition in the economy. Some believe that government victories in Section 7 cases are meaningless because the terms of final decisions are too loosely enforced or do not promote competition even if strictly enforced. Others conclude that antitrust agencies, through relief, can impose costly constraints on defendant firms in addition to the legal expenses and the interruption of productive activities associated with antitrust suits. Green et al. (1972), and Pfunder et al. (1972) criticize the enforcement record of antitrust agencies. Ellert (1976) casts doubts on the effectiveness of the Justice Department or the Federal Trade Commission in enhancing competition. Kummer (1975), Stillman (1983), and Eckbo (1983) examine the effects of antitrust complaint announcements on the stock returns of merging firms and conclude that relief is expected to be costly.

Wier (1983) tests the validity of the costly relief hypothesis by analyzing defendant's returns on equity around announcements of Section 7 final decisions. Her tests reveal that stockholders of firms that lose Section 7 cases suffer abnormal losses of 2% on average, while stockholders of firms forced to abandon proposed acquisitions suffer abnormal losses of 8%. These abnormal returns lead her to conclude that the enforcing agencies can impose costly constraints on defendant firms. Of course, constraints that are costly for a firm may be socially beneficial when they prevent the creation of monopoly power or costly for the aggregate economy when they impose an inefficient allocation of resources. She does not attempt to allocate the costs of relief among the loss of productive efficiency or the reduction in monopoly rents. Stillman (1983), Ellert (1976), and Eckbo (1983) address this issue.

Mergers and antitrust suits filed by the Justice Department or the Federal Trade Commission occur at different times for different firms. In order to draw conclusions about the general impact of mergers and suits from events scattered in time, Wier analyzes security returns in "event time," a technique pioneered by Fama et al. (1969) in a study of stock splits. Stock returns are indexed by event time with the Section 7 proceedings termination day assigned an index value of zero. By focusing on the termination day rather than the merger proposal or antitrust complaint announcement days, Wier can capture the market estimates of the costs or benefits associated with the terms of final judgments since litigation and other case-related expenses are capitalized prior to the final decision day. Wier's sample includes all firms whose stock returns are available from the daily returns file of the Center for Research in Security Prices and whose Section 7 violation cases occurred between December 1962 and March 1979. Her sample excludes all mergers regulated by agencies other than the Justice Department or the Federal Trade Commission. These include bank, railroad, and other transportation industry mergers.

Using firms involved in all attempted mergers in her sample, Weir forms five equally weighted portfolios containing (1) all targets, (2) bidders in canceled acquisitions, (3) dismissed, (4) convicted, and (5) settling bidders in completed acquisitions. This grouping of securities in portfolios enables her to identify the common effects of Section 7 decrees since unrelated specific events which also influence stock returns tend to average out in the grouping process.

The costly relief hypothesis predicts that the shares of firms subject to relief will show negative abnormal performance while the shares of dismissed firms will show positive abnormal returns when the final judgment is rendered.

In order to allow for the possibility that the merger alters the nature of the firm, α_i and β_i for each firm in the sample are estimated first over preevent days -308 to -109 and then over postevent days 61 to 261.

Daily portfolio average prediction errors (APE_t) are obtained by averaging the individual security prediction errors each day for days -108 to $+60$. The cumulative average prediction error over the interval (D_1, D_2) is

$$\mathrm{CAPE}_{D_1,D_2} = \sum_{D=D_1}^{D_2} \frac{1}{J} \sum_{j=1}^{J} \mathrm{PE}_{jD},$$

where J is the number of securities in the portfolio and PE_{jD} is the prediction error for security j on day D. The standardized CAPE is assumed to have a t distribution.

For proposed acquisitions that were later canceled following antitrust complaints, target firms have average abnormal losses of 10.85% between the proposal and the cancellation announcements, which includes an average -6.50% [standard error (se) = 0.49%] abnormal return on the eve of the publication of the cancelation in the *Wall Street Journal*. A large response on the eve of the assigned event date is consistent with the hypothesis that the event has an impact on stock prices because we allow for the possibility that news of the event reaches many market participants through news wires or other rapid media prior to the newspaper publication. Acquiring firms do not earn statistically significant abnormal returns at any time throughout the period of interest. From these results, Wier concludes that most if not all of the gains from canceled mergers are expected to accrue to target firms' shareholders. These results are consistent with much of the recent literature on corporate control. See, for example, Jensen and Ruback (1983), who find that "corporate takeovers generate positive gains, that target firm shareholders benefit, and that bidding firm shareholders do not lose." Bidders, who seem to anticipate only normal returns from their acquisitions, probably lose less by canceling their merger plans than by going through antitrust legal proceedings. Target firms show large average gains around the merger proposal announcements and then average losses at least as large in magnitude when mergers are canceled. This is consistent with both the proposition that enforcing agencies are successful in eliminating the potential monopolistic use of target assets and the proposition that enforcing agencies are successful in preventing a more efficient allocation of resources.

The sample of completed acquisitions also yields interesting insights. Convicted firms suffer statistically large average abnormal losses of 1.67% (se = 0.39%) on the complaint announcement day. They also experience sizable average losses of 0.53% on the eve of and 1.56% (se = 0.31%) on the day of the final judgment announcement. These losses, like those of target firms whose acquisitions are canceled, are consistent with the costly relief hypothesis, although they are not very large.

In the case of dismissed complaints, the filing of the complaint induces a small average abnormal loss of 1.08% (se = 0.48%) on the same day. There are no significant abnormal returns around the complaint dismissal announcement. This evidence is inconsistent with the costly relief hypothesis since unanticipated dismissals of legal proceedings that could lead to costly relief should be associated with upward share price adjustments. By disaggregating the data, Wier finds some weak evidence of positive abnormal returns (1.61%, se = 0.83%) for cases in which appeals were dropped by the Justice Department. She points out that the lack of significant responses may be due to the difficulty of identifying the relevant event date for dismissed complaints. Dismissals may occur as a result of the failure of the Justice Department to announce its intention to appeal within the prescribed 60-day period or as a result of a formal pullout by the antitrust agency. The failure to appeal is an especially diffuse event.

The last portfolio is formed from firms that settle with the antitrust agency after the complaint has been filed. These firms earn large negative average abnormal returns around the complaint announcement day: −1.23% (se = 0.27%) on the eve of the announcement and −1.04% and −0.66% on the next two days. Like dismissed firms, the abnormal performance of settling firms is negative, although not as clearly so, around the final settlement announcement: −0.20% (se = 0.27%) on the settlement announcement day. In the case of settling firms, Wier faces again the problem of elusive event dates, which may explain the lack of significant responses to the announcements.

In conclusion, the evidence bearing on the costly relief hypothesis is mixed. Complaint announcements results in abnormal losses for four of five portfolios of firms involved in Section 7 cases. Subsequently, shareholders of target firms whose acquisitions are canceled in reaction to antitrust complaints suffer large losses that wipe out, on average, the gains realized at the time of the acquisition proposal announcement. Convicted firms' share prices also perform as predicted by the costly relief hypothesis. However, dismissed and settling firms' share prices do not respond systematically to final decision announcements. This lack of measured responses could be due to problems in isolating the exact event dates relevant to the dismissals and settlements.

Another question of interest to researchers concerned with antitrust laws is whether the abnormal gains realized by the shareholders of merging firms are due to gains in productive efficiency made possible by the improvement in resource allocation or to the increased probability of successful industry cartelization. Eckbo (1983) asserts that the behavior of horizontal rivals to the merging firms ought to be indicative as to the nature of the abnormal gain.

Eckbo predicts that the value of rival firms should increase as a result of anticompetitive horizontal mergers. Because horizontal mergers reduce the number of firms in an industry, they also reduce the cost of achieving collusion among the remaining firms. Hence, any announcement that increases the probability of a successful merger should coincide with abnormally high returns on the stock of rivals to the merging firms. Likewise, negative announcements should coincide with abnormally low returns.

Eckbo's sample consists of 191 horizontal mergers in mining and manufacturing industries that occurred between 1963 and 1978. He uses the market model and continuously compounded daily stock returns to estimate abnormal returns for 101 days around each event. He computes cumulative average abnormal returns for bidders and targets in challenged and unchallenged mergers with respect to merger proposal and complaint announcements. His results are consistent with Wier's. Target firm shareholders typically earn relatively large abnormal gains in the period surrounding the merger proposal, whereas bidding firm shareholders earn much less.

Abnormal stock returns around antitrust complaint announcements show bidders experiencing insignificantly negative abnormal returns (−0.85% over days −10 to +5), whereas targets suffer large negative abnormal returns (−9.60% over the same time interval).

Eckbo then forms an equally weighted portfolio of horizontal rivals for each merger. Rival portfolio abnormal returns are estimated in the same fashion as merging firms.

Rivals of the 126 unchallenged horizontal mergers systematically earn small but significantly positive abnormal returns around the merger proposal (0.76% over days −3 to +3). The behavior of rivals' returns around the merger proposals of 65 challenged mergers is similar (0.90% over the same period). These findings are consistent with both hypotheses. Under the productive efficiency hypothesis, a merger proposal signals to all industry participants that the resources of merging firms can be more efficiently employed. Under the collusion hypothesis, the proposal increases the likelihood of successful collusion.

When focusing on the complaint announcement, Eckbo finds that, unlike bidding and target firm shareholders, horizontal rival shareholders do not suffer any loss in response to this unexpected increase in the cost of collusion. Based on the lack of response of horizontal rival stock prices to the filing of Section 7 complaints, Eckbo rejects the collusion hypothesis in favor of the productive efficiency hypothesis.

To substantiate his conclusion further, he also examines the abnormal performance of rivals of target firms in 68 vertical mergers. Since vertical mergers do not reduce the number of independent producers in an industry, there are few a priori reasons to expect them to yield results consistent with the collusion hypothesis. He finds small negative abnormal returns for rivals around 57 unchallenged vertical merger announcements (−0.01% over days −10 to +5). In the case of 11 challenged vertical mergers, rivals earn insignificant positive average abnormal returns around merger announcements (1.95% over days −10 to +5) and complaint announcements (1.93% over the same period). These returns are similar to the returns of horizontal merger rivals. The evidence from vertical mergers, which are unlikely to have collusive effects, clearly adds to the credibility of the productive efficiency hypothesis. Stillman (1983) also examines antitrust policy toward horizontal mergers. Based on the stock performance of rivals to 11 challenged horizontal mergers, he also concludes against the collusion hypothesis.

4. Individual Case Studies

The research studies reviewed above provide considerable information regarding the normal or average responses of capital values to mergers and events related to antitrust litigation. Because these studies demonstrate that asset prices generally respond rapidly to such events, they provide the empirical foundation for analyses of individual antitrust cases using market model tests. In studying a single case, however, application of the market model methodology must include consideration of the impacts on security prices of simultaneous events that are unrelated to the litigation since these effects cannot be expected to "average out" as they would in the analysis of many cases. Two approaches to handling this complication are illustrated in the case studies discussed here.

One of the most careful and elaborate applications of capital market methodology is Michael Jensen's study (Jensen, 1978) of the 1970 merger of H & B American Company and Teleprompter. This merger of two firms in the cable-television industry is somewhat unusual in that the alleged restraint of trade involved illegal agreements between one of the merging firms and third parties. The value of the Teleprompter (TPT) shares that H & B American (HBA) shareholders received in this merger were affected by these illegal activities which came to light shortly after the merger was consummated.

HBA and TPT began merger discussions in May 1969. The intent to merge was announced on 11 August 1969, approved by the shareholders of the two companies on 10 May 1970, and took effect on 17 September 1970. Shareholders of HBA received one share of TPT stock for each $3\frac{1}{8}$ shares of HBA that they held. At the time of shareholder approval, the transaction involved a premium of 20.6% to HBA shareholders, which is in the lower part of the range of premiums normally paid to selling shareholders in such mergers. Shortly after the merger, however, it came to light that TPT had engaged in illegal transactions with officials of Johnstown, Pennsylvania, during 1966, and it was alleged to have done so with Trenton, New Jersey, officials in 1967–1968. The plaintiff alleged that the market price of TPT at the time of the merger thus did not represent the true value of TPT. Jensen was asked to estimate what the market price of TPT would have been at the time the merger took effect if these actions had been known to the investing public at that time.

On 28 October 1970 the son of the mayor of Johnstown, Pennsylvania, received a grand jury supoena, and he testified on 6 November 1970 regarding alleged 1966 bribes by TPT to Johnstown officials. The mayor and other officials testified during November and December, as did the chief executive officer (CEO) of TPT. TPT and its CEO were indicted on 28 January 1971, and the *Wall Street Journal* reported their innocent pleas on 17 February 1971. The *Journal* reported during March 1971 that five TPT executives were supoenaed by the Trenton grand jury and that TPT and three of its officers were named as co-conspirators in that case.

Jensen estimated abnormal returns on TPT shares during the November 1970–March 1971 period using both monthly and daily data. In his monthly market model analysis, Jensen included an index of returns to nine firms in the cable-TV industry as well as the market index in order to account for events that had industry-wide impacts on security prices. He used 50 months of data (and excluded November 1970–April 1971) to estimate the normal returns equation, which explained 63% of the variance in monthly TPT returns. The estimated abnormal returns for the months of November 1970 through March 1971 were −4.3%, −10.2%, −14.8%, −18.5%, and −11.5%, respectively, resulting in a compound abnormal return of −47.2% over that period. The actual price of TPT shares declined by only 12.8% during this period but the market index rose 23.6% and the cable-TV industry index rose 27.8%.

Jensen's daily analysis also included the market index and a cable-TV index. The normal returns equation was estimated using 400 daily observations with data from November 1970–April 1971 excluded from the estimation. The use of daily data permitted Jensen to associate abnormal returns more precisely with events related to the illegal activity and enabled him to make adjustments for extraneous factors affecting TPT share prices (e.g., other acquisitions, earnings announcements).

During November and December 1970, the estimated abnormal return on TPT shares was −14.7%. Essentially all of these losses occurred at the time of or immediately following events pertaining to the Johnstown affair. TPT was indicted by the Johnstown grand jury on 28 January 1971, and this was reported in the *Journal* on 29 January and 2 February. In the seven trading days beginning 28 January, the estimates of TPT's abnormal returns are −1.9%, −13.5%, −6.3%, −5.3%, +4.2%, −0.9%, and −6.9%. The standard error for each of these prices changes is 2.7%, so the departure from normal returns is quite significant. Abnormal returns during February and March 1971 continued to be closely associated with events in the Johnstown and Trenton cases. On 1 April 1971 the resignation of TPT's chairman was announced. Abnormal returns for seven days beginning 1 April are +13.7%, +3.8%, +2.9%, −5.7%, −2.7%, −0.2%, and −3.6%. In fact, the TPT chairman did not leave his position at that time.

During the November 1970–April 1971 period there were also several other announcements relating to TPT but unrelated to the litigation issues. Using the same normal daily return model, Jensen estimated the abnormal returns associated with these events and adjusted TPT's cumulative abnormal performance measure to eliminate their effects. From 28 October 1970 to 12 April 1971 the cumulative abnormal return on TPT shares, after adjustment for unrelated abnormal returns, was −48%. Litigation-related events continued occasionally for another two years, but they had little further systematic impact on TPT's abnormal performance index.

Jensen's analysis played a major role in the disposition of this case. He gave extensive deposition testimony based on his affidavit and its appendixes. These include a discussion of the market model methodology and extensive numerical and graphical exhibits showing his analysis of TPT's abnormal returns due to the illegal activity and other factors. Jensen estimated the damages to HBA shareholders using the adjusted cumulative abnormal return of −48%. This estimate provide the basis for settlement of the case just before it was scheduled to go to trial.

Our second illustration of the use of capital market tests concerns an alleged price fixing conspiracy by several large firms in a major industry during the early 1970s. Litigation on various aspects of this matter is still continuing, so we cannot provide identifying information about the participants or give exact calendar dates here.

This industry was characterized by substantial output price fluctuations during the relevant time period. Of particular interest to the litigation is a short time period during which prices were stabilized at "list" prices following heavy discounting. Such price restorations had occurred previously but were typically followed relatively quickly by renewed discounting. On this occasion they remained high for an extended period. Contemporary public statements attributed the price increases both to high demand and to cost increases. As part of its investigative work, however, the plaintiff developed evidence suggesting the possibility of a meeting of industry executives at which a pricing agreement might have been formulated.

In this case capital market tests were used as one part of a larger econometric study to investigate the various explanations of the price increase. Multiple regression and Box–Jenkins models, which were developed to forecast product consumption, indicated that the period of several months prior to the price increase was characterized by generally high demand, but that demand was low just prior to the increase. The daily market model analysis of a portfolio of the firms in question revealed a positive abnormal return of nearly 10% during the week surrounding the price increase with no systematic patterns in returns during either the preceding or succeeding months. Further, a market model for smaller, less integrated firms in the industry showed much smaller abnormal returns during the week in question, and the *Wall Street Journal Index* did not report any other major events that might have explained this increase in security prices. The capital market tests thus lended credence to the price-fixing explanation of the price increase and proved helpful to attorneys in deciding their strategies for developing the case.

5. Conclusion

In this chapter we have reviewed applications of capital market methods in the analysis of the impacts of antitrust litigation. Capital market tests are theoretically based on the efficient markets hypothesis and the market model of equilibrium security returns. These tests have been successfully applied in economy-wide studies of mergers and antitrust policies. They are also coming into increasing use in analyses of individual firms and markets in the context of antitrust and other litigation. As the illustrative cases presented here show, capital market tests yield direct estimates of the impact of such events as antitrust violations by applying straightforward statistical techniques to capital market data.

References

Annual Report of the Director of the Administrative Office of the U.S. Courts (1983).

Brown, S. J., and Warner, J. B. (1980). "Measuring Security Price Performance," *Journal of Financial Economics,* **8,** 225–240.

Eckbo, B. E. (1983). "Horizontal Mergers, Collusion and Stockholder Wealth," *Journal of Financial Economics,* **11,** 241–274.

Ellert, J. C. (1976). "Mergers, Antitrust Law Enforcement and Stockholder Returns," *Journal of Finance,* **31,** 715–732.

Elzinga, K. G. (1969). "The Antimerger Law: Pyrrhic Victories?," *Journal of Law and Economics,* **12,** 43–78.

Evans, D. S. (ed.) (1982). *Breaking Up Bell,* Amsterdam: North Holland.

Fama, E. (1970). "Efficient Capital Markets: A Review of Theory and Empirical Work," *Journal of Finance,* **25,** 383–417.

Fama, E. (1976). *Foundations of Finance,* New York: Basic Books.

Fama, E., Fisher, L., Jensen, M. C., and Roll, R. (1969). "The Adjustment of Stock Prices to New Information," *International Economic Review,* **10,** 1–21.

Fisher, F., McGowan, J. J., and Greenwood, J. E. (1983). *Folded, Spindled, and Mutilated,* Cambridge, Mass.: MIT Press.

Green, M., Moore, B., and Wasserstein, B. (1972). *The Closed Enterprise System,* New York: Grossman.

Jensen, M. C. (ed.) (1972). *Studies in the Theory of Capital Markets,* New York: Praeger.

Jensen, M. C. (1978). "Affidavit in Leeds vs. Teleprompter," U.S. District Court for Central District of California. CV 71-2739.

Jensen, M. C., and Ruback, R. S. (1983). "The Market for Corporate Control: The Scientific Evidence," *Journal of Financial Economics,* **11,** 5–50.

King, B. F. (1966). "Market and Industry Factors in Stock Price Behavior," *Journal of Business,* **39,** 139–140.

Kummer, D. (1975). *Stock Price Reaction to Announcements of Forced Divestiture Proceedings,* Unpublished dissertation (University of Oregon, Eugene, Oregon.)

Pfunder, M., Plaine, D., and Whittemore, A. M. (1972). "Compliance With Divestiture Orders Under Section 7 of the Clayton Act: An Analysis of the Relief Obtained," *Antitrust Bulletin,* **17,** 19–120.

Posner, R. A. (1970). "A Statistical Study of Antitrust Enforcement," *Journal of Law and Economics,* **13,** 363–417.

Roll, R., and Ross, S. (1980). "An Empirical Investigation of the Arbitrage Pricing Theory," *Journal of Finance,* **35,** 1073–1103.

Ross, S. (1976). "The Arbitrage Theory of Capital Asset Pricing," *Journal of Economic Theory,* **13,** 341–360.

Schwert, G. W. (1981). "Using Financial Data to Measure Effects of Regulation," *Journal of Law and Economics,* **24,** 121–158.

Stillman, R. (1983). "Examining Antitrust Policy Towards Horizontal Mergers," *Journal of Financial Economics,* **11,** 225–240.

Wier, P. (1983). "The Costs of Antimerger Lawsuits: Evidence From The Stock Market," *Journal of Financial Economics,* **11,** 207–224.

The Use of Court-Appointed Statistical Experts
A Case Study

Robert F. Coulam
Department of Social Sciences
Carnegie-Mellon University
Pittsburgh, Pennsylvania[1]

Stephen E. Fienberg
Departments of Statistics and Social Sciences
Carnegie-Mellon University
Pittsburgh, Pennsylvania

ABSTRACT

The increasing use of statistical methods in legal proceedings has been accompanied by a variety of criticisms, some of which question the capacity of the normal adversary process to produce sound statistical judgments. One potential way to improve how courts handle statistical issues is through the use of court-appointed experts, under Rule 706 of the Federal Rules of Evidence. Since experience with court-appointed statistical experts is limited, this chapter analyzes in detail how a court-appointed expert was used in one actual case, a recent Title VII class action involving claims of sex discrimination. Based on an analysis of selected documents and interviews with many of the key participants, we conclude that the court-appointed expert in this case (a) helped the judge to set priorities for managing the resolution of statistical disputes; (b) provided authoritative judgment to resolve minor, but protracted, statistical disputes that had clearcut technical answers; (c) clarified the implications of the different statistical judgments proposed by the party experts, for statistical matters lacking clearcut technical answers; and (d) helped generally to educate the judge about statistical problems and to give the judge greater confidence in resolving statistical disputes. Meanwhile, the judge found Rule 706 to be a workable framework for appointing and overseeing the work of the court-appointed expert, and he appeared to be enthusias-

[1] Present address: Budget Office, Massachusetts Department of Public Works, 180 Tremont Street, Boston, Massachusetts 02111.

tic about the procedure. Party attorneys were more skeptical. Thus, this study suggests that a court-appointed expert can provide a judge with important assistance in resolving statistical disputes, although wider experience with the procedure would be necessary to confirm its value.

Legal proceedings are using statistical methods with increasing frequency. As a result, the courts are increasingly presented with complex issues of measurement, methodology, and statistical inference that they rarely confronted before (e.g., see Finkelstein, 1980; Kaye, 1982; Fienberg and Straf, 1982). As is noted in Fienberg (1982):

> Perhaps the most difficult task facing the courts is the evaluation and assessment of statistical analyses and opinions, especially in cases where there is conflicting statistical testimony. Courts are often confronted by conflicting statistical testimony based on different types of information, analyses, and judgments, because statisticians disagree but also because courts rely on the presentation of selected evidence favoring different points of view. . . . Yet the complexity of statistical issues raised in some cases will clearly put a resolution of conflicting expert testimony beyond the ken of even the most thoughtful and well-trained jurist.

One of the ways to address such situations is through the use of court-appointed statistical experts. Judges have long had the common-law power to appoint experts (e.g., see Learned Hand, 1902). The adoption in 1975 of the revised version of the Federal Rules of Evidence, however, provided an explicit procedure (and, to that extent, some encouragement) for the court to appoint expert witnesses of its own (see Rule 706). There are, however, few examples of the use of court-appointed *statistical* experts to date and no empirical literature on the subject. The effectiveness of this procedure thus remains undemonstrated to the judges, attorneys, and statisticians who may have the most to gain from it.

The chapter describes a recent sex discrimination case in which the judge invoked the provisions of Rule 706 and appointed his own statistical expert. We use the case as a vehicle to explore the practicalities of the rule and to suggest the potential value of court-appointed witnesses. In Section 1 we review Rule 706 and some discussion of it by others. In Section 2 we present a summary of one lawsuit in which a court-appointed statistical expert was used. We examine in detail the roles of the expert statistical witnesses and the uses made of the court-appointed expert. Finally, in Section 3 we discuss the implications of our case study for the use of court experts in litigation where complex statistical issues are involved.

1. Background: The Provisions of Rule 706

Judges are not mere umpires—as noted, they have long had the power to call and question witnesses. There have been suggestions in some cases that this power rises to the level of a *duty*. Efforts at reversal on this ground have been unavailing, however (e.g., see Cleary, 1972, Sec. 8, n. 38). Meanwhile, as noted in a discussion of the new federal rules:

> Excessive intervention by the judge is improper because it interferes with the jury's right to decide facts, is contrary to our tradition of adversary practice, undermines the party's right to an impartial arbiter, and results in prejudice to the parties. Flagrant abuse may amount to a denial of due process. (Weinstein and Berger, 1982a, Vol. 3, at 614-4, citations omitted)

Between this power (not quite a duty) and "excessive intervention" lies the range of judicial discretion to call witnesses. The range of discretion for the court is substantial—either to refrain from appointing, or to appoint and use, an expert.

The procedures for a judge to appoint an expert witness are set forth in Rule 706 of the Federal Rules of Evidence (Federal Judicial Center, 1975):

> (a) *Appointment.* The court may on its own motion or on the motion of any party enter an order to show cause why expert witnesses should not be appointed, and may request the parties to submit nominations. The court may appoint any expert witnesses agreed upon by the parties, and may appoint expert witnesses of its own selection. An expert witness shall not be appointed by the court unless he consents to act. A witness so appointed shall be informed of his duties by the court in writing, a copy of which shall be filed with the clerk, or at a conference in which the parties shall have opportunity to participate. A witness so appointed shall advise the parties of his findings, if any; his deposition may be taken by any party; and he may be called to testify by the court or any party. He shall be subject to cross-examination by each party, including a party calling him as a witness.
>
> (b) *Compensation.* Expert witnesses so appointed are entitled to reasonable compensation in whatever sum the court may allow. The compensation thus fixed is payable from funds which may be provided by law in criminal cases and civil actions and proceedings involving just compensation under the fifth amendment. In other civil actions and proceedings the compensation shall be paid by the parties in such proportion and at such time as the court directs, and thereafter charged in like manner as other costs.
>
> (c) *Disclosure of appointment.* In the exercise of its discretion,

the court may authorize disclosure to the jury of the fact that the court appointed the expert witness.

(d) *Parties' experts of own selection.* Nothing in this rule limits the parties in calling expert witnesses of their own selection.

These detailed requirements apply in civil and criminal cases. Although the rule does not so specify, it will ordinarily be invoked considerably before trial, since to comply with subdivision (a) there must be time for (1) a hearing on the order to show cause, (2) consent by the designated expert, (3) notification of the expert of his duties either in writing or at a conference, and (4) findings by the expert, which (5) must be communicated to the parties. Additional time may also be required if the judge exercises his option to request the parties to submit nominations, or if the parties exercise their right to depose the expert. Usually the process can be set in motion at a pretrial conference pursuant to Rule 16 of the Federal Rules of Civil Procedure or Rule 17.1 of the Federal Rules of Criminal Procedure (Weinstein and Berger, 1982a, Vol. 3, at 706-12.)

The judge has discretion to choose whether or not to appoint an expert the parties agree on. The judges may wish to follow the parties' agreement, since their agreement constitutes a consensual guarantee against abuses of the rule. However, the judge may feel that appointment of a particular expert is desirable, independent of any wishes of the parties.

The expert must consent to act—a reluctant expert is considered a hazard to the purposes of the rule. The trial judge may choose to inform the expert of his duties orally (at a conference the parties attend) or in writing. The parties are permitted to take the deposition of a court-appointed expert as a matter of *right*. A party is also expressly permitted to cross-examine the court-appointed expert, even if the party calls the witness at trial.

The compensation of the court-appointed expert is left substantially to the discretion of the judge and in general may be charged to the parties like any other litigation cost. The factors a judge may take into account include the nature of the case, why the need for a court-appointed expert arose, the status of the parties, and the decision and its consequences (Weinstein and Berger, 1982a, Vol. 3, at 706-22). The language about "just compensation" in 706(b) serves primarily to assure that the costs for the court-appointed expert will not be taxed against (and thereby reduce) an award of just compensation guaranteed by the constitution.

The final two provisions of Rule 706 [subdivisions (c) and (d)] are straightforward. Under 706(c), the judge is granted discretion about whether or not to inform the jury that the *court* appointed the expert. This provision gives a judge latitude to withhold the fact of court appointment when the judge thinks the jury will be unduly influenced by that fact.

Finally, 706(d) delimits the purposes of Rule 706 by expressly providing that nothing in 706 should be read to limit the parties' rights to call their own experts. This provision is in the nature of a reassurance about the purpose of the rule and a barrier against creative legal analysis that might read a limitation on the parties' rights by implication.

After Rule 706 was formally proposed, Congress made only minor changes in its wording. It was not the subject of floor debate. The Advisory Committee that proposed the rule did note that there were some concerns about the "aura of infallibility" that might cloak a court-appointed expert. And in Congressional testimony, the American Trial Lawyers Association expressed the fear that, in medical malpractice cases, judges would appoint locally prominent physicians—physicians unduly sympathetic to a local colleague who was being sued. But Rule 706 generally raised little controversy (Weinstein and Berger, 1982a, Vol. 3 at 706-2–706-4).

Since its passage in 1975, the rule has been little used, although authors often discuss its possible use. For example, in interviews with participants in nine carefully selected cases, Saks and Van Duizend (1983) elicited much enthusiasm and support for the use of court-appointed experts. But none of the judges interviewed had ever appointed a court expert, nor had the lawyers and experts interviewed ever participated in a case involving the use of such experts. In fact, the *Federal Rules of Evidence Digest* (Pike and Fischer, 1981, 1983) reports a total of 14 cases concerning the appointment of experts of all kinds under Rule 706(a). Using the Westlaw system, we performed a computer search of reported cases for federal courts of general jurisdiction, and we found approximately 250 cases with one of various forms of reference to "court-appointed experts." Of these, only 75 cases also mention at least one of a large variety of potential statistical terms. Out of the 75, only 57 cases actually resulted in the appointment of an expert. A selective reading of each of the 57 cases revealed that only 4 cases clearly involved the appointment of an expert with either statistical, econometric, or psychometric credentials. One of those four cases is discussed in this paper. The others are *In re Antibiotic Antitrust Action,* 410 F.Supp. 706 (D. Minn. 1975); *Officers for Justice v. Civil Service Commission,* 371 F.Supp. 1328 (N.D. Calif. 1973); and *Vanguard Justice Society, Inc., et al. v. Harry Hughes et al.,* 471 F.Supp. 670 (D. Md. 1979). The *Antibiotics* case involved the use of a court-appointed expert at the damage stage of the litigation, after liability had been determined.

This computer survey should be viewed as only a first approximation of the use of court-appointed statistical experts. Many pertinent cases may not have resulted in published opinions for the computer search to discover. Meanwhile, the opinions in the cases we examined do not provide much information about how or why the judges appointed statistical experts, about the kind of experts appointed and how they were used, or about the virtues and costs of the procedure. An example of these problems is *U.S. v. Reserve Mining Co.* 380 F.Supp. 11 (D. Minn. 1974). In that case, the judge appointed a biostatistician as his expert to evaluate the risk of cancer due to asbestiform fibers in the drinking water in Duluth, Minnesota. Yet the appeals court opinion (*Reserve Mining Co. v. EPA,* 514 F.2d 492 (8th Cir. 1975)) and most discussions of it refer to the statistical testimony of another court-appointed expert, a medical witness (e.g., see Yellin, 1983). Published opinions may thus only fortuitously reveal whether and how a court-appointed statistical expert was used. In any event, the category "statistical expert" is itself vague. Accordingly, the apprehension of the cases is selective, and the classification of the cases (as using statistical experts) is in important ways subjective.

Nonetheless, this rough count confirms the conventional wisdom: the use of court-appointed experts of any kind is rare, and the use of court-appointed statistical experts is rarer still. The question naturally arises: if the rule is so helpful, why is it so little used for cases involving statistical proof? There are no systematic data on the subject. However, Weinstein and Berger (1982a, Vol. 3, at 706-8–706-12) offer a thoughtful summary of possible reasons about why Rule 706 has been infrequently used in cases of all kinds (statistical and nonstatistical, civil and criminal). Their reasons include (a) uncertainty about the source and extent of compensation for the expert; (b) the fear that an expert who bears the imprimatur of the judge will unduly influence a jury and thereby tend to undermine the constitutional right to trial by jury; (c) the fear that a judge-appointed expert will become an inquisitor and thereby erode the traditional adversary process; (d) the difficulty of obtaining a truly "neutral" expert for some cases; (e) the concern that a court-appointed expert might do a more routine, less penetrating job than a party expert; (f) uncertainty about the mechanics of the rule; and (g) ignorance of the rule and its value.

We cannot speak with authority about which of these reasons (or others) lie behind the infrequent use of Rule 706. Most of the barriers outlined above, however, rest on ignorance of the rule or on *possibly* misplaced fears of its effects. The best way to address these problems is to test speculations about the effects of the rule with data about how the rule works in actual cases. There have been some studies of nonstatistical court experts [see, e.g., a study of the use of court-appointed medical experts, in "The Doctor in Court" (1967)]. We propose in this chapter to examine how a court-appointed statistical expert affected the conduct of one particular case. A case study cannot refute or confirm all of the prevalent worries about Rule 706, but it can provide concrete examples of the value of a court-appointed expert and some reassurance about the means at the judge's disposal to prevent any abuses.

2. Case Study: A Title VII Class Action

Introduction

This lawsuit was brought in Federal District Court under Title VII of the Civil Rights Act of 1964, as amended. Plaintiffs were the female employees of (and applicants to) a medium-sized company. The company's operations included management and professional positions (specifically in writing and editing), as well as sales, clerical, and basic production positions. The suit claimed that women faced discrimination by the company in hiring and initial job placement, promotions, wages, quality of work assignments, and personal treatment by male colleagues and superiors. The suit was brought as a class action. In its comprehensive allegations, it depicted a "pattern and practice" of sexual discrimination throughout defendant's operations.

The individual plaintiffs were a small group of named women, representing the class. They had some outside funding to support the litigation, but their resources were ultimately limited. The plaintiffs' apparent litigation strategy: to use extensive pretrial discovery and expert analysis to obtain information to corroborate and refine their charges. By contrast, while the litigation was costly and unwelcome, defendant was in a better position than plaintiffs to absorb the costs. The defendant's apparent litigation strategy: to fight plaintiffs every step of the way, conceding nothing on principle and taking tactical advantage of the asymmetry in resources between the parties.

Case Chronology

The lawsuit lasted approximately 8½ years, from late 1973 to the middle of 1982. But it never reached trial—it was settled by the parties just before trial was scheduled to begin. Broadly speaking, there were eight phases to the litigation:

1973–1974 Administrative Actions—Complaints were filed with the Equal Employment Opportunity Commission (EEOC) by 40–50 individuals. The EEOC concluded there was probable cause for complainants to file a civil lawsuit.

1975–mid 1976 Complaint Filed/Class Certified—The named plaintiffs filed a complaint on behalf of themselves and others similarly situated, alleging a pattern and practice of sex discrimination in defendant's hiring, work assignments, promotions, wages, and other areas. The defendant broadly

	and tenaciously opposed these claims. Discovery commenced. The issue of class certification was briefed and argued, with the judge deciding in mid-1976 that the lawsuit could be maintained as a class action.
Mid 1976–mid 1978	Discovery/Motion for Summary Judgment—During this two-year period, contentious discovery continued. The defendant moved in 1978 for summary judgment and modification of the class, a motion denied by the judge with leave to renew at trial.
Mid 1978–mid 1980	Party Experts Analyze and Report—During this two-year period, discovery continued. In mid-1979, plaintiffs' statistical expert filed his report, based on the extensive data plaintiffs had obtained in discovery. Defendant's experts, in turn, filed their reports in early 1980. Depositions of the experts by opposing parties began. Discussions also began at this time among the judge and attorneys on the schedule and procedure to be followed at trial.
Late 1980	Judge Appoints Expert—The judge scheduled completion of discovery. The parties prepared pretrial statements and negotiated the stipulations they were willing to make for trial. Near the end of the year, the judge issued an order to show cause why an impartial expert should not be appointed. In early 1981, he appointed an econometrics professor to be his expert.
1981	Judge's Expert Reports/Pretrial Order Negotiated—During the first eight months of the year, the court-appointed expert worked with the party experts to refine their conflicting data bases and to rerun certain analyses. In July and August, he filed his own reports. In September, the judge issued a 1000-page pretrial order, setting out the agreed stipulations of the parties and the issues and procedures for the forthcoming trial.
October 1981	Defendant's Motion Renewed—Defendant's earlier motion for partial summary judgment and modification of the class (denied in 1978, with

leave to renew) was renewed, briefed, and argued. In October, the judge granted a key part of the motion. The effect was to find for the defendant with respect to all hiring of editorial employees—the most important group of professional employees in defendant's operations. (Such questions as promotion and wages for these employees, once hired, remained.)

Late 1981–mid 1982 Settlement—Shortly after the judge's ruling, the case was settled on the major issues. Settlement negotiations ensued, to embody the terms of the tentative settlement agreement in writing acceptable to both sides. In March, the parties reached a formal settlement, whereby defendant agreed (a) to make good-faith efforts to achieve specified goals in the placement, training, and promotion of females across a broad range of positions in its management and operations; (b) to make payments of over $100,000 to female employees, according to certain rules of eligibility and priority; and (c) to pay up to approximately $200,000 in legal fees and disbursements incurred by plaintiffs. The parties could not agree on an allocation of the costs for the court-appointed expert. They left this matter to the judge. In June, the judge ratified the settlement and allocated costs to the parties to pay for the services of the court-appointed expert.

This summary does little to convey the complexities—to say nothing of the intensity—of the litigation. It may serve, however, as a background for the fairly narrow set of issues of interest in the examination of the role of court-appointed experts. The next subsection is really an embellishment on this background information. It provides a somewhat more detailed description of the major legal issues that were resolved in the course of the lawsuit, as an aid to understanding the problems of expert evidence in the case.

Some Major Legal Issues

The essence of plaintiffs' complaint was that defendant had engaged in patterns and practices of sex discrimination across its entire organization. These alleged discriminatory acts affected hiring, placement, promotions, wages, work assignments, and other practices of defendant. These claims affected both applicants to and employees of the defendant. The class certified by the court in 1976 included both groups, applicants and employees. The class was bounded in time as provided by statute. The time limit derived from the statute dated back 300 days from the original EEOC claims in late 1973. Thus, the originally certified class included all females who were either applicants to or employees of the defendant in or after early 1973.

In 1978, defendant moved for partial summary judgment and modification of the class. As noted in the chronology above, the motion was denied, with leave to renew at trial. In 1981, when the motion was renewed, the character of the evidentiary record had changed. Discovery was now complete. Experts for both sides had issued reports and had been deposed. The court-appointed expert had worked with the parties, performed his own critical analyses, and answered certain statistical questions posed by the judge. Finally, a 1000-page pretrial order had been negotiated by the parties. In other words, unlike in 1978 when the motion was first made, the factual record was essentially complete; it had been subjected to sophisticated analysis; there were many outstanding disagreements on statistical issues, but these had been to some degree moderated by the pretrial negotiations of the parties and by the efforts of the court-appointed expert; and the parties were on record (in the pretrial order) about the stipulations they were willing to make and the proofs they proposed to offer at trial.

For one key group—applicants to editorial positions—defendant could make a convincing empirical case that its hiring had been nondiscriminatory over the pertinent period. Since the question was presented as a motion for summary judgment, the burden was entirely on the defendant to show that there was no issue of material fact between the parties concerning defendant's claim. The judge ultimately agreed that defendant met that burden—he agreed with defendant's argument that none of the plaintiffs' data (a) demonstrated a statistically significant disparity between the percentage of females hired versus the percentage of females in defendant's applicant pool, or (b) otherwise demonstrated a pattern and practice of discrimination in editorial hiring. Regarding the statistical data, the disparities between hires and applicants that the judge found were less than a single standard deviation for the female editorial applicants. The judge found these deviations statistically "insignificant," especially with reference to the "two or three standard deviation" measure proposed by the Supreme Court in *Castaneda v. Partida,* 430 U.S. 482, 496 n.17 (1977). Meanwhile the judge found other, nonstatistical evidence advanced by plaintiffs (e.g., anecdotal evidence) insufficient to support an inference of sex-biased hiring of editorial applicants by defendant.

The judge therefore concluded that plaintiffs could not demonstrate a ''pattern and practice'' of discrimination by defendant against the editorial applicants. He accordingly found for the defendant on this part of the motion.

This ruling eliminated an important group of applicants from the plaintiff class. The judge's ruling in other respects (concerning class action requirements) complicated plaintiffs' pursuit of the lawsuit, but left defendant still exposed to liability for discrimination against noneditorial applicants and all female employees. (These other aspects of the judge's ruling were important to the lawsuit, but are not essential to an understanding of the use of experts and statistical evidence in the case. For simplicity's sake, we omit them from the discussion here.) A tentative settlement of the case was reached by the parties shortly thereafter, and definitively negotiated over the next few months.

Questions of discrimination with respect to other groups of female *applicants*—and with respect to promotions, wages, and other practices concerning female *employees*—remained at issue and would have been the main factual issues in contention at trial. These issues were objects of extended discovery and analysis by the statistical experts. They were never authoritatively judged, as the settlement by the parties preempted any formal judgment on them. Each party, however, sought in the course of the lawsuit to strengthen the arguments it could make on these issues. Much, though by no means all, of the evidence adduced on these issues was statistical. For the statistical evidence, plaintiffs attempted to demonstrate statistically significant patterns of difference in the treatment of males and females, while the defendant attempted to show that there was no such statistical evidence. The character of these statistical debates is set forth in the following subsections.

Description of the Statistical Experts

The lawsuit involved the use of four statistical experts, one for the plaintiff (hereafter denoted as Dr. P), two for the defendants (Drs. D1 and D2), and a court-appointed expert (Dr. C).

Dr. P is a statistician who teaches in the Department of Statistics at a large eastern state university and has served as department chairman. He received his Ph.D. a few years prior to the litigation and has been employed as statistical expert in a large number of Title VII class action discrimination cases, working both for plaintiffs and for defendants. Although he has not published widely, he has coauthored an introductory text on the use of statistics in business and has developed a solid professional reputation.

Dr. D1 is a distinguished statistician who has taught in the Department of Statistics at a private eastern university for 25 years. For about 12 of those, he served as department chairman. He received his Ph.D. before World War II and is one of the few statisticians who is a member of the National Academy of Sciences. Although widely viewed primarily as a mathematical statistician, in recent years he has coauthored papers on statistics and the law. He has appeared as an expert statistical witness in several lawsuits, including Title VII class action cases. His work for the defendant in this case focused primarily on the statistical analysis of employee compensation.

Dr. D2 is a distinguished economist with an extensive career in government and academia. His extensive list of publications includes several books on employment and unemployment and on labor market information. He has recently completed a decade as dean of the business school at the same university as Dr. P. He has also appeared as an expert witness in a variety of Title VII cases and has worked primarily with defendants. His work for the defendants in this case focused primarily on issues concerning the statistical data base and hiring and promotion.

Both the plaintiffs' expert and the defendant's experts were well-known, highly qualified statisticians with extensive experience in Title VII litigation. It is in exactly such situations, with strong statistical experts on both sides of the case, that the adversarial nature of court proceedings often makes it difficult for a judge to reconcile conflicting statistical evidence.

Finally, Dr. C, the court-appointed expert, is an econometrician with strong expertise in the use and development of statistical methodology. He received his Ph.D. a few years prior to the litigation. He is a professor of economics at a major private university in the northeast. He has published widely in statistical and economics journals. Although he had appeared as an expert witness in one previous case, he had no previous experience as a statistical expert in a Title VII discrimination case.

Major Statistical Issues

The analyses of each of the party experts are summarized below, followed by an overview of the court-appointed expert's own studies.

Plaintiffs' Case

The plaintiffs' expert, Dr. P, carried out a variety of statistical analyses on issues relating to hiring, initial placement, promotion, and compensation as they differentially affected female applicants and employees. His basic data base included (a) the defendant's employment master payroll tapes for 1973–1977, (b) census labor market data, (c) employees' educational attainment data as coded from company personnel files, and (d) manually generated employment statistics for 1965–1973.

The statistical report of Dr. P contains a variety of analyses. The report begins by comparing the 1701 full-time workers as of 31 December 1977, by job groups or EEO categories and by sex, with "comparable" Census Bureau labor market data. The basic conclusion Dr. P reaches is that there is a "highly statistically significant underrepresentation of females in the managerial, professional, and sales categories, and a statistical overrepresentation in the clerical category." In addition, he finds that women are essentially excluded from the production process at the company. Next he turns to a salary profile, by sex and job group or department, and finds that women earn less than men in similar job groups.

On the issue of hiring and initial placement, Dr. P analyzes hires by sex and position (for nonproduction jobs excluding clericals), and finds that women "are hired at a disproportionately low rate into the more desirable positions," when compared with labor market sex ratios. He focuses specific attention on disparities in writing and editorial jobs. Finally, he notes that while the rate of female hires in other professional jobs is "not statistically significantly less than expected," females are hired into the lowest paying of these jobs.

On the issue of promotions, the report attempts to show that for 1975–1977 the promotions to the higher-paying jobs in sales and in writing and editorial job groups went disproportionately to men. No formal statistical inferences are presented on this issue.

Finally on the issue of compensation, Dr. P uses regression analyses to control for differences in seniority, age (used as a proxy for prior experience), educational attainment, and EEO category of current job, while assessing the effect of sex on salary. His basic regression analysis uses all full-time employees as of 31 December 1977, and is summarized in Table 1A. Dr. P concludes: "The effect of the sex variable was highly statistically significant at well below the 0.00001 level . . . females with the same seniority, educational attainment, age, and EEO category as males would earn $61 a week or $3172 a year less than the similarly situated males." In addition he reports on separate regressions using the same basic independent variables for each EEO group, for the editorial department, for the writing staff, and for all employees hired since 1971. For each, the sex variable is statistically significant, with the coefficient varying from a low of $14 for the clerical EEO group to a high of $112 for the managerial group. (See Table 1B.)

The report of Dr. P, submitted in July 1979, was subsequently amended in October 1979, but his conclusions were basically unaltered.

Defendant's Response

As we noted above, the defendant's response to the statistical issues raised in Dr. P's report came in two parts, prepared by two different experts, Drs. D1 and D2. We describe these separately.

TABLE 1. Results of Regression Analyses by Dr. P, Using Wages as Response

(A) All Employees as of 31 December 1977 ($R^2 = 0.73$)

Variable	Estimated Coefficient	Significance Level
Sex		
Female/male	−61.00	0.000
Seniority		
Years	15.70	0.000
$(\text{Years})^2$	−0.88	0.000
$(\text{Years})^3$	0.01	0.000
Age		
Age	12.83	0.000
$(\text{Age})^2$	−0.14	0.000
Educational Attainment		
Bachelor's	8.75	0.719
Master's	7.37	0.785
High school	−22.51	0.341
Ph.D.	77.88	0.264
Unknown	−4.03	0.849
High school plus	−8.55	0.726
Some college	44.31	0.524
EEO Category		
Professional	160.54	0.000
Managerial	141.02	0.000
Clerical	−41.73	0.000
Technical	64.20	0.000

(B) Auxiliary Regression Results for Employees as of December 1977

Regression for	Estimated Coefficient for Sex	R^2
Editorial department	−45	0.77
Writing staff	−44	0.62
All employees hired since 1971	−45	0.72
Managerial	−112	0.26
Professional	−52	0.44
Sales	−68	0.68
Clerical	−14	0.54

Dr. D2 attacks how Dr. P structured the company's data. Dr. D2 cited errors in classifying job groups and gross errors in tabulating data extracted from the payroll tapes. He also deprecated Dr. P's use of labor market information based on census data. Dr. D2, working with company staff, prepared an alternative data base of hires and promotions for 1974–1978, especially those associated with editorial positions. He claims the application and hiring processes for these editorial positions were handled by the editorial department, outside the application and hiring processes for essentially all other parts of the company (a point not contested by plaintiffs). For the *editorial staff,* he notes that the data "show no statistical evidence of discrimination of [sic] hiring" relative to the applications. (Recall that Dr. P had examined hiring relative to labor market sex ratios, rather than actual applications.) Dr. D2 then goes on to note the superiority of the applicant data to the labor market data used by Dr. P. But he also attempts a "more refined" comparison with census data, using an approximation of the relevant labor pool for a highly select group such as the company's editorial staff. Dr. D2 again finds no evidence of discrimination.

For *noneditorial positions,* Dr. D2 then does an alternative comparison of hire rates by sex and job group with census occupational data for the local SMSA. He finds no statistical evidence of a pattern or practice of discrimination. The apparent difference between the two analyses is Dr. D2's use of very broad census job groupings that include narrow categories with relatively large percentages of female workers.

Dr. D2's promotional analysis is cast in terms that are, in our reading, difficult to understand and are not directly comparable with those of Dr. P's analysis. The conclusion is again one of no evidence of discrimination.

Finally, Dr. D2 alleges a number of statistical flaws in Dr. P's report and in particular is critical of the static (as opposed to flow) nature of much of the analyses.

Dr. D1, in his report for the defendant, takes on the issue of discrimination in compensation, reviewing Dr. P's use of regression analysis and presenting his own alternative regressions. The discussion at the beginning of his report adopts a stance similar to that used by defendants in other Title VII cases (e.g., that salary differentials by sex are attributable to differences in labor market pay rates for, and the differences in female representations in, various job categories; see Conway and Roberts, this volume). In each case he explains why he uses different explanatory variables in his regression analyses.

Dr. D1's basic criticism of Dr. P's regression analyses is that they "cut across different occupations with differing pay scales and in which there are widely varying female representations." More minor criticisms relate to some problems in the use of the data base (that Dr. P apparently corrected in his amended report) and to an 18% missing rate for data on educational attainment for the company's writing staff (46 of 249 employees). His own analyses focus on six job groups that are much more narrowly defined than the EEO groupings used by Dr. P and are based on payroll data as of 31 December 1978 (a year later than the data used by Dr. P). The total number of employees covered by these analyses is about 660, with almost 500 of these in two of the six groups. The resulting multiple regressions, summarized here in Table 2, show estimated sex regression coefficients that are not statistically significant at the 0.05 level, although two of the groups show substantially large coefficients interpretable as sex-related pay differentials of in excess of $30 per week.

TABLE 2. Results of Regression Analyses by Dr. D1, Based on Employees as of 31 December 1978

(A) Editorial, News, and Sports Staff ($R^2 = 0.67$)

Variable	Estimated Coefficient	Significance Level
Years of Seniority	14.51	0.001
$(\text{Years})^2$	−0.72	0.005
$(\text{Years})^3$	0.01	0.01
Age	21.47	0.001
$(\text{Age})^2$	−0.25	0.001
Prior experience	3.05	0.005
Merit increase	48.91	0.001
Promoted	38.63	0.001
Former manager	91.48	0.001
Sex	−4.78	0.60

(B) Editorial Staff Management ($R^2 = 0.50$)

Variable	Estimated Coefficient	Significance Level
Seniority in management	20.30	0.001
$(\text{Seniority in management})^2$	−0.65	0.05
Age	27.90	0.05
$(\text{Age})^2$	−0.33	0.03
Sex	−30.42	0.25

(C) Advertising Inside Sales Group ($R^2 = 0.69$)

Variable	Estimated Coefficient	Significance Level
Seniority	11.77	0.001
$(\text{Seniority})^2$	−0.40	0.005
Sex	7.92	0.38

TABLE 2. (Continued).

(D) Advertising Outside Sales Group ($R^2 = 0.77$)

Variable	Estimated Coefficient	Significance Level
Seniority	4.21	0.001
A.S.R.–Sr. Status	98.53	0.001
Sex	−33.12	0.15

(E) Clerical Employees ($R^2 = 0.43$)

Variable	Estimated Coefficient	Significance Level
Seniority	9.33	0.001
$(\text{Seniority})^2$	−0.17	0.001
Sex	−1.81	0.85

(F) Clerical Managers ($R^2 = 0.71$)

Variable	Estimated Coefficient	Significance Level
Seniority	9.21	0.001
Age	0.95	0.50
Sex	−21.46	0.40

A major factor in the regressions for several of the groups is the inclusion of special explanatory variables. For example, for Dr. D1's editorial job group, for which the estimated sex coefficient was equivalent to a differential of only $5 per week, the added variables were (a) whether the employee received a promotion since January 1975, (b) whether the employee received any merit increase since January 1975, and (c) whether the employee had previously served as a manager but was returned to the nonmanagerial group without a reduction in pay. The inclusion of these variables have a very substantial effect on the sex coefficient. (These additional variables were challenged by Dr. P in his rebuttal report.)

Dr. D1's summary conclusion on the basis of this analysis is "that no valid statistical inference can be drawn to substantiate the assertion of a pattern and practice of sex discrimination in compensation."

The Reports of the Court's Expert

The reports of the court's expert, Dr. C, were prepared in two parts. The first addressed the compensation analyses; the second addressed issues of hiring, placement, and promotion. In response to these reports, the judge prepared a series of questions about technical statistical terminology and Dr. C's conclusions regarding the conflicting claims of the party experts.

The answers to these questions were presented by Dr. C in an addendum to his reports.

From both procedural and statistical perspectives, Dr. C's report on compensation issues may be of greatest interest and so we describe it first. In it, he describes a series of additional regression analyses that he directed Dr. P and Dr. D1 to perform—these were intended to supplement the analyses described in their reports, and to allow Dr. C to assess the validity of the specifications used for the various regression equations. For example, Dr. P was asked to perform analyses to elucidate the impact of missing information for the education variable (594 missing values out of 1701 individuals), and various additional regressions to examine the impact of pooling across job groupings (i.e., aggregation bias). Dr. C concludes that the impact of the missing data on Dr. P's regressions was minimal, but that aggregation bias does exist in some of Dr. P's analyses. Nonetheless, Dr. C's summary appraisal seems to provide modest support for Dr. P's regressions: there remains "strong evidence of statistically significant results" in the coefficient of the sex variable for several large groupings of employees. The major exception to these findings is for editorial department employees.

Dr. D1 was also asked to carry out additional regression analyses (e.g., to exclude various "employer controlled variables" such as seniority). For the editorial employees, Dr. C appears on balance to find Dr. D1's analyses convincing, but he does raise a series of questions about the remaining regression models and the related conclusions regarding the lack of statistically significant, sex-related wage differentials. The first report concludes with Dr. C's attempt to examine the overall disparity in wages by sex in Dr. D1's results, using first a sign test and then Fisher's technique for combining p-values. For the sales groups, in particular, Dr. C finds "quite striking evidence of overall wage differentials by sex."

The judge in his questions on this part of the report focuses directly on Dr. C.'s conclusions regarding the editorial department and sales department employees. Dr. C responds in his addendum that, for the editorial workers in particular, he remains "unconvinced by [Dr. P's] results in the narrow sense of a finding of a statistically significant wage disparity." Despite a marginally significant result for editorial staff workers in Dr. P's analyses, he does *not* find the results with respect to sex "strong enough to make a conclusive inference of wage discrimination."

In his second report, Dr. C addresses Dr. P's and Dr. D2's analyses of the remaining statistical issues. Here he reports not on requested reanalyses done at his direction by the experts for the two sides, but rather on his own statistical tests. He also attempts to weigh the conflicting arguments and data. Without reviewing all of the details in his 34-page report, we note that Dr. C appears to be reaching conclusions that reflect a middle road between the claims of the experts for the two parties. On some issues, such as those relating to editorial employees, he confirms most (but not all) of the analyses and conclusions of the defendant's expert, Dr. D1. On others, he accepts the results of Dr. P. On yet others, he argues that neither expert has quite carried out convincing analyses.

The judge once again focuses several of his questions specifically on issues regarding the editorial department. In his responses, Dr. C again appears to suggest that evidence of discrimination in hiring, placement, and promotions for editorial employees is weak, and that Dr. D1's analyses for the defendant are, in general, convincing.

Procedural and Statistical Aspects of the Use of the Court-Appointed Expert

Beyond the written analyses described above, the court-appointed expert was used in a variety of different ways in the relatively short time he functioned in this case. Recall that this lawsuit was filed in early 1975. The judge did not appoint his expert until early 1981, after most pretrial matters had been completed. While there were many outstanding factual issues between the parties—hence the need for trial—the major preliminary matters that remained were the negotiation of pretrial stipulations by the parties and, as it happened, the resolution of defendant's renewed motion for partial summary judgment. The case was tentatively settled 10 months after the court's expert was appointed, 3 months after the court expert's written reports, and 1 month after the judge's decision on the motion for partial summary judgment (the latter apparently being the catalytic event prompting the settlement).

While the court's expert did not by himself resolve the major factual disagreements in the case, he did help to reduce the differences among the parties and to educate the judge about the statistical issues the case presented. In this section, we provide a procedural summary of how the court expert was actually used, focusing on how he was able to work with the parties and the judge amid the intense adversarial environment of this case. Rule 706 itself provides only the outlines of how a court-appointed expert is to be used, leaving much to the discretion of the judge and the accommodations of the parties. There is no broad conceptual scheme to our description of the role of the judge's expert, but rather a series of relationships and interactions which we summarize briefly.

1. The judge had appointed an expert in one prior case: a patent case, involving complex chemical data. He knew of other cases in which a 706 expert had been used to sort out medical testimony of party experts. His reason for appointing his own expert in this case was straightforward. The parties were far apart in their statistical arguments. He felt that having his own expert would enable him and his law clerks to sort out the contending statistical arguments, with a confidence they could not easily obtain from the piecemeal education that party experts would provide.

2. In late 1980, the judge on his own motion (and to the apparent surprise of the parties) issued an order to show cause why he should not appoint an expert. (Rule 706 establishes this procedure.) A hearing to show cause was held, and no objections were entered. The judge directed the parties to submit names for the position, an optional procedure under Rule 706. Each of the parties submitted different names.

3. The judge picked as his statistical expert an econometrician who was on neither plaintiffs' nor defendant's list. The judge issued a written order appointing the expert and defining the expert's duties in January 1981. The order is brief and general—only two legal pages long. After setting forth the need for a court expert, the order directs the parties to submit for the expert's examination "all material pertaining to this cause which [the expert] may request from time to time." The order goes on to provide that the expert may consult with the court on the nature of his duties and the scope and format of his reports "within the parameters of the adversary process." The expert was also authorized to consult with the employees, representatives, and experts of the plaintiffs and defendant as he might deem necessary. Finally, the order provides that costs for the expert's services shall be fixed by court order when appropriate. In such order, the court will allocate the expert's compensation between the parties, and "such allocation shall be taxable as costs in this cause within the court's discretion."

4. The judge began his use of Dr. C by holding a meeting with Dr. C and the attorneys. The judge made it clear to the attorneys that they were to provide Dr. C with whatever materials he requested. The judge's general strategy, in this meeting and thereafter, was to dramatize his impartiality, and that of Dr. C, by always dealing with Dr. C in the open. Thus, while the judge held informal conversations with Dr. C, he had the attorneys present whenever substantive matters were discussed. When Dr. C submitted his written analyses of the party experts' reports, the judge responded with a set of written questions to his expert, which Dr. C answered in writing. In these ways, both sides were kept informed about what the judge was doing with the court-appointed expert, and suspicions about the expert's power were allayed.

5. The lawyers in this case were aggressive and had continual disputes with each other and with the judge. Dr. C felt restricted by that atmosphere. He had to do considerable work with the parties' experts (getting them to rerun analyses, etc.). While he occasionally contacted the experts directly to set dates or clarify straightforward questions, he did not discuss matters of substance with the experts without including the attorneys or by making conference calls, sending letter copies, etc.

6. Dr. C had a practical concern to protect himself in this work. Specifically, he considered retaining a personal attorney to avoid being browbeaten during any of the proceedings. The judge reassured Dr. C that he (the judge) would be present at any questioning and would prevent any abuse. Meanwhile, there were procedural devices readily available to the judge to protect the court expert. For example, in the fall of 1981, when Dr. C's deposition was being scheduled, the judge required attorneys for the parties to submit in advance the areas and questions on which they intended to depose Dr. C.

7. Notwithstanding the atmosphere of contention, the party attorneys and experts were cooperative with Dr. C, particularly when the judge (a) made it clear that he would consider and weigh Dr. C's work with that of the other experts, and (b) indicated that he would back up Dr. C's efforts by the threat of potential sanctions (e.g., the potential exclusion of evidence at trial). In any event, it was in neither party's interest to impede Dr. C in his work.

8. One main function of Dr. C was to educate the judge on statistical matters. Indeed, Dr. C feared that the judge might listen to him too much, taking his word as definitive. (There is no evidence that this in fact happened.) As a check on abuses, the parties did have the right to depose Dr. C and to cross-examine him at trial, under Rule 706. However, it was clear that Dr. C could do a lot of damage to one side or the other, prior to trial, if he were not careful.

9. Broadly speaking, there were two classes of statistical problems that Dr. C confronted in this case:

a. Many statistical/data problems were unresolved and protracted before the court expert was appointed, but were in fact resolvable by the insertion of a "neutral," authoritative expert.

- For example, when Dr. C first began his work, there was a longstanding argument between the parties about how to classify particular employees for statistical purposes. (There were 100 or so discrepancies in how employees were classified in the data sets the two sides were using.) The first thing Dr. C said to the judge was that the parties would "have to get the data straight before doing any statistics." Otherwise, it would be difficult to tell whether the different conclusions of the parties were based on differences in statistical inference or differences in the data sets. This led the judge to put pressure on the parties to resolve the classification problems. The problems were resolved in about three months, at which point statistical analyses could move forward from common base data. *The presence of a court-appointed expert enabled the judge to establish appropriate priorities for managing the conflict between the two parties, better to clarify the real sources of difference between the party experts' views.*

- A second example: One argument by one of defendant's experts, not discussed above, was that one could not do binomial tests on small categories of employees. (Defendant's argument was based on the use of asymptotic normal approximations to the binomial.) Dr. C responded that one need not use normal approximations. Dr. C then did exact calculations using tables of binomial probabilities. That ended the dispute. *The court-appointed expert was thus able to serve, for some issues, as a source of authoritative judgment as to the validity of competing statistical claims of the parties. By himself, the Judge could not have settled these issues with confidence.*

b. A second group of statistical problems were at some point irreducible—they turned on matters of statistical judgment, on which a group of neutral experts might differ. *The court-appointed expert could not resolve these issues, but he could clarify some of the tradeoffs involved.*

- For example, as discussed earlier, there was an extended dispute between the party experts about how regression equations to estimate wage differentials should be specified. There was an inescapable tradeoff between the size of the employee groups analyzed and the homogeneity of the groups. Dr. C developed certain pooling tests to describe more accurately the statistical implications of different specification choices. These tests would have permitted the judge to make a more informed decision about the competing claims of the two parties, had the case come to trial. The influence of these tests on the out-of-court settlement negotiations is necessarily speculative. While both parties came in for some criticism, the tests on balance may have reinforced plaintiffs' claims.

10. One important issue created by the presence of a court-appointed expert was expense. To do his work, the court expert required the parties to do reanalyses of certain data. These reanalyses were expensive, but there was no practical way for a party to object to them if the party were to win critical statistical arguments. For plaintiffs, the costs were particularly important. Plaintiffs not only had to buy the time of their expert and other personnel, they also had to pay for computer services. None of these costs had been budgeted by plaintiffs, since, necessarily, they had had no idea that the court would appoint an expert. Defendant, meanwhile, had its own computer and was probably in a better position to absorb this, as other, unexpected expenses.

11. Another issue of expense: the court-appointed expert himself had to be paid by the parties. There was a short, but intense, dispute between the parties about who should pay the expert as part of the overall settlement of the case. The parties could not agree on an allocation of these costs, but they ultimately agreed to leave their detailed settlement agreement silent on the issue—in effect, to leave the matter to the judge. At defendant's request, Dr. C was required to submit a detailed breakdown of the time he had worked on the case. In the end, the judge allocated these expenses 90% and 10% between defendant and plaintiffs, respectively.

12. In our view, the court-appointed expert helped to resolve a key statistical issue in the case: whether the data could support plaintiffs' claim that defendant had engaged in discriminatory hiring of editorial applicants. The judge ruled on this issue in response to the defendant's motion for partial summary judgment. The judge found for the defendant, thus eliminating this claim and excluding these applicants from the plaintiff class. He expressly emphasized in his ruling that his judgment was not affected by Dr. C's views. Nonetheless, as we note above, statistical issues surrounding the appropriateness of the claims for the editorial applicants were a focus of the judge's official exchange with Dr. C. At the same time, given the commitment of attorneys and judges to the traditional adversary process, and given possible questions that might be raised on appeal, a judge might be cautious in explicitly acknowledging the contributions of a court-appointed expert. (As a federal judge unrelated to this case commented to us, "Even if the judge did rely on [the court-appointed expert], he couldn't say so.") It is our view that Dr. C's clear response to the judge's questions paved the way for the judge's ruling.

Summary Views of the Participants

The subsections above constitute our interpretation of the case, based on part of the written record, the experts' reports, and interviews with most of the major participants. In this subsection, we shift the focus slightly, to give our impression of how the participants themselves evaluate the role of the court-appointed expert in this case.

The Judge

It is our assessment that having a court-appointed expert permitted the judge to feel bolder, more confident, in making decisions that required the resolution of statistical issues (ranging from the informal minutiae of pretrial negotiations to the formal decision, in late 1981, on the defendant's motion for partial summary judgment).

For the judge, having a court-appointed expert was not like having a statistician on his staff—too much of the interaction had to be formal and in the open, with party attorneys present—but the procedural requirements were not unduly cumbersome.

The judge is enthusiastic about the possibilities of using a court-appointed expert in any subsequent litigation where he faces complex statistical disputes among party experts.

Dr. C

Dr. C also came away from the case feeling that the procedure had worked well and that he had made a useful contribution to the resolution of the case. His worries about personal vulnerability were forthrightly resolved by the judge's efforts to prevent any abuse. The need for Dr. C to be careful in how he fulfilled his role—dramatized by the contentiousness of the parties—was an important requirement in any event.

The "most difficult" problem Dr. C faced in fact had little to do with Rule 706 itself and a lot to do with the more general problem of statistical evidence and the law: There is a hodgepodge of case law about law and statistics (although there are some helpful books and treatises). It is difficult to know what guidelines a statistician (party- or court-appointed) should follow. Dr. C feels that the area could benefit from a study by lawyers and statisticians to develop guidelines for the statistician to follow. As to Rule 706 itself, however, Dr. C sees it as a reasonably practical way for a judge intelligently to sort the competing claims of party experts.

The Parties' Experts

Only one of the parties' experts was prepared to make an assessment for us regarding the use of the court-appointed expert. He did not recall that there was any special need for a court-appointed expert, noting that it was a typical Title VII case, in which the two sides had somewhat different but overlapping data bases. He noted: "The plaintiff's expert made a selection from his data base and used certain variables in a series of regression analyses. The defendant's experts made a different data selection and used an overlapping but different set of variables in their regressions. As a consequence, the two sides disagreed."

Attorneys for Plaintiffs

It is an everyday matter for these attorneys to have to educate a judge on the complex facts at issue in the lawsuits they bring. From their point of view, the disputes that remained with the defendant were essentially garden-variety disputes based on differing views of the law and the facts. At a price—uncertainty and added expense—the court-appointed expert did help to clarify the statistical issues in the case and was able to resolve a number of data problems that had impeded the litigation. However, the court-appointed expert did not in any major way alter the task these attorneys foresaw to convince the judge of their case.

Meanwhile, a number of other facts in the case limit the significance of the case as an exemplar of Rule 706. First, since the case never came to trial, the utility of the rule in a trial setting remains undemonstrated. Second, in his one formal ruling on the merits of the case (the motion for partial summary judgment), the judge expressly discounted the relevance of the findings of the court-appointed expert. Third, while the court-appointed expert may nonetheless have influenced the judge's decision with respect to editorial applicants, the claims advanced for editorial applicants were the weakest part of plaintiffs' case. Thus, the judge's ruling may be important not for how it resolved the claims of the editorial applicants (with the court expert's assistance). Defendant won, as expected, on those claims. The ruling may rather be important for how it sustained the balance of plaintiffs' claims for trial, substantially unaided (these attorneys claim) by the court-appointed expert.

Accordingly, in the view of plaintiffs' attorneys, the court-appointed expert was helpful on some issues, but did not have an important effect on the course or outcome of the litigation. In an abbreviated discussion, counsel for defendant seemed to share this summary appraisal.

3. Implications of the Study

This study involves only one case, and the value of the court-appointed expert in the case is a matter of opinion—the conclusions to be drawn are not undisputed. Our judgment of the facts of this case, as we know them, leads us to suggest the following implications:

1. In the pretrial period, at least—and even in a highly contentious case—the judge found the use of a court-appointed expert to be fairly straightforward as a procedural matter. The court-appointed expert himself found the procedures and protections adequate to do the necessary work. Thus, while this one case did not test Rule 706 in all its potential complexity, it at the same time did not reveal any notable inadequacies in the rule or in related court procedures.

2. The court-appointed expert aided the resolution of the disputes between the parties in four principal ways:

- **a.** case management—he could suggest priorities for the judge to follow (e.g., cleaning up the data) to isolate the true statistical disagreements between the parties.

b. authoritative judgment—some of the statistical issues raised in the course of litigation were tactically useful but scientifically weak; the court-appointed expert was in a position to help resolve these issues without extended debate.

c. clarification of trade-offs—for matters of statistical judgment, the expert could clarify the advantages and disadvantages of different statistical judgments proposed by party experts.

d. education of the judge—in a rather general way, the court-appointed expert was able to give the judge added confidence in appraising various statistical disputes and resolving them.

To be sure, the normal adversary process can bring about each of these results. But the results are not automatic, nor are they necessarily reached in an expeditious way, given the inevitable mixture of tactics and statistics that accompanies the dialectics of debate between the parties.

3. Attorneys have a powerful commitment to the adversary process, and they do not necessarily segregate statistical issues from the other complex issues on which they have to educate judges. They did not see particular harm in the use of a court-appointed expert in this case, but they were the least convinced that it made an important difference.

4. The restrained attitude of the attorneys stands in notable contrast to the relative satisfaction of the judge with the use of a court-appointed expert in this case. Perhaps this difference is not surprising, as *the difficulties the court-appointed expert helps to solve are most acutely felt by the judge, not the attorneys*. The assistance the court-appointed expert provided in this case—in terms of case management, authoritative judgment, clarification of trade-offs, and general education—helped the judge directly. It helped the party attorneys indirectly and unevenly (even as it created problems of expense and uncertainty for them). In other cases, the party attorneys might be enthusiastic. But in this case, the judge was enthusiastic and the party attorneys were not, a disparity that seems best explained by the more direct benefits the judge received.

5. This case did not really test one of the fears expressed when Rule 706 was passed: namely, that the fact-finder would be unduly influenced by the court expert. There is a potential for abuse, compensated somewhat by the protections the rule provides (e.g., deposition as a matter of right, authority to call and cross-examine the court expert at trial, and discretion in the judge to withhold the fact of court appointment from a jury).

6. The problem of who pays for the court's expert was a matter of contention here, but the parties seemed satisfied in the end to leave the problem to the discretion of the judge.

7. Rule 706 does not eliminate a central problem that any statistical expert faces: what guidelines to follow. Court-appointed experts can help to smooth disagreements within the particular facts of specific cases. But they cannot provide the clarity and predictability needed, more generally, in the use of statistical methods by the judicial process.

Acknowledgments

An earlier draft of this paper was prepared as background material for the National Research Council's Panel on Statistical Assessments as Evidence in the Courts. Members of that panel, as well as Morris H. DeGroot, Joseph B. Kadane, and Robert E. Klitgaard, provided helpful critiques of the paper. The paper is based in part on documents from the case, including copies of the experts' reports, kindly provided by the presiding judge. We were also privileged to interview and to receive comments on this paper from many of the participants involved, including the judge, counsel for plaintiffs, and the court-appointed expert. We appreciate their generosity in assisting our research effort. The views expressed here and the conclusions we draw are our own and do not necessarily reflect the views of any of the parties who sponsored or assisted us.

References

Cleary, E. W. (ed.) (1972). *McCormick's Handbook of the Law of Evidence*. 2nd ed. St. Paul, Minn.: West.

"The Doctor in Court: Impartial Medical Testimony" (1967). *Southern California Law Review,* **40,** 728–741.

Federal Judicial Center (1975). *Federal Rules of Evidence Annotated.* New York: Matthew Bender.

Fienberg, S. E. (1982). "The Increasing Sophistication of Statistical Assessments as Evidence in Discrimination Litigation—A Comment on Kaye," *Journal of the American Statistical Association,* **68,** 784–787.

Fienberg, S. E., and Straf, M. L. (1982). "Statistical Assessment as Evidence (with Discussion)," *Journal of the Royal Statistical Society* [*A*], **145,** 410–421.

Finkelstein, M. O. (1980). "The Judicial Reception of Multiple Regression Studies in Race and Sex Discrimination Cases," *Columbia Law Review,* **80,** 737–754.

Hand, L. (1902). "Historical and Practical Considerations Regarding Expert Testimony," *Harvard Law Review,* **15,** 40–58.

Kaye, D. (1982). "Statistical Evidence of Discrimination," *Journal of the American Statistical Association,* **77,** 773–783.

Pike and Fischer, Inc. (ed.) (1981). *Federal Rules of Evidence Digest.* Wilmette, Ill.: Callaghan.

Pike and Fischer, Inc. (ed.) (1983). *Federal Rules of Evidence Digest: Cumulative Supplement.* Wilmette, Ill.: Callaghan.

Saks, M. J., and Van Duizend, R. (1983). *The Use of Scientific Evidence in Litigation.* Denver: National Center for State Courts.

Weinstein, J. B., and Berger, M. A. (1982a). *Weinstein's Evidence,* Vol. 3. New York: Matthew Bender.

Weinstein, J. B., and Berger, M. A. (1982b). *Weinstein's Evidence,* Vol. 3 supplement. New York: Matthew Bender.

Yellin, J. (1983). "Who Shall Make Environmental Decisions (with Discussion)," *American Statistician,* **37,** 362–373.

Does Electronic Draw Poker Require Skill to Play?

Joseph B. Kadane
Departments of Statistics
and of Social Sciences
Carnegie-Mellon University
Pittsburgh, Pennsylvania

ABSTRACT

This paper records the author's experience in preparing testimony and testifying in *Commonwealth of Pennsylvania v. Electro-Sport Draw Poker Machine,* Serial No. 258. The legal background of the issue is discussed, and alternative ways of presenting evidence on the issue are addressed. The decisions of the Allegheny County Court of Common Pleas, the Pennsylvania Superior Court, and the Pennsylvania Supreme Court are all reported. This case presents an example of the kinds of choices an expert witness has in preparing testimony for cases of this kind.

1. Introduction

On 11 March 1980, a Pennsylvania State Trooper seized a draw poker machine in Allen's Grill in Neville Island Township in Allegheny County, Pennsylvania, charging it was a gambling device per se. I was hired by the machine's defenders to testify on the question of whether these machines require skill to play. I begin with a description of how a draw poker machine is played, and a brief description of the law of gambling devices per se. I then describe my testimony and the decisions of the courts. Ultimately the Pennsylvania Supreme Court decided that chance predominates over skill in playing draw poker machines.

2. What Is a Draw Poker Machine?

A draw poker machine plays a form of solitaire poker with the customer. It is activated by a coin (typically a quarter), which causes a point to appear in the "points" counter. By pushing a "play credit" button, one may then choose any number of points, up to 10, to be transferred to the "bet" column. Pushing the "deal" button causes five card images to appear on the screen. These are electronic pictures of ordinary playing cards, from ace to two, and the ordinary four suits. Under each card is a "discard" button. By pushing that button, the image of the card disappears. If one wishes not to discard at all, one may push a "stand" button. Otherwise, when one has finished discarding, one pushes the draw button, which causes new cards to appear in place of those discarded. If one feels one has made a mistake in pushing discard buttons, a "cancel" button may be pushed, which restores the initial five cards. "Cancel" will work only if pushed before the "draw" button.

Hands are evaluated in "skill" points according to the system shown in Table 1.

The number of skill points awarded corresponds to the most valuable type of hand it qualifies for. Some machines award 250 skill points for a royal flush. The number of skill points, multiplied by the "bet," is added to the "points" counter, allowing the player to play more games.

All the machines I saw had a counter measuring the number of quarters that had been spent in the machine. In addition, some had a "knock-off button," permitting all remaining points to be removed from the "points" counter (i.e., restoring it to zero). These machines also had a meter recording the number of points knocked off.

3. What Is a Gambling Device Per Se?

Nearly anything (including the *World Almanac* and a television set) may be used in the course of making bets. However, some devices (like roulette wheels and slot machines) have no other reasonable uses, and these are called gambling devices per se. A gambling device per se may be confiscated by the Commonwealth of Pennsylvania without any evidence of its being used for gambling.

The common law of Pennsylvania provides three elements that must all be present in order to establish that a device is a gambling device per se: consideration, a result determined by chance rather than skill, and a reward. "Consideration" is whether one must pay to play. This element is conceded by all parties to be present in the Electro-Sport Draw Poker Machine. "Reward" is whether there is a valuable prize for winning. In

TABLE 1. Evaluation of Hands in Solitaire Draw Poker

Type of Hand	Number of Skill Points
Straight flush	50
Four of a kind	25
Full house	10
Flush	8
Straight	5
Three of a kind	3
Two pairs	2
Pair of aces	1

this case, one can win only further free games, which in other cases (*In re Wigton,* 1943) had been held not to constitute reward. While "reward" has many aspects, the court in this case ruled that an Electro-Sport draw poker machine with a knock-off button and meter satisfies the element of reward, and that one without it does not. A machine with a knock-off button and meter that is used for gambling is easy for absentee owners to monitor, since the difference between the number of quarters that were put into the machine and the number of games knocked-off by the bartender and paid for by him (at 25 cents per game) is the net take of the machine. The Pennsylvania Supreme Court ruled in this case that an Electro-Sport draw poker machine without a knock-off button and meter does not satisfy the element of reward, and hence is not a gambling device per se. For those with the button and meter, however, the remaining question is the issue of chance and skill, which I was asked to address.

4. My Testimony

Using an Electro-Sport machine like the one that had been seized, I wanted to make dramatic the role that skill plays in the game. To do this, I compared two strategies, a "dumb" strategy consisting of automatically standing pat with every hand, and a "smart" strategy in which I did my best to win, knowing what I do about probability. I played 128 games on the machine, and found that the smart strategy won 159 skill points, whereas the dumb strategy won only 34 skill points. Thus I found a factor of about $4\frac{1}{2}$ between my smart and dumb strategies, and proposed this as a measure of how much skill is involved in the game.

In the hands that I played, I never threw away a winning hand. There are, however, some winning hands I would have thrown away. For example, if I had been dealt the aces of hearts and spades and the king, queen, and jack of spades, I would have discarded the ace of hearts. While this does break up a winning hand (the pair of aces is worth one point), it gives a better expected number of points. Of course, the probability is greater than $\frac{1}{2}$ that this move would have resulted in zero points for the hand. Consequently it was not inevitable (but it was highly likely) that the skill ratio I based my testimony on would come out to be greater than one.

A larger factor could have been found, of course, with a "dumber" strategy that actively sought to minimize the expected number of points scored. It was my decision that such a comparison would not have been useful to the court, and I made no such comparisons. Conversely, a smaller factor would have resulted if I had compared the smart strategy to a "smarter" dumb strategy, a point made in cross-examination in this case, and by an opposing expert witness in a similar case in Michigan.

Another decision I made was whether to report plays of the machine at all. Under the assumption that each hand is equally likely, one could in principle compute the ratio of the expected winnings of the strategy that maximizes expected winnings to the expected winnings of the stand-pat strategy. However, to present such evidence would have highlighted the assumption. Verifying the assumption empirically looked at least as difficult a task as to play the machine and report the results. Thus although I played the machine using my "smart" strategy with the assumption of equally likely hands, my statement about skill did not depend on such an assumption. An electrical engineer who had participated in the design of the Electro-Sport machine testified that it was designed with a pseudorandom number generator that started manipulating a Boolean string when the machine is turned on. Thus to get nonrandom results from the machine would have required timing in microseconds, which is beyond human capability. I found no empirical evidence that suggested the machine's random device had been tampered with.

As the number of games played goes to infinity, the factor I used will presumably settle down to a number that reflects the ratio of the expected winnings of my "smart" strategy to that of the stand-pat strategy. However, I have not specified my "smart" strategy for all possible hands (it is even possible that, by whim, I would play the same hand differently on different occasions). Thus theoretical calculation of this ratio is not possible.

Under the assumption of independent and identically distributed trials, I could have calculated a variance for the estimated ratio. This would have quantified the obvious problem that would have occurred with very small numbers of plays. But in this instance it seemed unnecessary, and no one asked me for it.

One might ask why I played 128 games. As I recall, my playing and recording of hands were interrupted by a telephone call at that point. I decided that I had enough data to analyze, and that became the data set used in the trial. One might argue that because I knew the results when I chose to stop, I might have biased the outcome. In fact, the ratio had been essentially constant for some time, so I doubt that any such bias occurred.

I testified that both skill and chance played a role in solitaire draw poker, and that skill and chance are not necessarily opposed elements in a game. I compared draw poker to the game of matching pennies, which in a single play is entirely a matter of luck; to chess, which is entirely a matter of skill; and to backgammon, which has both chance and skill. In my judgment, draw poker is more like backgammon, as it involves both chance and skill.

Apparently the courts were looking for a statement that goes something like "success at playing electronic draw poker is determined 70% by skill and 30% by chance." What might be the meaning of such numbers? Why would not 30% skill and 70% chance be equally plausible? If one could encode full strategies, specifying what cards one would discard in every possible hand, it would be possible to compute (in principle) the probability, in k hands, that a player with one strategy would win more skill points than a player with another strategy (assuming independent draws from equally likely hands). The strategy with the higher expected number of skill points would be nearly certain to win more skill points as the number of plays k gets large, as is guaranteed by the central limit theorem. But real players find it hard to say how they would discard in each of the $\binom{52}{5}$ possible hands. Even if they could, would these calculations get us any closer to skill versus chance? In fact, skill can mean many things. It might mean whether one can identify players who are particularly good at the game. It might mean whether one can, by practice or by thinking about it, find ways to improve. But none of these theories appears to get one any closer to the desired form of statement. The interpretation I chose emphasized that there are decisions under the control of the player that greatly influence how many skill points the player wins. I had no knowledge of, and no way to measure, how "skillful" typical players in bars are. Hence I was not able to quantify the extent to which skill, as opposed to chance, matters in draw poker.

The opposing expert witness was an FBI agent who testified that solitaire draw poker had no element of skill, apparently defining skill in the sense of manual dexterity. My profession and curriculum vitae were contrary to that position—how could it be that one could be a professor and grant Ph.D.'s in a subject that was not a skill? None of the courts to review this case apparently took the FBI agent's testimony seriously.

A similar problem was faced by Solomon and several of his colleagues in relation to pinball in the early 1960s. Solomon (1966, pp. 322, 323) writes:

> About four years ago I appeared as an expert witness in several trials in San Francisco. Some colleagues at Stanford and Berkeley also appeared there in this role in several similar trials. A number of shopkeepers had been arrested on the charge that they had paid a prize in money to the players (San Francisco policemen in mufti) of pinball machines which the State labeled a game of chance, that is, skill did not predominate in achieving a high score in the game. In those trials in which I participated, some experimentation preceded the day in court. A pinball machine was obtained, of the same model as the one to be produced in court (the machine in question being held in storage by the Police Department until the day of the trial). For one thousand games, the plunger was pulled back and released to propel each of the metal balls (usually five) on their way to drop into various holes and thus produce a score for the game. The ratio of the number of wins—score greater than a critical value—out of 1000 games was assumed to be the probability of winning by chance alone, that is, the probability of win which was built into the machine. On this same machine an expert then tried his prowess. The distance the plunger is pulled back and the manner in which it is released, plus the application of hand-pressure to the body of the machine, influence the score. Too much pressure causes the game to terminate abruptly and this is recorded as a defeat. The expert, who happened to be a house-painter with a neighborhood reputation for proficiency in this activity, ran through 1000 games. As before, the ratio of number of wins to number of games was assumed to be the expert's probability of win. The sample size of 1000 trials is large enough to make sampling error negligible (in these situations a standard deviation less than 0.01), and yet not wear out the expert, or the graduate student who pulled the plunger without ever checking to see the flight of the metallic ball. Over several pinball machines which differed in structure, the expert's performance was just about twice as good as could be attained without the use of any human skill. However, the ratios of winning performance without skill went from 0.05 to 0.10, and correspondingly, the winning performance ratios of the expert from 0.10 to 0.20.

It is notable that Solomon's measure of skill is very similar to mine. See also Solomon (1982). The California pinball cases do not seem to provide a basis for how to decide whether a game is predominantly skill or predominantly chance.

5. Court Response

The hearing was held before Judge Dauer of the Allegheny County Court of Common Pleas, who ruled orally at the time that "the court finds as a matter of fact that skill is a definite factor in playing this game and winning this game; certainly the evidence presented [showed that] a knowledge of the use of probabilities and statistics was certainly involved" [transcript at pp. 120–121]. His written opinion repeats this idea: "A player with a high degree of skill will produce a better outcome than a player with a low degree of skill . . . To constitute a gambling device 'per se,' the game must be purely one of chance with no skill involved in reaching the desired result" (Opinion, p. 6).

The case was appealed by the prosecution to the Pennsylvania Superior Court, which wrote:

> A device or game does not involve gambling per se merely because an element of chance is involved in its play or because it may be the subject of a bet. Football, baseball and golf, as well as bridge, ping pong, billiards or, for that matter, tiddledywinks, all involve an element of chance, yet the mere playing thereof is not gambling; betting on them is. *See Commonwealth v. Mihalow,* 142 Pa.Super.Ct. 433, 16 A.2d 656 (1940). Admittedly, someone might attempt to bet on the outcome of games played with the Electro-Sport Draw Poker Machine. This possibility, however, does not make the machine a gambling device per se.
>
> In support of its argument that no skill is involved in the play of draw poker, the Commonwealth compares it to a slot machine where the object of the game, to match a variety of objects on a visual display, depends entirely on a random spin. We find the analogy to be inapplicable. Draw poker involves far more than a mere glance at the face of cards, whether they appear on pieces of pasteboard, or on a viewing screen. Skill is exercised in choosing which cards to hold, in deciding which to discard, in considering whether to stand pat and in weighing the probabilities of drawing the desired card.[5] It is disingen-

[5] Illustrations of the kind of choices available to Draw Poker Machine players are found in R. F. Foster, *Foster's Complete Hoyle,* at 174 (1927):

> There is no rule to prevent his throwing away a pair of aces and keeping three clubs if he is so inclined; but the general practice is for the player to retain whatever pairs or triplets he may have, and to draw to them. Four cards of a straight or a flush may be drawn to in the same way, and some make a practice of drawing to one or two high cards, such as an ace and a king, when they have no other choice. Some hands offer opportunities to vary the draw. For instance: A player has dealt to him a small pair; but finds he has also four cards of a straight. He can discard three cards and draw to the pair; or one card, and draw to the straight; or two cards, keeping his ace in the hope of making two good pairs, aces up.

uous to compare the Draw Poker Machine to a game in which the player merely inserts a coin and waits for bells, bars, cherries, oranges, etc., to match up. In playing the Draw Poker Machine, there are a variety of "hands" which can be sought and a varying range of possibilities of obtaining them through a draw. A player faces similar odds to any game of draw poker; he has a measure of control over the outcome and he is similarly dependent upon his individual skill. To successfully play the Draw Poker Machine, as in regular poker, the player must be familiar with the relative value of the various combinations which may be held and must also know the odds against improving any given combination by drawing to it. *See Foster's Complete Hoyle, supra* at 169. An expert witness, testifying in Mr. Allen's behalf, Mr. Kadane, Professor of Statistics at Carnegie Mellon University, found, after over a hundred plays, that it was possible to win four times as many games playing a "smart" strategy as opposed to a "dumb" strategy. Professor Kadane therefore concluded that success with Electro-Sport Draw Poker depends "a great deal" on the player's skill.

The assertion by the Commonwealth that the "maximum score that a player can achieve is preordained upon the deal" is correct, but irrelevant; it is equally applicable to any game of cards. Once the cards are shuffled and cut, they are aligned so that the maximum hand is predetermined, unless someone engages in improper dealing, but the extent to which that hand is subsequently achieved is dependent upon the skill of each player as manifested by his subsequent play. In the Draw Poker Machine, in particular, the element of skill comes into operation when the player calculates the odds and determines whether to draw cards, and if so, how many.

The argument that a player may occasionally win against the law of probabilities does not mean that the game does not [differ] in substance from a card game of poker is to ignore contemporary reality.[6]

Thus the Pennsylvania Superior Court also found that the skill element is present in the Electro-Sport Draw Poker Machine, and hence that it is not a gambling device per se. Finally, the case was appealed by the prosecution to the Pennsylvania Supreme Court, which wrote:

[6] Appellant also contends that the brief duration of the game, approximately fifteen seconds, is further evidence that the machine is a gambling device. *Laris Enterprises, Inc. Appeal,* 201 Pa. Super. Ct. 28, 192 A.2d 240 (1963). While in a proper case, brevity of play may tend to prove one of the elements required for a finding that a machine is a gambling device per se, we conclude that since there is ample evidence that skill is required to play Electro-Sport Poker, the fact that the playing time is brief provides little, if any, support for Appellant's position.

Whether the result is determined by chance poses a far more difficult question. Superior Court, citing *Nu-Ken, supra,* and *In re Wigton,* 151 Pa. Superior Ct. 337, 30 A.2d 352 (1943), stated the standard to be that:

In order to conclude that a machine is a gambling device per se, *it is necessary to find that successful play is* entirely a matter of chance as opposed to skill.

Electro-Sport, supra, at 58, 443 A.2d at 297 (emphasis added). Initially, we note that the cited cases do not stand for the proposition that success be *entirely* a matter of chance. Rather, they hold that the "mere fact that a machine involves a substantial element of chance is insufficient" to find the machine a gambling device per se. *Nu-Ken, supra,* at 433, 288 A.2d at 920, citing *In re Wigton, supra.* Thus a showing of a large element of chance, without more, is not sufficient. Nor must the outcome of a game be wholly determined by skill in order for the machine to fall outside the per se category. As Superior Court pointed out:

A peculiar combination of luck and skill is the sine qua non *of almost all games common to modern life. It is hard to imagine a competition or a contest which does not depend in part on serendipity. It cannot be disputed that football, baseball and golf require substantial skill, training and finesse, yet the result of each game turns in part upon luck or chance.*

Electro-sport, supra, at 60, 443 A.2d at 298. We are thus left with the task of determining in each case the relative amounts of skill and chance present in the play of each machine and the extent to which skill or chance determines the outcome.

The expert witness who testified on behalf of the tavern owner is a professor of statistics at Carnegie-Mellon University. He stated that he could win at a rate four and one half ($4\frac{1}{2}$) times greater using a "smart" strategy (*i.e.*, employing his knowledge of statistics) than by using a "dumb" strategy (*i.e.*, always "standing pat" on the initial hand dealt by the machine). Based on his play of the Electro-Sport machine, he concluded that skill was a "definite factor" in playing the game. RR-130a. On cross-examination, however, he conceded that chance was also a factor both in determining the initial hand dealt and the cards from which one can draw, concluding that there was a "random element" present. *Id.* He could not say how to apportion the amounts of skill and chance. The expert witness for

> the Commonwealth testified that no skill was involved in playing the game.
>
> While appellee has demonstrated that some skill is involved in the playing of Electro-Sport, we believe that the element of chance predominates and the outcome is largely determined by chance. While skill, in the form of knowledge of probabilities, can improve a player's chances of winning and can maximize the size of the winnings, chance ultimately determines the outcome because chance determines the cards dealt and the cards from which one can draw—in short, a large random element is always present. *See Commonwealth v. 9 Mills Mechanical Slot Machines,* 62 Pa.CommonwealthCt. 397, 437 A.2d 67 (1981). That the skill involved in Electro-Sport is not the same skill which can indeed determine the outcome in a game of poker between human players can be appreciated when it is realized that holding, folding, bluffing, and raising have no role to play in Electro-Sport poker. Skill can improve the outcome in Electro-Sport; it cannot determine it.

Thus the Pennsylvania Supreme Court found that skill was not as important as chance in determining the outcome in Electro-Sport Draw Poker, and thus that the chance element is present. As explained above, this led them to determine that draw poker machines equipped with knock-down buttons and meters are gambling devices per se and that those without them are not.

Part of the difference between the decisions of Judge Dauer of the Allegheny County Court of Common Pleas and the Pennsylvania Superior Court, on the one hand, and the Pennsylvania Supreme Court on the other, can be explained by their different understandings of the law. Both Judge Dauer and the Pennsylvania Superior Court interpreted "skill versus chance" to mean that if any skill had been shown, this is sufficient to remove the chance element, and hence sufficient to declare the machines not gambling devices per se. They also both found, on the facts that I had shown, that skill was involved, and thus ruled that these are not gambling devices per se.

The Pennsylvania Supreme Court, however, interpreted the requirement as being whether skill or chance predominates. Thus it was not sufficient for the Supreme Court that I had shown that skill was definitely involved. In its last few words quoted above, the court advances a theory based on "determining the outcome." While those words are vague, no doubt they will be scrutinized by counsel and courts as they argue other cases on this point, until another case goes to the Supreme Court and, perhaps, further clarification is obtained.

Cases Cited

In re Wigton (1943), 151 Pa.Super.Ct. 337, 30 A.2d 352.

Commonwealth v. 9 Mills Mechanical Slot Machines (1981), 62 Pa.CommonwealthCt. 397, 437 A.2d 67.

Commonwealth v. Mihalow (1940), 142 Pa.Super.Ct. 433, 16 A.2d 656.

References

Solomon, H. (1966). Jurimetrics. In F. N. David (ed.), *Research Papers in Statistics,* New York: John Wiley & Sons, pp. 319–350.

Solomon, H. (1982). Measurement and burden of evidence. In J. Tiago de Oliveira and B. Epstein (eds.), *Some Recent Advances in Statistics,* London: Academic, pp. 1–22.

Comment

John Lehoczky
Department of Statistics
Carnegie-Mellon University
Pittsburgh, Pennsylvania

1. Introduction

In his chapter, Kadane raises a variety of provocative issues associated with the quantification and measurement of skill in a game involving both skill and chance. Kadane considered these issues in the context of the game of draw poker as played on certain electronic machines. In this game, an individual is "dealt" five cards and can decide which, if any, of the cards to discard. Any discarded cards are replaced. The resulting hand is evaluated according to a table of values, and the player earns a certain number of skill points on each play of the game. The table of values is given by Kadane. Throughout this paper, we assume that a royal flush earns 50 points. The cost of playing is assumed to be $1.00, as is the value of a skill point. The legal issue involves the determination of whether electronic draw poker can be considered to be a game of skill.

Kadane attacks the question of skill in the following way. He considers two strategies, one called "dumb" and one called "smart." The dumb strategy consists of never drawing any cards. It was reported that this strategy earned 34 points in 128 plays. Unfortunately, the smart strategy was never defined by Kadane. Indeed, the only indication of its definition is that when dealt AKQJ of one suit and a second A, the smart strategy discards the second A. The performance of the smart strategy was reported as yielding 159 points in the same 128 hands. Kadane evaluates the performance of smart relative to dumb by computing the skill ratio, $\frac{159}{34}$ =

4.68. He feels that this ratio is a measure of the "amount" of skill in the game and in this case is sufficiently large to allow one to consider electronic draw poker to be considered a game of skill.

This comment addresses a variety of issues raised by the Kadane paper including:

1. The assumption of randomness in evaluating draw poker strategies,
2. The number of trials needed to accurately measure the skill ratio,
3. The use of a "skill ratio" as a measure of the amount of skill in a game,
4. The introduction and evaluation of strategies for playing draw poker.

2. The Randomness Assumption

Kadane introduces two strategies for draw poker, smart and dumb, and evaluates the performance of each by playing each over 128 trials. He indicates that his statement of skill did not depend on any assumption of the randomness of machine generated hands. It is likely he was implicitly assuming that all hands are equally likely in developing his smart strategy. For example, he indicates that he would discard an extra ace (thus risking earning 0 points instead of earning 1 or more for certain) in order to draw for a royal flush. This strategy is smart with respect to a model of randomness in which all cards are equally likely to be dealt, but need not be at all smart for other models of randomness. Moreover, in using a skill ratio as the sole measure of skill in the game, he is implicitly assuming that hands are independent of each other and identically distributed. If one did not make this assumption, then far more sophisticated experimentation would be required to determine if the smart strategy really was superior to dumb. Under the standard model of randomness, one can calculate the expected return from a variety of different strategies, although neither the optimal strategy nor its return are known. Throughout the remainder of this comment, we assume a total randomness model and evaluate all strategies with respect to it.

3. The Sample Size Required

In the Kadane chapter, a skill ratio approach was taken as a basis for evaluation of the degree of skill involved in the game. This ratio was estimated using the empirical skill ratio, namely, the number of points earned under the smart strategy divided by the number of points earned under the dumb strategy, assuming a common number of trials. Generally, owing to the volatility of ratio estimators and the skewness in the point scale, this approach requires a large number of trials to provide sufficient accuracy. Furthermore, in Section 4, we argue that when dealing with unfair games, the average payoffs (excluding entry fees) of strategies are not an adequate measure of their efficacy. Let us assume, however, that we want to estimate the skill ratio, namely, the ratio of the payoff (excluding entry fee) using the smart strategy to the payoff using the dumb strategy. Indeed, one can use standard combinatorial techniques to evaluate the average payoff of the dumb strategy (it is 0.246); thus one need only estimate the average payoff using the smart strategy. If one were to play independent trials and let X denote the average payoff of the smart strategy, then $X/0.246$ would estimate the skill ratio. This statistic would be approximately normally distributed about the true mean and would have variance in the order of $64/n$. The variance of the optimal strategy is not known, nor is it estimable from the data in Kadane's chapter. Nevertheless, theoretical calculations for enhanced unsophisticated strategies suggest that this variance should be at least 4. If one uses the observed average payoff for the dumb strategy in the denominator rather than the known theoretical value, then the variance of the empirical skill ratio increases substantially. One can approximate the situation using asymptotic methods. The empirical skill ratio X/Y is approximately asymptotically normal with mean μ_X/μ_Y and variance $(\sigma_X^2 + \sigma_Y^2/\mu_Y^2)/n\mu_Y^2$. For the draw poker game, we can get a rough order of magnitude by letting $\mu_X = 1$, $\mu_Y = 0.25$, $\sigma_X = 2$, and $\sigma_Y = 1$. This gives a variance of $544/n$ or roughly 8 times the variance arising from using the known value in the denominator. Indeed, with $n = 128$, we find that the empirical skill ratio has variance about 4, surely too large for accurate estimation of the true skill ratio. This analysis is very rough. The numerator and denominator used by Kadane are actually positively correlated, a fact that serves to reduce the variance somewhat. Nevertheless, using the known performance of the dumb strategy (which gives a variance of 0.5 for 128 trials), one can conclude that the skill ratio is at least 3. An approximate 95% confidence interval for the payoff of the smart strategy is (0.900, 1.59). It will be seen in Section 6 that there is an unsophisticated strategy that offers a payoff of 0.857. The skill ratio for this strategy versus the dumb strategy is nearly 3.5, so it is clear that Kadane's smart strategy and the optimal strategy must have skill ratios at least this large.

4. The Use of the Skill Ratio as a Measure of Skill

Section 3 indicated that the number of trials used by Kadane was not adequate to accurately estimate a skill ratio. Suppose, however, one were able to carry out an unlimited number of trials or calculate the skill ratio exactly using theory. The skill ratio itself, even if estimated exactly, cannot determine whether the game is one of sufficient skill to be legal in Pennsylvania. There are two problems:

1. The skill ratio obscures the absolute rates of return of the strategies used.
2. A strategy with a large skill ratio relative to another may still be inferior to the other.

The skill ratio measures the relative rates of return of two strategies. If one strategy (smart) has an expected payoff given by $r\theta$, while another (dumb) has payoff θ, then the ratio is r, independent of θ. As θ decreases, the payoff from the smart strategy decreases. If θ is very small, then the ratio r obscures the fact that the difference in rates of return, $\theta(r - 1)$, is small and that the game involves little skill.

A second problem is that of the general evaluation of the performance of one strategy versus another. It is important to restrict our analysis to unfair games (supermartingales). This means that we shall consider only the case in which the expected payoff for any strategy is less than the $1 entry fee. We assume throughout that electronic draw poker forms a supermartingale. This means that the expected payoff of the optimal strategy is less than $1. Kadane's data suggest that his smart strategy may have a positive rate of return; however, we believe this is due solely to sampling variability. When dealing with supermartingales arising from this sort of game (i.e., sums of i.i.d. bounded random variables) a player's fortune will become arbitrarily small if he plays the game a sufficiently long time. That is, he will lose an arbitrarily large amount, limited only by his initial capital. Ruin is certain, although the time until ruin occurs is random and will vary with each distinct strategy.

A simple example may help to clarify these points. Suppose instead of draw poker, we play a game with an entry fee of $1 and the possibility of winning $2, $1, or $0 on each play of the game. Suppose using one strategy (smart) the player wins $1 with probability 0.99 or $0 with probability 0.01. The expected payoff is $0.99 for this strategy. Consider now a second strategy (dumb) which results in winning $2 with probability 0.125 and $0 with probability 0.875. This strategy has a rate of return of 0.25. The resulting skill ratio is approximately 4, thus showing a strong superiority in favor of smart. In fact, the first strategy is inferior to the second. Indeed, using the first, the player can never win on any trial and may lose $1. Its only virtue is that one gets to play many trials before ruin occurs. The second strategy at least affords a positive probability of winning and of increasing an initial fortune. For example, if one began with $1 and quit after the first trial, then the average final fortune would be 0.25, whereas using the smart strategy, it would be 0.99. If, however, we played until we have won or lost $1, then the expected final fortunes would be 0.25 and 0, respectively. It follows that unless the duration of the trials is important, the relative rates of return are insufficient by themselves to determine if a game is one of skill. One must make a more detailed analysis of the performance of the strategies in question.

5. Characterizing a Game of Skill

In Section 4 we argued that the skill ratio is not a sufficient measure of the degree of skill in a game. Rather, one must make a more detailed analysis of the particular strategies. In this section, we argue that one must be very careful in defining a "no-skill" strategy to use as a basis of comparison with a "skillful" or "smart" strategy. When defining a no-skill strategy, one should restrict attention to strategies which satisfy three conditions:

1. The strategy is well-intentioned (i.e., it does not attempt to deliberately do poorly).
2. The strategy assumes a full knowledge of the rules and scoring system of the game.
3. The strategy can be based on a few simple heuristics that a child could apply after a few minutes of instruction.

The third requirement merits some explanation. As an illustration, we consider the game of video tic-tac-toe. In this game, the player deposits $1 for each play. The player is allowed to choose whether or not to go first and then plays a standard game of tic-tac-toe against a machine that is programmed never to lose and to win if possible. If the player wins, he wins $100,000! Of course, this payoff never occurs. If the game is tied, the $1 is returned, whereas it is lost if the player loses. We now must determine whether this is a game of pure chance or a game of sufficient skill to be legal in Pennsylvania. Using the Kadane approach, we would introduce an ignorance or no-skill strategy and compare it with a smart or optimal strategy. In this case, the optimal strategy is the one that leads to a tie game on every trial. What would be a no-skill strategy? It seems that a dumb (but not suicidal) strategy would be a random-number-generated strategy. The player would pick at random whether to go first or second, and would select boxes at random from the remaining boxes. It is fairly easy to compute that the player can never win, has a $\frac{1}{96}$ probability of a tie game when he is second, and a probability of 0.05 of a tie if he goes first. This game offers a rate of return of 0.03 using the no-skill strategy. The resulting skill ratio is more than 30. Nevertheless, there is near universal agreement that tic-tac-toe is not a game of skill. The reason is that one need only apply one or two simple rules in order to never lose the game (e.g., play first in the center, play in a corner, block the opponent.). When one can play optimally or nearly optimally using a few simple rules that can be implemented by a child, one generally considers the game to involve no skill. We argue in Section 6 that this is also the case with electronic draw poker.

6. The Evaluation of Skill in Video Draw Poker

In this section, we deal specifically with the question of whether video draw poker can be considered to be a game of skill. We argue first that the dumb strategy used by Kadane is an unacceptable representative of a no-skill strategy and hence is an unacceptable basis of comparison. In particular, this strategy ignores the possibility of improving hands at no risk. For example, if one were dealt a hand valued at one point (pair of aces), the Kadane dumb strategy would draw no cards. A player with no skill but knowing the scoring rules of poker would keep the two aces and discard the three other cards. This offers the possibility of obtaining a hand of value 25, 10, 2, or 1 point. A more sophisticated strategy is, of course, possible. We assert that playing to enhance the value of a hand is not a skillful strategy; rather it is one that assumes only knowledge of the scoring system and no knowledge of probability theory.

There are 2,598,960 distinct poker hands that can be dealt. Among them, there are 2,317,320 that do not score any points. Of the 281,640 that do score points, 18,696 cannot be enhanced without risk, while 262,944 can be enhanced without risk. We thus define the *enhancement strategy* s_e as follows: Draw no cards if dealt a straight flush, four of a kind, a full house, a flush, or a straight; draw two cards with three of a kind; draw one card with two pairs; draw three cards with a pair of aces; and draw five cards with a hand of no points.

The return from the enhancement strategy can be calculated by theory. It has a rate of return of 0.570, compared with 0.246 from the Kadane dumb strategy. It embodies no skill in or knowledge of probability theory, and hence cannot be thought of as a skillful strategy. Furthermore, this strategy dominates the dumb strategy in all cases. No matter what cards are dealt, the enhancement strategy is guaranteed to do at least as well as the dumb strategy and better in many cases. We regard this enhancement strategy as being a better representative of the no-skill strategies.

We next introduce a series of strategies that use some simple rules for discarding. These strategies are analogous to the optimal tic-tac-toe strategy, which was characterized by a few simple playing rules. The strategies we consider consist of a sequence of discarding rules that are to be applied sequentially, using the first applicable rule. If none applies, then all five cards should be discarded. The rules are

R_0: Keep and enhance any hands worth at least one point.
R_1: Keep a pair; draw three cards.
R_2: Keep a four-card flush; draw one card.
R_3: Keep a four-card straight (consecutive or broken); draw one card.
R_4: Keep an ace; draw four cards.

We create a sequence of strategies s_1, s_2, s_3, and s_4. Strategy s_i is defined by applying rules R_j, $j = 0, \ldots, i$, in that order. This sequence was suggested by Dorian Feldman.

In each case, the expected rate of return can be calculated using combinatorial arguments. The results are given in the following, where s_d represents the Kadane dumb strategy:

s_d: 0.246
s_e: 0.570
s_1: 0.768
s_2: 0.809
s_3: 0.836
s_4: 0.857

We have argued earlier that s_e can be considered a no-skill strategy. Furthermore, we have argued that $s_1, \cdots s_4$ are specified by a few simple heuristics, and hence do not involve skill. That each of these strategies has an expected return of nearly 0.8 (or more) suggests that they must be near the optimal. Indeed, I would estimate that the optimal strategy (the one that maximizes expected return) has a rate of return only slightly above 0.9. Furthermore, the gains of the optimal over the simple heuristic strategies are in large part from events of small probability, for example, discarding an ace from AKQJ and a second ace. While such a strategy serves to increase the overall rate of return, it will not increase by much the expected final fortune when the gambler's ruin formulation is used. Consequently, it would seem that there is very little room for skill in electronic draw poker over and above the simple heuristic rules cited earlier. A final resolution of this issue will, however, have to wait until either Kadane clearly describes his smart strategy so that its performance can be determined exactly or the optimal strategy (in the sense of highest rate of return) and its performance are determined. These performances can then be compared with that of the heuristic strategies defined earlier in this section. Only then can a final determination concerning skills be made.

7. Conclusion

Kadane has presented an analysis of video draw poker that shows that it is possible to make a profit playing the game using a smart strategy and that a representative no-skill strategy has a rate of return far below the rate of return of the smart strategy. We have shown that the empirical evidence pointing to video draw poker having an optimal expected payoff larger than the entry fee can be easily explained by sampling variability. Although the dumb strategy is very inferior to the optimal strategy, calculations presented in Section 6 indicate that simple heuristic strategies can perform very well and possibly at a near-optimal level. Indeed, it appears that the difference between the best heuristic strategy listed in Section 6 and the optimum arises mostly from the inclusion of low-probability, high-payoff results. If one focuses on the final fortune of a player, the differences are likely to be further narrowed. This brings us to the conclusion that video draw poker cannot be considered to be a game of skill, even though there are unskillful ways to play it.

Rejoinder

Joseph B. Kadane

Lehoczky's comment makes important contributions to the scientific study of electronic draw poker, but neglects, at times, the legal context of my paper. I thank him for the former, and will defend my work from the latter.

1. The randomness assumption: Had I presented calculations that assumed randomness, I would have been vulnerable on cross-examination to questions about how I knew that every hand was equally likely. Since I did not have explicit knowledge of that, it seemed to me legally wise to base my testimony on experimentation rather than calculations. I still think this was wise, Lehoczky's comment notwithstanding.

2. Sample size and accurate approximation of the skill ratio: Lehoczky's rough calculations are quite interesting, and will advance our understanding of electronic draw poker under the randomness assumption. His conclusion, however, that "the number of trials used by Kadane was not adequate to accurately estimate a skill ratio," is unwarranted in the legal context. All three courts found that I had shown that skill is a factor in playing electronic draw poker. None complained that they were worried about sampling error in my "skill ratio." (This is not too strong an argument, since it is not clear how much they understood about sampling error). Nonetheless, for this specific legal purpose, I believe that my sample size was adequate.

3. The use of the skill ratio: Lehoczky's first point, about the skill ratio being a ratio, is of course correct, but the use of the ratio was not likely to mislead the court. I did report the number of games played and won under each strategy, but the skill ratio did (and does) seem to be a convenient dimensionless quantity.

His second point has to do with the appropriateness of the expected return as a utility function. Of course other utility functions are possible and interesting (e.g., see Dubins and Savage, 1965), but as a first mode of analysis, the expected return is quite reasonable. In any event, the ratio of expected utilities is one way of describing the relative worth of two strategies, although it is not invariant to adding a constant to the utility function. I do consider it likely, but not certain, that electronic draw poker is subfair (i.e., that the expected payoff of the optimal strategy is less than one). If it is, then the expected return determines the duration of the game. Since the defendant's position is that the game is "for amusement only," one gets more amusement the longer the game, and, hence, the higher the expected return. Thus larger expected returns are arguably in the interest of the player.

4. Skill: This is indeed the most difficult of the questions raised by my paper. I do not know the expected worth of the optimal strategy, nor do I know a simple set of heuristics whose expected worth approaches that of the optimal strategy. But suppose after further work and computer study such a set is found. Suppose, in addition, that it can be taught to a child. Could one say that, having applied all our tools of combinational analysis, supermartingales and fast computation, that the result is a skill-less strategy? Perhaps *ex post* one could, but before the work is done, I must say I still think that electronic draw poker requires skill to play well. As I remarked, the *legal* question of how to define skill is still very obscure. But perhaps with further work, solitaire draw poker will become a well-understood game, and hence skill-less, at least for those who read and write technical papers.

Additional References

Dubins, L., and Savage, L. J. (1965). *How to Gamble If You Must: Inequalities for Stochastic Processes,* New York: McGraw-Hill.

Inference in Cases of Disputed Paternity

Donald A. Berry **Seymour Geisser**
School of Statistics
University of Minnesota
Minneapolis, Minnesota

ABSTRACT

Blood testing for hereditary factors is being used increasingly in paternity cases to infer that a particular man is the father. Geneticists calculate a probability of paternity using Bayes' theorem, making various assumptions about the genetic factors used and the other evidence in the case. These assumptions are criticized, and the role of Bayes' theorem in a legal setting is discussed. The role of the forensic statistician in helping a court combine quantitative genetic evidence with nongenetic evidence is described. The effects of statistical errors and laboratory errors are discussed.

1. Introduction

There are two important kinds of evidence in cases of disputed paternity: blood tests for hereditary factors and testimony concerning sexual intercourse between the mother and alleged father. If blood tests exclude an alleged father, then the case usually proceeds no further. If an alleged father is not excluded, then the question of sexual intercourse becomes central. In many courts the particulars of blood tests play an increasingly important role in this setting.

Given the genetic makeup of parents, that of an offspring has a particular probability distribution. The problem of inference in cases of disputed paternity is to decide which genetic structure produced a given result. So this is a typical problem in statistical inference.

The basic statistical tool for problems involving "inverse probabilities" is Bayes' theorem. While its use for scientific inference is controversial among statisticians, it has been readily adopted by geneticists for purposes of genetic counseling and by some in cases of disputed paternity. Its application will be discussed in Sections 3 and 4. The approach we describe is used by many authors and, in particular, Salmon and Salmon (1980).

The procedure followed varies considerably from one laboratory to another. The following is a scenario that some facilities follow in obtaining and reporting results of blood tests. First, a blood sample is taken from the alleged father. At the same time he is identified and photographed; his thumbprint may be taken. When the samples are drawn from the mother and child, the mother is asked to verify that the man in the photograph is indeed the alleged father.

The child, mother, and alleged father are compared with respect to various blood-group genetic systems. The systems chosen vary greatly from one laboratory to another, and we have records of cases that show that different systems can be chosen by the same laboratory. Most states employ tissue-typing for human leukocyte antigen (HLA). This latter possibility will be discussed in Section 6, but our examples deal with red-blood-cell antigens and enzymes and serum proteins.

The testing laboratory prepares a report with its findings. This report (which is similar to Table 9 in Section 4) lists the various blood factors tested and gives measures of "likelihood of paternity." A main purpose of this paper is to elucidate and critically examine the assumptions and ensuing calculations in such a report. This is done in Sections 4 and 5. Another purpose is to describe an appropriate method of presentation for quantitative evidence in the presence of other kinds of evidence.

The focus of this chapter is the role of the forensic statistician in paternity cases, particularly as it applies to educating lawyers and communicating with juries. This role revolves around the use of Bayes' theorem: How are likelihoods calculated? Are blood-group factors and other genetic polymorphisms independent? What is the effect of classification errors? How are prior probabilities assessed and interpreted? How should nongenetic evidence be combined with genetic polymorphisms? What are appropriate reference populations? The chapter is written mainly for statisticians, but the discussion is kept at a level appropriate for many nonstatisticians. Some of the ideas and criticisms of the usual approach are similar to those of Ellman and Kaye (1979) and Aickin and Kaye (1982).

A short summary of the necessary genetical background is given in Section 2. The interested reader is referred to Elandt-Johnson (1971) for a much more extensive presentation.

2. Background Genetics

Until recently, genetic testing was used in paternity suits to exclude an alleged father. If the testing procedure did not exclude, it was regarded as largely irrelevant. These tests involved mainly the red-blood cell antigens.

These antigens are inherited substances, present on the surface of the cells, that have the capacity to induce the production of other substances, termed *antibodies*. Antibodies in turn react with antigens; it is assumed that the serum of an individual in which the red blood cells are suspended does not possess nor can produce antibodies against its own antigens. So a blood group system is a property of the individual's serum by which the antigen is recognized.

Consider a hypothetical system, say GH. An individual having blood group G then will not produce anti-G (antibody), but can possess or produce anti-H, say, an antibody to an alternative blood factor H belonging to the same system. A battery of diagnostic tests are available to determine an individual's blood group.

The logical basis for an exclusionary result depends on a relatively simple genetic construct. An individual's genotype, or hereditary configuration, for a particular inherited antigen consists of two of a number of alternative forms of the hereditary unit, called *alleles*, one for each parent. For example, for the simplest systems with only two possible alleles, say, G and H, an individual's genotype will be one of GG, GH, and HH. If the parents are GG and HH, then all offspring must be GH, a G from the first parent and an H from the second. On the other hand, if both parents are GH, then the child can be any of the three genotypes. Since each child inherits one of its two letters (alleles) from each parent, certain men can be logically excluded if the mother's and child's genotypes are known. Table 1 lists these exclusions. No alleged father of genotype GH can be logically excluded under any genotypic combination of mother and child.

TABLE 1. GH Paternal Exclusion

Child's Genotype	Mother's Genotype	Excluded Father's Genotype
GG	GG	HH
GH	GG	GG
HH	GG	Mother excluded!
GG	GH	HH
GH	GH	None
HH	GH	GG
GG	HH	Mother excluded!
GH	HH	HH
HH	HH	GG

When the alleles are codominant, as in the MN red-blood-cell antigen case, it is possible to establish the genotype of any subject. When one allele dominates another it is only possible for the test to establish the phenotype—the physical expression of a genotype, which may be influenced by environmental conditions, in this case the presence or absence of the dominant allele.

The ABO red-blood-cell antigen system is basically a three-allele system (although modern methods are able to discern at least two variants of the A allele, A_1 and A_2). It is a mixed system, since A and B are codominant and both dominate O. Hence an individual whose phenotype is A has genotype either AO or AA; similarly for B. However, the phenotype of the codominant case AB and also that of the recessive case O completely determine the corresponding genotypes AB and OO.

As an example consider the following phenotypic frequencies for the ABO system in a sample of $n = 6004$ white Californians reported by Grunbaum et al. (1978):

$$n_{\mathrm{O}} = 2891, \qquad n_{\mathrm{A}} = 2149, \qquad n_{\mathrm{B}} = 724, \qquad n_{\mathrm{AB}} = 240$$

(They actually report relative frequencies to three decimals—the frequencies we give are approximated from their figures.) Let $p_{\mathrm{O}}, p_{\mathrm{A}}, p_{\mathrm{B}}, p_{\mathrm{AB}}$ be the population phenotypic relative frequencies and $g_{\mathrm{O}}, g_{\mathrm{A}}, g_{\mathrm{B}}$ the allelic frequencies. Then if we use the Hardy–Weinberg law,

$$(g_{\mathrm{O}} + g_{\mathrm{A}} + g_{\mathrm{B}})^2 = p_{\mathrm{O}} + p_{\mathrm{A}} + p_{\mathrm{B}} + p_{\mathrm{AB}},$$

it follows that

$$\begin{aligned} p_{\mathrm{O}} &= g_{\mathrm{O}}^2 \\ p_{\mathrm{A}} &= g_{\mathrm{A}}^2 + 2g_{\mathrm{O}}g_{\mathrm{A}} \\ p_{\mathrm{B}} &= g_{\mathrm{B}}^2 + 2g_{\mathrm{O}}g_{\mathrm{B}} \\ p_{\mathrm{AB}} &= 2g_{\mathrm{A}}g_{\mathrm{B}}. \end{aligned}$$

The Hardy–Weinberg law depends on the assumption of random mating, which, strictly speaking, assigns to each individual of one sex in a population an equal chance of being a partner of a given mate of the opposite sex. That is interpreted as a mating between unrelated individuals. The transmission of the inherited units, one from each parent, is then presumed to be statistically independent of the particular blood systems to which the law is applied.

One estimate of p_{O} is n_{O}/n and in turn g_{O} can be estimated to be $\sqrt{n_{\mathrm{O}}/n}$. But these estimates are not efficient. Assuming the sample is random, the likelihood function of $g_{\mathrm{O}}, g_{\mathrm{A}}, g_{\mathrm{B}}$ (where $g_{\mathrm{O}} + g_{\mathrm{A}} + g_{\mathrm{B}} = 1$) is proportional to

$$g_{\mathrm{O}}^{2n_{\mathrm{O}}}(g_{\mathrm{A}}^2 + 2g_{\mathrm{O}}g_{\mathrm{A}})^{n_{\mathrm{A}}}(g_{\mathrm{B}}^2 + 2g_{\mathrm{O}}g_{\mathrm{B}})^{n_{\mathrm{B}}}(g_{\mathrm{A}}g_{\mathrm{B}})^{n_{\mathrm{AB}}}.$$

Methods for numerically determining the maximum likelihood estimates of the various allelic, genotypic, and phenotypic frequencies are readily available (see Elandt-Johnson, 1971). The maximum likelihood estimates of the allelic relative frequencies for the above data are $\hat{g}_O = 0.692$, $\hat{g}_A = 0.224$, and $\hat{g}_B = 0.084$. The corresponding estimates of genotypic and phenotypic relative frequencies are given in Table 2. [These phenotypic estimates differ from those of Grunbaum et al. (1978), who apparently used maximum likelihood estimates for the allelic frequencies, but the sample proportions for the phenotypic frequencies.]

Logical phenotypic exclusions for the ABO system are given in Table 3. This table also gives estimates of the proportion of men who are excluded from paternity on the basis of ABO blood type.

TABLE 2. ABO System Proportions (Estimates for a White California Population)

Phenotype:	O	A		B		AB
Genotype:	OO	AO	AA	BO	BB	AB
Genotypic frequency:	g_O^2 (0.479)	$2g_Og_A$ (0.310)	g_A^2 (0.050)	$2g_Og_B$ (0.116)	g_B^2 (0.007)	$2g_Ag_B$ (0.038)
Phenotypic frequency:	0.479	0.360		0.123		0.038

TABLE 3. ABO System Paternal Exclusion

Child's Phenotype	Mother's Phenotype	Excluded Paternal Phenotypes	Proportion of Excluded Males	Estimate for White Californians
O	O	AB	p_{AB}	0.038
A	O	O, B	$p_O + p_B$	0.062
B	O	O, A	$p_O + p_A$	0.839
AB	O	Mother excluded!	—	—
O	A	AB	p_{AB}	0.038
A	A	None	0	0.000
B	A	O, A	$p_O + p_A$	0.839
AB	A	O, A	$p_O + p_A$	0.839
O	B	AB	p_{AB}	0.038
A	B	O, B	$p_O + p_B$	0.602
B	B	None	0	0.000
AB	B	O, B	$p_O + p_B$	0.602
O	AB	Mother excluded!	—	—
A	AB	None	0	0.000
B	AB	None	0	0.000
AB	AB	O	p_O	0.479

The estimated probability of exclusion for the ABO system in a white population is the average of these estimates. This average is with respect to the probabilities of the various child–mother phenotype combinations. These are given in Table 4. For example, for "child: O; mother: O," the calculation of g_O^3 proceeds as follows: The probability of "mother: O" is g_O^2; if the mother is O, the child inherits an O allele from her, and the probability that the other allele is O is simply g_O. For the population of white Californians, the estimated probability of exclusion on the basis of ABO is 0.181. Salmon and Salmon (1980) suggest that only systems with high exclusion probability be used in paternity cases.

There are many known systems of red-blood-cell antigens. The first system for exclusion of paternity based on blood-group evidence was used more than 40 years ago, although early tests were used only to exclude putative fathers. As the number of red-blood-cell groups used in such tests increased it was realized that continual nonexclusion enhances the possibility of paternity. As still other genetic polymorphic (multiple allelic forms) systems, including serum protein groups and white-blood-cell antigens such as human leukocyte antigen (HLA), were introduced into this legal enterprise, systematic efforts were made to determine a canonical measure of the "likelihood of paternity" of an alleged father. In

TABLE 4. Proportion of Mother–Child Phenotype Combinations

Child's Phenotype	Mother's Phenotype	Combination Proportion	Estimate for White Californians
O	O	g_O^3	0.331
A	O	$g_O^2 g_A$	0.107
B	O	$g_O^2 g_B$	0.040
AB	O	0	0
O	A	$g_O^2 g_A$	0.107
A	A	$g_A^3 + 3g_O g_A^2 + g_O^2 g_A$	0.223
B	A	$g_O g_A g_B$	0.013
AB	A	$g_A^2 g_B + g_O g_A g_B$	0.017
O	B	$g_O^2 g_B$	0.040
A	B	$g_O g_A g_B$	0.013
B	B	$g_B^3 + 3g_O g_B^2 + g_O^2 g_B$	0.055
AB	B	$g_A g_B^2 + g_O g_A g_B$	0.015
O	AB	0	0
A	AB	$g_A^2 g_B + g_O g_A g_B$	0.017
B	AB	$g_A g_B^2 + g_O g_A g_B$	0.015
AB	AB	$g_A^2 g_B + g_A g_B^2$	0.006

many European and U.S. courts it has now become standard practice to accept genetic evidence in terms of a "probability" that an alleged father is indeed the father. This concept is developed in Section 3.

3. From Evidence to Inference

Either an alleged father is the true father or not, and a court may ultimately be called on to render a verdict. If genetic testing excludes the man, then, barring a gene mutation and errors in testing and transcribing, he is not the father. (Presumably a modern court would reach a conclusion different from the one that decided against Charlie Chaplin in a famous case from the 1940s, *Berry v. Chaplin,* 74 Cal.App.2d 652, 664–65, 169 P.2d 442, 450-451 (1946)). If the man is not excluded, then the evidence is not decisive—one possibility is to quantify "degree of paternity": How likely is it that the man is the father on the basis of the quantitative evidence? Such questions are statistical in nature and can be addressed by either Bayesian or classical methods.

The Bayesian approach is ideal for this problem in the sense that Bayes' theorem gives the relation between "inverse probabilities": the probability of guilt given evidence and the probability of evidence given guilt. The usual objection to a Bayesian approach is that the decision maker must assess prior information before, or independently from, the evidence at hand. This in turn leads to a subjective interpretation of probability. There is a natural "subject" or "decision maker" when a paternity case is brought to court: the individual juror (or the judge in nonjury trials). The ability (or willingness!) of a juror to make a probability assessment is another matter—this will be discussed in Section 5.

The problem of disputed paternity can also be addressed from a classical hypothesis-testing point of view. The null hypothesis is that the alleged father is the father, and the alternative is that he is not. A test based on a series of blood-group determinations that rejects the null only when the alleged father is excluded has significant level 0 but indeterminate power. (Any other nonrandomized test with $\alpha < 1$ will be arbitrary and difficult to describe. Randomized tests are inappropriate in this and similar contexts and would be disallowed by courts—see Section 5 for related discussion.) There is no natural alternative (unless there are two potential fathers, with the blood types of both available) at which to evaluate power, and averaging power over the population seems reasonable. This of course is a Bayesian notion. To our knowledge classical approaches have not been used in paternity cases. The remainder of this article deals with the Bayesian approach.

Bayes' Theorem

Bayes' theorem is an immediate consequence of the definition of conditional probability. Stated simply, it says that the probability of a statement being true given some new evidence E is proportional to the probability that it was considered true before obtaining E times the probability that E would obtain if the statement were true.

For example, we suspect that a coin that has just been tossed five times yielding five heads is not a fair coin. The probability that a fair coin would show five heads in any five independent tosses is $\frac{1}{32}$. There is a nonsensical tendency among naive users of significance tests to say that $\frac{1}{32}$ is the probability that the coin is a fair coin. The probability that the coin is fair is of course related to the evidence at hand, but the problem cannot be addressed unless alternative hypotheses are specified. Suppose that prior to tossing a coin we consider it to be fair with probability 0.95 and, say, two-headed with probability 0.05. Given the new evidence, the probability that the coin is fair is proportional to $(0.95)(\frac{1}{32})$ and the probability that the coin is two-headed is proportional to $(0.05)(1)$; since the corresponding probabilities must sum to 1 they are approximately 0.37 and 0.63, changed from 0.95 and 0.05. The output of Bayes' theorem is a "posterior" probability assessment of the truth of the opposing statements—posterior to the new evidence.

This example is obviously simplistic. There is seldom a single clear-cut alternative (such as "two-headed coin") to the statement under consideration. It is unlikely in a legal case that all parties would agree that one of two particular persons is guilty. Usually there are a large number of alternatives. The probability of obtaining E, the evidence at hand, must be assessed under each alternative. In addition, the probability of each of the possible alternatives must be assessed.

Let $S_1, S_2, \ldots$ stand for the possible true statements and $\Pr(S_i)$ the prior probability of S_i. The likelihood of S_i on the basis of evidence E is $\Pr(E \mid S_i)$. Then the posterior probability of S_i is given by Bayes' theorem:

$$\Pr(S_i \mid E) = \Pr(E \mid S_i) \cdot \Pr(S_i)/K.$$

The constant K is determined by the requirement that the total probability is 1:

$$K = \Pr(E) = \Pr(E \mid S_1) \cdot \Pr(S_1) + \Pr(E \mid S_2) \cdot \Pr(S_2) + \ldots.$$

An Example

To illustrate with a case of disputed paternity and very simple genetic evidence, consider ABO system phenotypes for white Californians discussed by Grunbaum et al. (1978), with frequencies given in Table 2. The least complicated application of Bayes' theorem is for a set of possible fathers with their phenotypes given (a more realistic assumption will be made in Section 4).

Six men (Mr. 1, . . . , Mr. 6) are the only possibilities as the father of the child in question, and they are deemed equally likely on the basis of other evidence. That is, $\Pr(S_i) = \frac{1}{6}$, where S_i is the statement "Mr. i is the child's father." Evidence E consists of the blood-type information given in Table 5. The problem is to incorporate this new evidence.

Since parental genotypes AB and OO cannot produce a type O child, $\Pr(E \mid S_4) = 0$. Since type O crossed with type O always gives rise to type O,

$$\Pr(E \mid S_1) = \Pr(E \mid S_5) = 1.$$

The other likelihoods are complicated by the fact that the genotypes of types A and B are not known. Consider S_2, or equivalently, S_6. If Mr. 2 is the father, then he must be genotype AO—this has probability $2g_O g_A / p_A = \frac{31}{36}$ given he is type A. Further, the probability of AO (father) and OO (mother) giving rise to type O (child) is $\frac{1}{2}$: the child is type O if and only if the father's O allele is passed on. So

$$\Pr(E \mid S_2) = \frac{1}{2} \cdot \frac{31}{36} = 0.431.$$

Similarly,

$$\Pr(E \mid S_3) = \frac{1}{2} \cdot \frac{116}{123} = 0.472.$$

The value of K is 0.556. The required likelihoods and posterior probabilities are given in Table 6.

TABLE 5. Evidence E—ABO System

Person:	Child	Mother	Mr. 1	Mr. 2	Mr. 3	Mr. 4	Mr. 5	Mr. 6
Phenotype:	O	O	O	A	B	AB	O	A

TABLE 6. From Prior to Posterior via Evidence E

Statement:	S_1	S_2	S_3	S_4	S_5	S_6
Probability prior to E:	$\frac{1}{6}$	$\frac{1}{6}$	$\frac{1}{6}$	$\frac{1}{6}$	$\frac{1}{6}$	$\frac{1}{6}$
Likelihood $\Pr(E \mid S_i)$:	1	0.431	0.472	0	1	0.431
Probability posterior to E:	0.300	0.129	0.141	0	0.300	0.129

TABLE 7. **Evidence E'—Gc System**

Person:	Child	Mother	Mr. 1	Mr. 2	Mr. 3	Mr. 4	Mr. 5	Mr. 6
Phenotype:	12	11	22	12	12	12	11	11

TABLE 8. **From Posterior to E to Posterior to E'**

Statement:	S_1	S_2	S_3	S_4	S_5	S_6
Probability prior to E':	0.300	0.129	0.141	0	0.300	0.129
Likelihood $\Pr(E' \mid S_i)$:	1	$\frac{1}{2}$	$\frac{1}{2}$	$\frac{1}{2}$	0	0
Probability posterior to E':	0.690	0.148	0.162	0	0	0

Among the six candidates, only Mr. 4 is exonerated by the evidence. Every other man with type AB would also be exonerated. On the other hand, men with type O have the highest likelihood.

Now suppose new evidence E' involving the Gc serum protein system is introduced. For the same population, estimates of the frequencies of alleles 1 and 2 are 0.710 and 0.290 (Grunbaum et al., 1978). Since these are codominant, the genotypic frequencies are the same as the phenotypic frequencies; these are 0.504, 0.084, 0.412 for genotypes 11, 22, 12 respectively. Evidence E' is given in Table 7.

Assuming E' and E are independent, Bayes' theorem can be applied again. The likelihoods of the S_i for these new data are given in Table 8 along with the probabilities posterior to both E' and E. The value of K, the probability of E' given E, is 0.435. Mr. 5 and Mr. 6 are now exonerated and, of course, Mr. 4 is still excluded.

As more genetic systems are included, more potential fathers are eliminated, and the probability of paternity for those not excluded tends to increase.

The way in which Bayes' theorem is applied in many courts is presented in Section 4, which can properly be regarded as an extension of this section. This application will be discussed and carefully scrutinized in Section 4 and, especially, in Section 5.

4. Current Use of Blood Testing in Law

The example in Section 4 is not very realistic. First, it is unusual for all possible candidates to be known with certainty—mothers conceal the number of possible fathers in at least 48% of paternity cases (Arthur and Reid, 1954). Second, many more genetic polymorphisms than that provided by the ABO and Gc systems are available. We shall reconsider the example of the previous section making the more realistic assumption that one man, say, Mr. 1, has been accused of being the child's father. For expository purposes, we shall consider only evidence from the ABO system.

Likelihood Ratios and Prior Odds

Let S_i continue to stand for "Mr. i is the father," and let S_1^c be the complement of S_1. Mr. 1 plays a central role in the current discussion, so it is convenient to write, for $i \geq 2$,

$$\Pr(S_i) = \Pr(S_i \mid S_1^c)\Pr(S_1^c).$$

Rewriting Bayes' theorem,

$$\Pr(S_1 \mid E) = \left[1 + \frac{\Pr(S_1^c)}{\Pr(S_1)} \cdot \frac{\Pr(E \mid S_1^c)}{\Pr(E \mid S_1)}\right]^{-1}$$

where the "likelihood" of S_1^c is

$$\Pr(E \mid S_1^c) = \sum_{i \geq 2} \Pr(E \mid S_i) \cdot \Pr(S_i \mid S_1^c),$$

really an average or integrated likelihood. Expressed equivalently,

$$\frac{\Pr(S_1^c \mid E)}{\Pr(S_1 \mid E)} = \frac{\Pr(S_1^c)}{\Pr(S_1)} \cdot \frac{\Pr(E \mid S_1^c)}{\Pr(E \mid S_1)};$$

the posterior odds ratio is the product of the prior odds ratio and the likelihood ratio.

Evidence E is the child's, mother's, and alleged father's ABO system blood types. The likelihood of S_i depends on Mr. i's blood type, and is given in Section 4. It may be appropriate to restrict laboratories from supplying any more than the various $\Pr(E \mid S_i)$ to court. But the current practice of many laboratories goes further, and their calculations are usually admitted, if not well understood!

First, it is assumed that the true father is a randomly selected man from some population (we return to this in Section 5) if the alleged father is not the father. Since all men with the same genotype have the same likelihood and conditional (on S_1^c) prior probability, they can be grouped together. The conditional prior probability of each group is the proportion of the corresponding genotype in the population. So

$$\begin{aligned}\Pr(E \mid S_1^c) = {} & g_O^2\Pr(E \mid \text{Father is OO}) + 2g_Og_A\Pr(E \mid \text{Father is AO}) \\ & + g_A^2\Pr(E \mid \text{Father is AA}) + 2g_Og_B\Pr(E \mid \text{Father is BO}) \\ & + g_B^2\Pr(E \mid \text{Father is BB}) + 2g_Ag_B\Pr(E \mid \text{Father is AB}).\end{aligned}$$

In the example of Section 4 in which both mother and child are type O, there is a much easier route. Namely, this evidence will result if and only if an O allele is selected (randomly) from the population; so $\Pr(E \mid S_1^c) = g_O$, or about 0.692 in the example.

That assumption is innocuous when compared to the next one! Suppose that Mr. 1 and one other man of unknown blood type are equally likely to be the father; that is, $\Pr(S_1) = \Pr(S_1^c) = 0.5$. Then the posterior odds equal the likelihood ratio. If, as in Section 3, Mr. 1 is type O, then

$$\Pr(S_1 \mid E) = \frac{1}{1 + 0.692} = 0.591,$$

increased somewhat from the prior probability since Mr. 1 is in a group of men—those with type O—who have the highest likelihood.

The inverse of the above likelihood ratio, $\Pr(E \mid S_1)/\Pr(E \mid S_1^c)$, plays an important role in some courtroom presentations. It is called the *paternity index,* or P.I., by Salmon and Salmon (1980). The higher the paternity index, the greater the relative likelihood of S_1.

One problem with converting a paternity index into a probability via Bayes' theorem is assessing the prior probability $\Pr(S_1)$. It is artificial to suppose, as we essentially did above, that exactly two men, including the alleged father, had intercourse with the mother near the time of conception, each the same number of times. The number of men who could be the father is usually a point of contention between the two sides. Another point of contention may be whether intercourse with the alleged father ever took place, or if it did, the timing of such intercourse relative to the child's birthdate.

The other problem with this paternity index is that it assumes a "random man" is the alternative to the alleged father. The question is, random from what population? Averaging with respect to different relative frequencies to obtain the likelihood $\Pr(E \mid S_1^c)$ can substantially affect it.

These two issues are related; both will be returned to in Section 5.

Likelihoods for Multiple Gene Systems: Independence

In the above example, suppose the Gc phenotypes of child, mother, and Mr. 1 are known to be 12, 11, and 22, respectively; call this evidence E'. Now the likelihood of S_1 for E' is

$$\Pr(E' \mid S_1) = 1.$$

Assuming that the true father is selected randomly from the hypothesized population when S_1 is false implies that the (average) likelihood of S_1^c for E' is

$$\Pr(E' \mid S_1^c) = g_1^2\Pr(E' \mid \text{Father is } 11) + 2g_1g_2\Pr(E' \mid \text{Father is } 12) + g_2^2\Pr(E' \mid \text{Father is } 22)$$

Again the analysis is simpler. Evidence E' obtains when a 2 allele is selected randomly from the population; this has probability $g_2 \approx 0.290$. Hence, using probabilities posterior to E as prior to E',

$$\Pr(S_1 \mid E, E') = 1 \Big/ \left(1 + \frac{0.290}{1} \cdot \frac{0.409}{0.591}\right) = 0.833.$$

This still assumes that the probability of S_1 apart from blood data is 0.5.

Bringing more and more evidence to bear in this way will tend to increase the posterior probability of Mr. 1 if he (or his identical twin!) is the father and tend to decrease it—perhaps make it 0—if he is not. This repeated application of Bayes' theorem is not only appropriate for blood-test data or other genetic polymorphisms but applies whenever information can be quantified using probabilities. But there is a proviso. It would be incorrect to use the same data twice. More generally, the individual pieces of information should be *independent*.

Instead of applying Bayes' theorem separately for E and then for E', a more direct path, in terms of odds, is as follows:

$$\frac{\Pr(S_1^c \mid E, E')}{\Pr(S_1 \mid E, E')} = \frac{\Pr(E \mid S_1^c)}{\Pr(E \mid S_1)} \cdot \frac{\Pr(E' \mid S_1^c)}{\Pr(E' \mid S_1)} \cdot \frac{\Pr(S_1^c)}{P(S_1)}.$$

The paternity index is now

$$\frac{\Pr(E \mid S_1)}{\Pr(E \mid S_1^c)} \cdot \frac{\Pr(E' \mid S_1)}{\Pr(E' \mid S_1^c)},$$

the product of individual likelihood ratios. Multiplication of probabilities is appropriate only if E and E' are statistically independent. For example, it must be that frequencies 0.504, 0.084, 0.412 for Gc phenotypes 11, 22, 12 hold for each blood type. If almost every type O has Gc-22, say, then the above calculations are inappropriate. Bayes' theorem would still apply, but the ABO and Gc systems would have to be considered jointly, with frequencies given for the various combinations of ABO and Gc phenotypes.

In paternity cases, calculations of an index and the posterior probability of paternity assume independence of the blood factors tested. There is some justification for this assumption. Many geneticists and pathologists claim that this is not an assumption, but a fact. Indeed, it is standard practice to assume independence of genetic polymorphisms in legal settings without stating the assumption. For example, the two-volume work by Schatkin (1984) never mentions it as an assumption, although hundreds of genetics articles are referenced. Also, Schatkin (1984, Chapters 5–9) cites many experts who casually multiply probabilities for up to 100 genetic polymorphisms (not all of which are specified), assuming independence without saying so. Independence of genetic traits is attributed to the loci controlling these traits being located on presumably different chromosomes or far enough apart on the same chromosome. This "assumption" is based either on theoretical considerations or on experimental data. It is therefore incumbent upon the forensic statistician to determine which in fact is the case and on what evidence it rests. Grunbaum et al. (1978) present data for over 10,000 individuals to show that 12 factors—including ABO and Gc—are either pairwise independent or, perhaps, negligibly dependent. While pairwise independence is weaker than independence, this result does lend credence to a calculation in which likelihoods for these 12 factors are multiplied.

A hypothetical example is shown in Table 9. None of the tests excludes the alleged father. Estimates of genetic frequencies for a white and a black California population were taken from Grunbaum et al. (1978). (The appropriate "population" to be used is that of the *true* father, not that of the *alleged* father, as used by all laboratories we know about—see Section 5.) For certain factors, these frequencies vary considerably by race. In this case the paternity index is 65 times larger in the black population than in the white one! Obviously, some of these blood factors are racially dependent.

TABLE 9. Gene System Likelihood Ratios, White Versus Black

Gene System	Child	Mother	Mr. 1	Likelihood Ratio (White)	Likelihood Ratio (Black)
ABO	O	O	O	0.692	0.690
Rhesus	—	—	—	0.208	0.074
PGM	11	11	12	1.534	1.619
AK	12	11	22	0.037	0.008
ADA	11	11	12	1.898	1.796
EAP	AA	CA	BA	0.670	0.428
EsD	12	12	12	1.807	1.840
G-6-PD	B	B	B	0.995	0.730
Hb	A	A	A	0.999	0.956
Hp	12	12	12	1.507	1.517
Gc	12	11	22	0.290	0.129
PGD	A	A	A	0.981	0.964
Product				8.00×10^{-3}	1.23×10^{-4}
Paternity index (=1/product)				125	8100
"Plausibility of paternity" (=1/(1 + product))				99.2%	99.99%

"Plausibility of paternity" is also given in the table. Although the black paternity index is 65 times that of the white, both plausibilities are close to 1. [Schatkin (1984, p. 8-37) cites a case in which the "plausibility of paternity" is 98.5% when the reference population is white and 54% when it is black.] This term is used by some to mean the posterior probability of paternity assuming the prior probability $\Pr(S_1)$ is 0.5 and the true father is selected randomly from the population if S_1 is false. While the term is misleading, it is better than the alternatives that are also used: "probability of paternity" and "likelihood of paternity."

Some laboratories take the liberty of transforming this "plausibility of paternity" into an assessment of the truth of the ultimate question by providing Hummel's likelihood of paternity, (Hummel, 1971), given in Table 10.

In practice, factors are tested that are not among the 12 given in Table 9. But if additional factors cannot be shown to be independent of all other factors tested, their use in calculating these indices should be criticized and should be disallowed in court (unless they serve to exclude a putative father).

Some of the factors given in Table 9 are not tested in practice. The reason is clear: several of them (e.g., AK, G-6-PD, Hb, PGD) are poor discriminators and a greater number of tests allows more room for misclassification errors (Chakraborty et al., 1974). These tests have low (average) probabilities of exclusion, but there are measures for use in selecting tests that are somewhat more appropriate for the analysis we have described. For a man selected randomly from the population, it is a trivial calculation to show that the expected value of his paternity index is 1 regardless of the number of gene systems involved. This is true whether or not the child–mother phenotypic combination is given. (A corollary is that the expected posterior odds of paternity are the prior odds.) Systems should be chosen if for that system the paternity index of a randomly selected man has substantial variability—measured, say, by its standard deviation.

TABLE 10. Hummel's Likelihood of Paternity

Plausibility of Paternity	Likelihood of Paternity
0.9980 – 0.9990	Practically proved
0.9910 – 0.9979	Extremely likely
0.9500 – 0.9909	Very likely
0.9000 – 0.9499	Likely
0.8000 – 0.8999	Undecided
Less than 0.8000	Not useful

The standard deviation of the system P.I. depends on the child's and mother's phenotypes. Table 11 provides an example using the ABO system and the white California population discussed in Section 2. Since these phenotypes are, of course, not available before deciding which systems to test, the unconditional standard deviation is required. The variance of P.I. can be calculated by averaging the conditional variances over the distribution of the various child–mother combinations since the mean P.I. is constant. For the example of Table 11, the requisite distribution is given in Table 4; the (average) variance is 0.700 and standard deviation is 0.837. Only systems with sufficiently large standard deviations of their P.I., say at least 0.5, should be tested. If, say, 12 blood-group systems are used and all have standard deviation 0.5, then the standard deviation of the P.I. is only $\sqrt{(1 + (0.5)^2)^{12} - 1} = 3.7$. On the other hand, if the individual standard deviations are 1, then the overall standard deviation is 64. The latter case not only provides a much better chance for exclusion, but also will yield a much larger probability of paternity for a man who is not excluded.

Probability of Paternity Based on Nonexclusion

An analysis using Bayes' theorem applied to exclusion and nonexclusion has been suggested by many authors (e.g., Lee, 1975; Wiener, 1976). For example, given that a man has not been excluded but 90% of all men would be excluded, the posterior odds of paternity are nine times the prior odds. Such inferences are weaker than the approach described earlier because they are based on a reduction of the data.

Sometimes this reduction is substantial. Suppose mother and child have type A blood. Then there is no information in the fact that an alleged father is not excluded: the prior odds of paternity would be unchanged

TABLE 11. Estimates of Standard Deviation of Random Paternity Index (White California Population)

Child–Mother Phenotypes	Standard Deviation of Random P.I.	Child–Mother Phenotypes	Standard Deviation of Random P.I.
O/O	0.443	O/B	0.445
A/O	1.230	A/B	1.229
B/O	2.282	B/B	0.410
AB/O	—	AB/B	1.235
O/A	0.445	O/AB	—
A/A	0.283	A/AB	0.209
B/A	0.297	B/AB	0.356
AB/A	2.277	AB/AB	0.997

because no men are excluded. However, the P.I. varies in white Californians from 0.44 (odds of paternity decreased) for type B men to 1.32 (odds increased) for type A men. For a nonexcluded man, the odds of paternity would usually increase if full information is used; it would be a mistake for the counsel of the alleged father to object to this approach in favor of the one described earlier!

5. Discussion and Recommendations

There are a number of important issues that we have not resolved. How accurate are blood tests? What effect do inaccuracies have? How are prior probabilities assessed? Who does the assessing? How does one combine genetic and other evidence? These are among the questions considered in this section. In addition, questions for expert witnesses and effective communication of these issues to a jury are discussed.

The Role of Probabilities in Law

Should courts be guided to a posterior probability of paternity, or probability of guilt in criminal cases? Fairley (1973) presents arguments for both sides. He also describes a study that shows unaided intuition to be inept in learning from probabilistic evidence. While the case for formal analysis, administered with appropriate qualifiers, seems strong, it assumes that probability does have a role in law.

Ellman and Kaye (1979) argue for the appropriate use of probabilistic evidence in legal cases. They indicate that many people view such evidence as being comparable to Rabelais's Judge Bridlegoose who rolled dice to decide cases, the higher roll winning. We agree with them that a decision based on an assessed probability is not a randomized decision. The distinction is between constructing dice to specifications and actually rolling them.

Courts frequently decide cases in which they are uncertain about the correct disposition. In giving his reasons for rolling dice to his peers, Judge Bridlegoose repeatedly says that he throws dice "just like you other gentlemen." This repetition—the phrase occurs in practically every sentence—suggests that Rabelais was convinced that an element of randomness is present in all court decisions. Judge Bridlegoose had the advantage over his peers in knowing which dice he used in each case! "It took the testimony of a sage, an oracle, a drunken party goer, a messenger, a sheepherder and his own wife before Oedipus could figure out who his father was" (*New York Times,* June 1981, as quoted by Schatkin,

1984, p. 8-29). Modern paternity cases offer little good evidence of a nature other than probabilistic.

Blood Tests and Other Evidence: Assessing Priors

Defining a probability of paternity on the basis of genetic testing alone is like assigning a probability to the proposition that a coin is fair using the results of several tosses of the coin. There is no logical foundation for an assignment based solely on the data, but people try to do it nonetheless. The following quote from Schatkin (1984, p. 809) is given without comment:

> As a rule, one hundred blood tests will result in some fifteen exclusions. Multiplying 15 by 2 (because the blood test potential for exonerating an innocent man is about 55 per cent.) in that series of one hundred blood tests carried out, 30 per cent. are actually not the fathers. Falling back on the analogy of a woman putting her hand into an urn and making her selection of whom to accuse, the thirty innocent men in that series represent a wrong guess on her part and a gamble on her part that failed. Blood test exclusions, therefore, demonstrate those cases where the woman "guessed wrong."
>
> As stated, 15 exclusions result from 100 blood tests carried out, and of those 100 men, 30 are actually not the fathers. And, 85 are not excluded. And of those 85 not excluded, we know that 15 are not the father. So that, what are the probabilities of one of those 85 actually being the father? We divide 85 by 15. Therefore the chances of one of those men not excluded being the actual father, is in the proportion of $5\frac{1}{2}$ to 1. We conclude, therefore, that if a man is not excluded, the chances are 5 to 1 that he is the father of the child.

A mistake made in the literature and practice of genetic testing for paternity is easy to identify. The entire discipline recognizes that Bayes' theorem must be applied, but some paternity testers [apparently dating to Essen-Möller (1938)] want to use the same prior probability in every case! Namely, they assume that, aside from the genetic data, the alleged father is the true father with probability $\frac{1}{2}$. From our personal experience, we know of one pathologist who testified in court that the probability, or "plausibility," of paternity does not depend on the number of men who had sexual intercourse with the mother near the time of conception. In fact, a candidate for the prior probability for an alleged father is the number of times he had intercourse with the mother near the time of conception divided by the total number of times she had intercourse in that period. (This could be refined to take into account more likely times for conception, viability of the man's sperm, etc.) But this information is known only to the mother, and even she may have forgotten. In assessing a prior probability a juror must digest a variety of conflicting testimony concerning this and other issues.

There may be cases in which some jurors actually have a prior probability of $\frac{1}{2}$. But introducing it in court under the guise of blood typing is grossly misleading unless the implications are made clear. Ideally, each juror should appraise the information, other than the blood-typing data, in the case at hand and assess a prior probability on that basis. They can then be told how to transform it into posterior probabilities and in turn use it to reach a verdict.

The posterior probability of paternity, say π', is a function of the prior, say π: namely,

$$\pi' = \left[1 + \frac{1}{\text{P.I.}}\frac{1-\pi}{\pi}\right]^{-1},$$

Of course, judges and juries will have trouble with such a formula. But it can be tabled, and the table can include the traditional $\pi = \frac{1}{2}$. Table 12 provides an example using Mr. 1's P.I. = 125 found in Table 9. It also gives the prior and posterior odds against paternity since some people think in those terms rather than in probabilities.

There are several substantive problems with this "ideal." One is that the juror may refuse to quantify nongenetic information in terms of a probability or even a range of probabilities. The juror can be asked to interpret probabilities in terms of small-stake bets or betting odds. And many jurors will go along willingly. But some may have an aversion, moral or otherwise, to betting. It may help to indicate that these are only thought bets or preferences designed to quantify strength of belief. (For related ideas, see DeGroot, 1970, Chapter 6.)

Some care in advising a jury in the matter of assigning a prior probability is necessary because the best mode of elicitation—dialogue—is not available. On the other hand, a high degree of precision is not necessary; only a gross assessment of magnitude is needed. Bets or lotteries can be

TABLE 12. Prior to Posterior Probabilities of Paternity for P.I. = 125

Prior probability π:	0	$\frac{1}{1{,}000{,}000}$	$\frac{1}{1{,}000}$	$\frac{1}{100}$	$\frac{1}{10}$	$\frac{1}{2}$	$\frac{9}{10}$	1
Posterior probability π':	0	0.00012	0.111	0.558	0.933	0.992	0.999	1
Prior odds against:	∞	999,999 : 1	999 : 1	99 : 1	9 : 1	1 : 1	1 : 9	0
Posterior odds against:	∞	8000 : 1	8 : 1	4 : 5	1 : 14	1 : 124	1 : 999	0

described, which will help a juror decide on a range of prior probabilities. For example, the jury can be told: "If you would prefer being paid $1 if this man is not the father to $1 if he is, then your prior probability of paternity is less than $\frac{1}{2}$. If, in addition, you would prefer $1 if this man is the father to $1 if a "1" is rolled using a fair die, then it is greater than $\frac{1}{6}$." A range of prior probabilities corresponds to a range of posterior probabilities as exemplified in Table 12. These bounds on the posterior probability may be sufficient to determine the juror's vote—in any event it will help. This suggests a method for setting up an analog of Table 12. The prior probabilities tabled can be calculated to correspond to some interesting posteriors probabilities (e.g., 0.99, 0.95, 0.90, 0.50).

The discussion given above assumes that jurors are willing to assess prior probability distributions. While little can be done if they have no feeling for randomness, there are various devices that may help. One is as follows. They can be asked to suppose that the proportion of the time the mother had sexual intercourse with the alleged father, as opposed to other men, during the time in which conception was possible is known. Then this could serve as a prior probability. Since it is not known, relevant testimony can be weighed. If a juror can be made to assess a probability distribution on this proportion ("How likely do you feel this proportion is less than 10%?"), then Table 12 can be used with π equal to the mean of this distribution. The values $\pi = 0$ and $\pi = 1$ are important in this regard, for they correspond to frequently heard testimony: namely, "never had intercourse with her" and "no other man." For example, if these latter are the only two possibilities, and are given equal weight by the juror, then the column $\pi = \frac{1}{2}$ in the analog of Table 12 is appropriate.

The most serious difficulty in assessing a prior probability of paternity is setting the genetic information aside to ensure that it is, in effect, not used twice in evaluating a posterior probability. It would help if blood-group and other genetic data were presented subsequent to all other evidence, but it can never be kept completely separate! That a case comes to trial almost always implies that the alleged father was not excluded, so "nonexclusion" is, in the broad sense, used twice. The double usage of this evidence can have an enormous impact on a jury. [Our experience is that cases that go to court are almost always decided for the plaintiff when the alleged father has a high paternity index and acknowledges intercourse with the mother *at some time,* although not necessarily during the time that conception was possible—Schatkin (1984) gives many case histories.]

One remedy is to calculate a paternity index conditionally on the fact that the man was not excluded. This would be easy to do and would result in substantially lower P.I.'s for nonexcluded men. Such an adjustment is appropriate and seems essential if justice is to be served.

The following recommendation by Wiener (1976) is related to this double usage: "The value of [the a priori probability of paternity] depends on the experience of the courts—e.g., if 75% of the defendants have been found innocent of the charge, then the a priori probability of paternity . . . is 0.75 [sic; he meant 0.25]." There are a number of serious objections to this proposal. One is that the nonexclusion of the alleged father is used twice in an obviously formal way (while true for the use of Bayes' theorem with a paternity index, this is especially clear in Wiener's context because he goes on to use Bayes' theorem conditioning on nonexclusion). As stated previously, prior probabilities should depend only on the particulars of the case at hand but not on any blood-typing data that will be presented in evidence.

What Reference Population?

The paternity index depends heavily on the reference population used. A common practice of laboratories is to use the genetic frequencies in the race of the alleged father to calculate the paternity index. The logic for this is difficult to comprehend; the calculation is appropriate only if the *true* father is of the same race. As indicated in the example of Section 4, the resulting bias can be substantial. While not perfect, it would be much more appropriate to use the race of the *child*. If the race of a true father is an issue in a particular case, then the P.I. should be calculated by averaging over the local population. Alternatively, if the alleged father is white, say, and the defense claims the true father is black, then different sets of calculations should be made. The jury can decide which to believe, or how to weight them.

An obvious difficulty in making calculations for the "correct" reference population is the lack of appropriate data. Suppose a woman becomes pregnant in a small secluded town in which there are few families, some inbreeding, and little genetic variation. Using the population of the entire country as the reference set is obviously inappropriate. In particular, the proportion of the town's population excluded by blood tests would likely be much smaller than the proportion of the larger population that would be excluded.

Blood-group data in a case alleging incest need not be handled differently from any other paternity case. Calculating the likelihood of the alleged father depends only on the three blood samples, and the fact that the mother's and alleged father's genetic structures are similar is of no additional consequence. But the likelihood of the "random man" is greatly affected if a suspected alternative to the alleged father is related to the mother or the alleged father, whether or not the latter two are themselves related. This information would be easy to incorporate if the blood groups of any such alternatives are known, and extremely difficult if not. Men whose identities are known can be regarded as "random," but not if they are related to the mother or alleged father.

Errors in Testing; Other Realities

The analysis of the preceding section assumed that tests for genetic factors are error-free. Chakraborty et al. (1974) cite studies showing that "misclassifications insofar as the blood groups are concerned are more common than generally acknowledged and even in highly reputable laboratories may involve 2%–3% of all determinations." This has an effect on the paternity index and in turn on the probability of paternity. While not done in practice, error rates could easily be incorporated: the genetic evidence can be given in terms of a probability distribution that incorporates the possibility of error, and likelihoods can be calculated on this basis. Not explicitly considering misclassification errors encourages a court to lend more credence to the report from a laboratory than is justified.

The possibility of misclassification errors means, of course, that no man is excluded with certainty, only with high probability. For example, assuming a 2% error rate in ABO classification, Mr. 4 would not have been excluded in Section 3. Table 13 revises Table 6 assuming a 2% error rate in the blood tests of the possible fathers (mother and child are both assumed to be type O—the error possibilities in their tests would make for still less change in the probabilities from prior to posterior).

In general, the greater the possibility of classification error, the closer a paternity index is to 1. This means that men who are not excluded as father will tend to have smaller probabilities of paternity, and men who are "excluded" will tend to have greater probabilities.

In the words of Wiener (1976), "researchers have had occasion to retest and have detected errors in more than a score of cases. . . . If the newer tests [new blood groups] are included, the possibility of error will, of course, be multiplied, especially because many of the newer tests are not

TABLE 13. Table 6 Modified by 2% Classification Error Rate

Statement:	S_1	S_2	S_3	S_4	S_5	S_6
Probability prior to *E*:	$\frac{1}{6}$	$\frac{1}{6}$	$\frac{1}{6}$	$\frac{1}{6}$	$\frac{1}{6}$	$\frac{1}{6}$
Likelihood $\Pr(E \mid S_i)$:	0.988	0.439	0.476	0.014	0.988	0.439
Probability posterior to *E*:	0.295	0.131	0.142	0.004	0.295	0.131

perfected.'' And, ''as the number of tests increase . . . a point will be reached at which the chances of exclusion increase more slowly even than the chances of error and where further testing is extravagantly costly. . . .''

The better laboratories recognize the possibility of error and do all tests in duplicate and some in triplicate (Polesky, 1975). Assuming the individual tests are independent—even though technicians are blinded, this assumption may not be entirely valid—the error rate is substantially reduced. Presumably they report the mode in case of disagreement. The actual policy in these matters should be made public. But more important, all results (three, if triplicate) should be presented in court. Requesting all the data would be an appropriate tool for a lawyer whose case is suffering. While this information may tend to obscure matters for a judge or jury, it is necessary for completeness. Withholding such information creates an illusion of precision that may be unwarranted. (The situation is analogous to that of an experimenter who makes observations in triplicate and uses only their means or medians in a regression problem, an all too common practice—R^2 will be artificially inflated.)

Blood tests are highly respected as evidence by the legal profession. For example, in a chapter entitled ''The Unerring Accuracy of Blood Tests,'' Schatkin (1984, p. 11-1) displays a rather curious argument to come to the conclusion that blood tests are accurate:

> Verification of the accuracy of blood tests came not long after their inception. During the ten-year period March 22, 1935 to March 22, 1945, 656 blood tests carried out in affiliation cases by order of the Court of Special Sessions in New York City resulted in 65 exclusions. The question naturally arises, Were those exclusions accurate? The answer is Yes, because each and every one of those 65 exclusions was followed by the mother's subsequent confession, for the first time, of sexual relations with another man about the time she became pregnant.

Actually, the information given is also consistent with *every* blood test being wrong!

There are other errors that enter into the calculation of a paternity index that should be mentioned. One form of error is statistical. The phenotypic frequencies used are based on samples and not complete population counts. Some laboratories use published frequencies while some others, notably blood banks, keep records of previous blood samples tested. For example, the estimates used in Table 9 were reported by Grunbaum et al. (1978) and were based on blood samples of 6004 white and 1024 black Californians that were collected from blood banks around the state. Assuming these are representative of Californians (a dubious assumption), the standard errors in estimating the percentage of O alleles in the white and black population to be 0.692 and 0.690 are 0.6% and 1.4%. So a 95% confidence interval for the proportion of O alleles among black Californians is 66.2% to 71.8%. Of course, the smaller the sample size, the larger the standard error of the estimate.

The next point seems minor and it seems difficult to deal with in the courtroom. More than one laboratory in the United States carries out calculations of likelihood ratios, and presents them in court, using six-digit accuracy; for example, a paternity index of 51.3204 (translating to a plausibility of paternity of 98.089%). In view of the presence of these errors, calculations reported to more than two digits are suspect. They do not deceive educated observers, but, again, they can create an illusion of greater accuracy in court. A lawyer may question the accuracy of the numbers, but even if the paternity tester recants and, in the example given above, says P.I. = 51, the plausibility of paternity is essentially unchanged. The endeavor can be perceived as inconsequential and unnecessary carping to a jury.

Even the three-digit accuracy reported in Table 9 is misleading. (We note that the frequency of AK in blacks was reported to one-digit accuracy. This alone makes the reported paternity index of 8100 subject to an error of up to 500.) Suppose the relative error in each of the 12 gene system likelihood ratios is 2%. Then the relative error in the product of the 12 is

$$\sqrt{(1 + 0.02^2)^{12} - 1} = 6.9\%$$

or about $\sqrt{12}$ as large, so an error of 15% is quite possible.

Additional Genetic Evidence

A posterior probability of paternity can change in two ways: through the prior and through the paternity index. We have discussed the prior and the denominator of the paternity index (which is affected by the choice of reference population) previously. The P.I. also depends on the probability of the child's phenotypic structure given the mother's and assuming the alleged father is the father. This likelihood can be changed—perhaps to exclude a previously nonexcluded man—by gathering more data.

There is no universal standard that indicates which blood-group systems to use in calculating a particular paternity index. Some limit on the number used is necessary to help ensure independence, minimize errors, and keep costs down. But clearly an alleged father who knows he cannot be the father should ask for replicate testing and testing on further systems until he is excluded. On the other hand, a mother who is certain that the father is a man who has been excluded should ask for retesting because either a mistake has been made or a "silent" gene is involved. In certain codominant systems, when it happens that the mother and child are both homozygous for the same allele and apparently the father lacks that allele and so is judged to be homozygous for the second dominant allele, he is excluded. In this case the man may not be homozygous but actually have a "silent" recessive gene that is not detectable by the standard test, so that the exclusion is false. The frequency of such a gene in the serum protein haptoglobin system is reported by Cook et al. (1969). Competent facilities take this into account in their reports (Dodd and Lincoln, 1981).

Another way of gathering additional relevant data without testing more factors is to test relatives of the mother and alleged father. For example, in Section 3, Mr. 2 could father a type O child with a type O mother only if his genotype is AO. So if it were determined that his true parents were both AB, then he would be exonerated. Still relevant, though not conclusive, would be evidence that his other n children with a type O mother were type A. The likelihood that he is AA for this latter set of data is 1, and the likelihood of AO is only $(\frac{1}{2})^n$. On the other hand, if one of his parents or one of his other children were type O, then his genotype is in fact AO and his P.I. would increase somewhat.

Other kinds of genetic evidence that have been used in paternity cases are less accurate measures of heredity but are much better understood by lay people: hair color, eye color, "family resemblance," and similar traits. Incorporating this information into the P.I. is difficult but incorporating it into the prior is possible. For example, if both the mother and the alleged father are blue-eyed and devoid of any brown pigment in their irises but the child is definitely brown-eyed, the alleged father is excludable. On the other hand, a devastating impact can be made on a jury when a mother presents her red-headed child in court and the nonexcluded alleged father also has red hair. Still, cautious lawyers will often balk at exhibiting a child in court because of the poor understanding people have of these easily observed hereditary traits.

Questions for a Geneticist or Pathologist in Court

A number of substantive issues have been raised in this paper concerning the way in which genetic information is used in paternity cases. Many of these issues should be exposed in court. We present a few sample questions for a geneticist here. These may aid an attorney in preparing a case or a geneticist in critically rethinking an analysis; they also serve as a review for this paper.

Question: How were the blood factors you analyzed chosen?

Discussion: Presumably, factors were chosen if their phenotypes could be classified reliably and if they were discriminatory (having a high exclusion frequency). Also, tests are time-consuming, so some limit is necessary.

Question: Do you always analyze these same factors?

Discussion: A negative answer can be embarrassing, especially if some factors were not tested in the present case that had been in others. For then the obvious question is,

Question: Could the alleged father have been excluded had you tested these other factors?

Answer: Yes.

Question: Are there still other factors that you have never used that might have excluded him?

Answer: Yes.

Question: Do the calculations you made assume that the various blood factors are independent?

Answer: Yes.

Question: Are they independent?

Discussion: There is no way for someone to know the answer a priori. The basis for an affirmative answer, whether theoretical reasoning or experimental data, should be requested. The paper by Grunbaum et al. (1978) mentioned previously provides limited documentation on the 12 factors given in Table 9—''limited'' to pairwise independence and Californians.

Question: What is the rate of classification error in your methods?

Discussion: Possible answer: less than 3% for an individual test, but we do all tests in triplicate with three different analysts so the chance for error is negligible. (Some geneticists claim a very small error rate; documentation should be requested.)

Question: What do you do when the analysts disagree?

Discussion: If the answer is ''we take the mode,'' then there is still a substantial chance for error. Possible answer: We retest until we are virtually certain of the result.

Question: What effect does the possibility of error have on the paternity index?

Discussion: The issue is very complicated. If substantial retesting is done to eliminate errors, then the effect may be negligible.

Question: Were there any disagreements in any of the tests in the current case?

Discussion: Probable answer: I don't know.

Question: How accurate are the estimates of the phenotypic frequencies of the factors you used?

Answer: They are based on thousands of samples and so are accurate to within 2%.

Question: What effect can these errors have on the paternity index?

Answer: It could change it by 10–20%, but it could not have excluded an alleged father whose P.I. is positive.

Question: How did you calculate the probability (or plausibility) of paternity in this case?

Discussion: The answer is bound to be long, involve references to formulas and computers, and include phrases like "standard practice in the field." It is quite unlikely that any juror will understand the answer. The answer may be couched in terms of frequency-based probabilities ("Take the ratio of 125, the P.I., to 125 plus 1 for a random man and express it as a percentage"), which will also suggest that the respondent does not really understand the meaning of a probability of paternity.

Question: Does this probability depend on whether or not the mother and alleged father had sexual intercourse proximate to the time of conception?

Discussion: Seemingly a silly question, but it is difficult to give a correct, extended answer. Obviously, it is not possible for the alleged father to be the father unless intercourse occurred near the time of conception. But a juror's prior probability should weigh the various possibilities in this regard. Then the posterior probability also weighs these possibilities—as it should. Put another way, a juror's probability of paternity is an average of two conditional probabilities, one assumes intercourse and the other does not (obviously, the latter is zero).

Question: Would this probability change if it were known that the alleged father and a number of other men had intercourse with the mother near the time of conception?

Answer: Yes. If the proportion of the time intercourse involved the alleged father is $1/n$, then on the basis of this information the odds against the alleged father are increased by a factor of $n - 1$ and there is a corresponding decrease in the probability of paternity.

Question: Under these conditions, what value of n would make the alleged father an even bet to be the father?

Answer: n = P.I., the paternity index.

Question: What would be the effect on the posterior probability if it were known that one of the other men who had intercourse with the mother was a relative of the alleged father?

Answer: Then the calculation of the posterior probability is wrong. If, for example, the relative was his identical twin, then both would have the same P.I. and, assuming equal frequency of intercourse, the same posterior probability, which means it can be no greater than 50%.

6. Other Available Tests and the Horizon

One of the great advantages of genetic testing for paternity is the potential savings in time, effort, and money—especially in regard to litigation. Excluded fathers are rarely if ever brought to trial by a plaintiff these days; on the other hand, when the true father is confronted with overwhelming genetical evidence as to his paternity, he might well accept the responsibility. Hence, as a practical matter, genetic testing becomes a reasonable and relatively inexpensive way to clear court dockets. It also serves to reduce welfare costs for the increasing number of children born out of wedlock since the father can be compelled to pay child support.

Few courts have been hesitant to accept red-blood-cell antigen testing (Schatkin, 1984). When this series fails to exclude, a resolution may be attempted using red-blood-cell and serum proteins or HLA typing; which one is used depends on state laws and the usual procedure of the laboratory concerned.

In discussing the admissibility of HLA testing, the Kansas Court of Appeals, *Tice v. Richardson*, 8 F.L.R. 1113 (see Schatkin 1984, p. 3-28), notes:

> Since HLA testing is a relatively new test insofar as its use in the courtroom is concerned, it has been dealt with by only a few appellate courts. Several courts have refused to admit the test to show probability of paternity. In so doing, however these courts have in general acknowledged the test as reliable but nevertheless rejected the evidence under specific statutes which limit admissibility of blood test results to those which exclude the alleged father.

The value of an HLA test, though considerably more expensive to administer, lies in the large number of alleles, resulting in a much higher exclusion probability. It is also claimed (Perdue et al., 1977) that the error rate of HLA typing classifications in pairs of replicate typing tests is less than 0.35% when performed under very carefully controlled conditions. The probability of exclusion for the HLA system alone is claimed to be 0.95 (Sussman and Gilja, 1981). They also report that a total probability of exclusion of 0.9995 is available when red-blood-cell antigens and enzymes and serum proteins are used in conjunction with HLA.

A new potentially rich source of polymorphisms based on recombinant DNA technology is described by Botstein et al. (1980). Suggestions are made that could eventually lead to a human genetic linkage map that would considerably elucidate modes of inheritance. The new markers are called restriction fragment length polymorphisms (RFLP) and can be assayed from small amounts of peripheral blood and, according to some, appear to be inherited as simple Mendelian codominant alleles. We may expect that the development, perfection, and use of this rich new source of genotypic differences could eventually lead to a probability of excluding an innocent man that, for all intents and purposes, is one. Of course it is not clear when such a goal will become a reality. In the meantime, since juridicial decisions cannot be put off, forensic statisticians can be of much service to litigants and the court by constantly probing assumptions, scrutinizing techniques, assessing the accuracy of results, examining the data closely, and providing a coherent framework for decision making.

Acknowledgment

The work by the authors was supported in part by NSF grant MCS 8102471 and NIH grant GM25271.

References

Aickin, M., and Kaye, D. (1982). Some mathematical and legal considerations in using serological tests to prove paternity. In R. H. Walker (ed.), *Inclusion Probabilities in Percentage Testing,* Arlington, Va.: American Association of Blood Banks, pp. 155–168.

Arthur, R. O., and Reid, J. E. (1954). "Utilizing the Lie Detector Technique to Determine the Truth in Disputed Paternity Cases," *Journal of Criminal Law, Criminology and Police Science,* **45,** 213–221.

Botstein, D., White, R.L., Skolnick, M., Davis, R. W. (1980). "Construction of a Genetic Linkage Map in Man Using Restriction Fragment Length Polymorphisms," *American Journal of Human Genetics,* **32,** 314–331.

Chakraborty, R., Shaw, M., and Schull, W. J. (1974). "Exclusion of Paternity: The Current State of the Art," *American Journal of Human Genetics,* **26,** 477–488.

Cook, P. L., Gray, J. E., Brock, R. A., Robson, E. B., and Howlett, R. M. (1969). "Data on Haptoglobin and D Group Chromosomes," *Annals of Human Genetics,* **33,** 125.

DeGroot, M. H. (1970). *Optimal Statistical Decisions,* New York: McGraw-Hill.

Dodd, B. E., and Lincoln, P. J. (1981). The application of blood groups and other genetic polymorphisms to problems of parentage. In D. Tills (ed.), *Biotest Bulletin,* Vol. 5, 16 pages. Fairfield, N.J.: Folex-Biotest Schleussner Inc.

Elandt-Johnson, R. C. (1971). *Probability Models and Statistical Methods in Genetics,* New York: John Wiley & Sons.

Ellman, I. M., and Kaye, D. (1979). "Probabilities and Proof: Can HLA and Blood Group Testing Prove Paternity?" *New York University Law School Journal,* **54,** 1131–1162.

Essen-Möller, E. (1938). "Die Beweiskraft der Ähnlichkeit im Vater Schaftsnachweis; Theoretische Grundlagen," *Mitt. Anthorop. Ges.* (Wein), **68,** 598.

Fairley, W. B. (1973). "Probabilistic Analysis of Identification Evidence," *The Journal of Legal Studies,* **11,** 493–513.

Grunbaum, B. W., Selvin, S., Pace, N., and Block, D. M. (1978). "Frequency Distribution and Discrimination Probability of Twelve Protein Genetic Variants in Human Blood as Functions of Race, Sex and Age," *Journal of Forensic Sciences,* **23,** 577–587.

Hummel, K. (1971). Biostatistical opinion of parentage based upon the results of blood group tests. In Ihm P. Schmidt (ed.), *Biostatistische Abstammungsbegutachtung mit Blutgruppenbefunden.* Stuttgart: Gustav Fisher. (Quoted in *Family Law Quarterly,* **10,** 262 (1976).)

Lee, Chang L. (1975). Estimation of the likelihood of paternity. In H. F. Polesky (ed.), *Paternity Testing.* Chicago: The Division of Educational Media Services, pp. 28–29.

Perdue, S., Terasaki, P. I., Honig, C. R., and Estrin, T. A. (1977). "Reduction of Error Rates in the Microlymphocitotoxicity Test," *Tissue Antigens,* **9,** 259–266.

Polesky, H. F. (1975). Medical-legal immunohematology problem cases. In H. F. Polesky (ed.), *Paternity Testing.* Chicago: The Division of Educational Media Services, pp. 89–106.

Salmon, D., and Salmon, C. (1980). "Blood Groups and Genetic Markers Polymorphisms and Probability of Paternity," *Transfusion,* **20,** 684–694.

Schatkin, S. B. (1984). *Disputed Paternity Proceedings, Vol. I and II,* 4th Ed., rev., New York: Matthew Bender.

Sussman, L. N., and Gilja, B. K. (1981). "Blood Grouping Tests for Paternity and Nonpaternity," *New York State Journal of Medicine,* **81,** 343–346.

Wiener, A. S. (1976). Likelihood of parentage. In L. M. Seideman (ed.), *Paternity Testing by Blood Grouping,* 2nd ed., Springfield: Charles C Thomas, pp. 124–131.

Comment: On Blood Test Reports in Paternity Cases

Donald Ylvisaker
Department of Mathematics
University of California
Los Angeles, California

1. Introduction

The use of blood tests in paternity cases has increased dramatically over the past few years, and a growing number of principals, attorneys, judges, and jury members require an understanding of the positive evidence of paternity provided when the alleged father has not been excluded by tests made on one or more genetic systems. Statisticians have a natural interest in quantification here, but many of them are not current with developments in this area. The chapter by Berry and Geisser should therefore be welcome on several levels. The chapter does a special service by clearly setting forth the issues and calculations involved in a Bayesian analysis of this kind of situation and by bringing out the special difficulties that attend such requirements as a prior probability, a reference population, and the independence of tests.

This comment goes further than Berry and Geisser's work in criticizing the reports now offered to the principals and the courts by the blood banks. As these reports not only list test results but go on to assess relevance and weight of evidence as well, the impression is created that "Science" speaks directly to these questions with little regard for the

Research supported by NSF Grant MCS 83-01587.

individuality of cases. The contention here is that blood-test evidence consists of the test results themselves, and that it devolves on attorneys and judges to argue in the appropriate context what weight should be attached to them. The probative value of blood testing under nonexclusion is not at issue, but it is misleading to have it be assigned an explicit magnitude by a laboratory whose information is limited to population statistics.

After a brief background is sketched, separate sections will deal with the modeling difficulties that surround the probability of paternity and the paternity index. Some comments and conclusions are given in the final section.

2. Paternity Disputes

Legal procedures vary by state, according to statute or court ruling, so there is no typical case to bring forward. Perhaps the future is represented by the Uniform Parentage Act approved by the American Bar Association in 1974 and adopted by 15 states as of February, 1985. The act provides for an informal hearing at the first stage, and then if no resolution is obtained through the recommendation of the referee, the matter is set for trial. Blood tests may be called for, and nonexclusionary evidence is allowed. Broun and Krause (1982) provide a most informative account of how various courts deal with blood testing issues, and relevant sections of the Uniform Parentage Act may be found there as well.

Test reports vary in form and language, but they are adequately exemplified by Table 9 in Berry and Geisser. In that example the paternity index of 125 may be read directly as an estimate that it is 125 times more probable that the putative father would produce, with the mother, a child as exhibited than that a man chosen at random from the population of whites in California would do so. If accuracy is granted, this statement carries its own weight in much the same way as the observation that a given man is, say, 10 times as likely to be left-handed as a man chosen at random from this population. The translation of the paternity index to a probability or plausibility of paternity, 99.2% in the example, conveys rather more. This will be discussed in Section 3.

It should be noted that blood banks are now in general agreement on how to express the conclusions of their testing. Evidence of this is the experiment conducted in connection with the Airlie Conference and described fully in Walker (1982): 14 blood-testing experts were asked to analyze separately a case in which nonexclusion resulted from the testing of six genetic systems. With the use of a common set of frequency tables provided them, they found paternity indices between 64.903 and 66.239, with corresponding probabilities of paternity ranging from 0.9848 to 0.9851. The uniformity in these numbers, following involved analyses, is an indication that underlying assumptions are not controversial and that the calculation machinery is well in place.

3. From Test Results to the Probability of Paternity

How should one assess the positive evidence of paternity when blood tests do not exclude the putative father? In Berry and Geisser it is observed that "the Bayesian approach is ideal for this problem . . ."; in a discussion session devoted to this question and reported in Lee (1982), M. R. Mickey brings out the complete class argument for a Bayesian analysis; Bias et al. (1982) assert that "the preferred analysis, beyond any rational dispute, should be Bayesian." A practical note is sounded by Aickin and Kaye (1982): "To force one item of evidence into a Bayesian mold while leaving all the others to be handled intuitively and without reference to their impact on the probability of paternity seems unlikely to assist the jury in fairly assessing the quantitative evidence together with the other information in the case."

In any event, arguments that the testing facilities themselves should pursue a full-fledged Bayesian analysis seem less than compelling. What is clear is that when they do so, it is with little knowledge of the particulars of the case. Thus their choice of a prior probability of $\frac{1}{2}$ appears to be a necessary one on the grounds of ignorance or fairness (although arguments have been advanced that this is favorable to the putative father; see, e.g., Finkelstein and Fairley, 1970; Hummel, 1982). In conjunction with an estimated paternity index, one has available a probability of paternity. The consequences of this deserve further comment.

From the empirical perspective of the laboratory, a well-chosen probability of paternity is presented to the court. Such a stylized computation will fix its own meaning and sense of scale in time, but its legal relevance is not at all clear. In particular the $\frac{1}{2}$ prior probability basis requires justification in the case at hand, according to an appellate court finding that "where statistical evidence is derived from a formula which relies upon certain factual assumptions, the accuracy of those assumptions must be determined by the jury as a preliminary fact before the statistical evidence may be accorded any weight. In the case of the HLA probability of paternity results, the 94.67% probability of paternity is irrelevant if the prior probability of 50% that is employed in the formula is not correct. In order for the jury to properly assess such statistical evidence, it had to first evaluate the correctness of the 50% assumption inherent in the formula. If it then found that assumption to be accurate, it could give full weight to the 94.67% probability of paternity accordingly; but, if it found the assumption to be invalid, it could accord less weight or disregard altogether the probability of paternity results.''[1] Thus, rather than being a direct aid in the understanding of test results, the introduction of a probability of paternity brings about the unnatural situation in which jurors debate the merits of a common prior.

The conclusion here is the one reached by Aickin and Kaye: ''Finally, we oppose giving the W-value calculated according to the Essen-Möller equation as if it were the probability of paternity. Any single number served up as the probability of paternity connotes more than the biostatistical evidence can establish.'' This view is plainly echoed by a California appellate court: ''evidence [of the 98.95% probability of paternity] was presented by the medical technologist from the university laboratory whose knowledge of the case was limited to the blood typing. She did not know the extent or nature of the other evidence to be introduced at trial and she was not in a position to weigh the evidence and make a mathematical determination of the probability of paternity.''[2]

4. The Paternity Index

Valuable understanding of blood-test results might be gained through knowledge of a paternity index, to the extent that it is capable of exact expression and is obtained through defensible calculations. Exactness is stressed by Nijenhuis (1982): ''The paternity index should fulfill the conditions of a likelihood quotient. It must be the quotient of two real chances; both chances should concern the same event under alternative conditions.'' While reliability of estimation procedures in use deserves and receives continuing attention, the discussion here will center on the choice of a reference population and the assumption of independence of test results.

What population should be used for comparison purposes? Race is an acknowledged factor. Availability of data is an operational constraint, and one will not soon have the luxury of making sophisticated choices. Blood banks must choose on the basis of race and location alone, and availability of data, which is known, takes precedence over relevance, which is unknown.

A case[3] involving Hawaiians of Chinese descent affords an illustration

[1] *Everett v. Everett,* 150 Cal.App.3d 1053, 1069-70; 198 Cal.Rptr. 391 (1984).

[2] *Alinda V. v. Alfredo V.,* 125 Cal.App. 3d98, 101; 177 Cal.Rptr. 839 (1981).

[3] The case is pending at this writing and is not to be identified further. Contact the author at a later date for more information.

of this point. Sixteen systems were tested, and nonexclusion was found. A first report analyzed results on 12 of these systems, omitting 4 for lack of data, and made extensive use of a study of 1605 Asians from Hawaii (86%) and northern California. The paternity index was found to be 144.19. A reevaluation was made at the time of trial using all 16 systems. Data came from seven studies in all including, as other examples, one of 373 Chinese from Taiwan and another of 277 Japanese from the northern part of Japan. The index was then reported to be 10,204.

Given this description, we are in no better position to judge the relevance of possible comparison populations (Asians, Chinese in Hawaii, or others) than was the laboratory itself. Whatever the explicit comparison intended, however, the precision with which the index is stated is not at all consistent with its quite uncertain accuracy as an estimate for that population.

Consider now the independence of tests. There are two components to such an assumption: for the calculation of the index as it depends on the putative father, independence of systems is presumably a matter of genetics[4]; with regard to the random male on the other hand, it is effectively the independence of traits in the population and is subject to statistical study. Berry and Geisser cite Grunbaum et al. (1978) on the establishment of pairwise independence of 12 systems in black and white Californians. They go on to assert, "But if additional factors cannot be shown to be independent of all other factors tested—and this requires multivariate data, not simply a geneticist's impression—their use in calculating these indices should be severely criticized and should be disallowed in court; unless they serve to exclude a putative father."

Independence of test results goes largely unquestioned in blood-bank computations. The Airlie case mentioned earlier involved a white trio and a set of assumptions in parentage testing was listed in Walker (1982) in connection with it. Regarding independence one finds, "Classical Mendelian principles are applied in the analysis of results—i.e., alleles segregate, nonalleles assort independently, and expected ratios of offspring are utilized." In that test case, 2 of the 6 systems are among the 12 in Table 9 of Berry and Geisser—ABO and Rh—while four are not, including HLA. In the case referred to at note 4, 5 of 16 systems used independently are in Table 9—ABO, Rh, Gc, Hp, and EsD—while 11 others are not, including HLA.

[4] Population frequencies may be used, however, as in Assumption 10 of Walker (1982), which allows that the alleged father and mother are "random" individuals within their populations and therefore the Hardy–Weinberg principle can be used in estimating genotype frequencies from their phenotypes unless the genotypes of the mother and/or child are apparent from inspection of their phenotypes.

Technically, of course, the assumption of independence of traits is not valid in any human population for circumstances such as those mentioned above. One need only compute that for the six systems in the first study, the population must support more than 500,000 categories of individuals with widely varying frequencies; there are over 200 billion categories for the 16 systems tested in the second case. A human population at Hardy–Weinberg equilibrium with respect to several genetic systems is a theoretical ideal, not a practical reality.

Is the assumption of independence justified nonetheless? Kiefer (1979) is well worth reading on just this point. The context of that work has close parallels with the present one: a proposed technique in taxonomy assigns a lower-level taxon to one of several higher-level taxa by estimating the probability of observing it under alternative hypotheses, using statistical independence of character traits in each competing taxon. Kiefer gives a lengthy criticism of the free use of independence and his arguments are quite pertinent here. He suggests discussion to reach an understanding of "whether it is preferable to use an inappropriate mathematical model and accompanying mechanical routine that yields precise odds, rather than use intelligent but less formal data analysis in settings that cannot even approximately support such a mathematical structure." And again: "An obvious scientific principle, exercised in any careful statistical study, is that one should not assert bold inferences that are based on a particular model whose validity is unknown, without also testing the adequacy of the model."

Testing for independence has not received much attention and indeed the problem of verifying its approximate truth for several systems is a formidable one, even in the most accommodating of populations. It appears that real progress in identifying biological fathers stems from genetic advances in the isolation of sophisticated single systems, not from statistical justifications of the product rule.

5. Comments and Conclusions

It is natural to call on blood testing to provide relief from simple conflicting testimonies in paternity disputes. Exclusion now has this effect in roughly 20% of all cases. When a matter goes to court following nonexclusion, widespread feeling exists that blood-test evidence should be accorded special status, for example, by having it presented after all other information has been given to the jurors. While this passes over the possibility of further quantifiable facts,[5] it may be that the formal separation of

[5] For example, the information in the alleged time of conception relative to the onset of menstruation as in *Everett v. Everett*, note 2 at 1064.

quantitative and other evidence is a sensible first step in assisting jurors to a decision that must incorporate disparate elements. The debate on such questions will not be joined here. What is clear is that as test results are a focus of attention, special care must be exercised in their presentation.

Current blood-bank reports carry with them the implicit claim that the meaning of blood tests can be directly and unambiguously measured. This cannot be supported. In particular, the precision with which the paternity index is typically given belies its real accuracy, and the meaning of the index is often diluted by computational reliance on different data bases and on the assumption of independence of traits; the additional reporting of a probability of paternity introduces unfortunate language and with little prospect of assisting jurors. In short, laboratory reports do not always meet the standards suggested by a California court in connection with blood-test evidence: "The HLA test interpretations are not based on arbitrarily assigned numerical probability values or on a statistical theory unsupported by evidence. Instead, they are based upon objectively ascertainable data and a statistical theory based upon scientific research and experiment."[6]

Uncomplicated yet faithful communication of the meaning of nonexclusion following extensive blood testing is more than one should expect, especially from an agency remote from the facts of a case. It is in court that a proper focus can be chosen to permit the weighing of such evidence in the case at hand.

Additional References

Bias, W. B., Meyers, D. A., and Murphy, E. A. (1982). Theoretical underpinning of paternity testing. *Proceedings of the American Blood Banks International Conference on Inclusion Probabilities in Parentage Testing,* 51–61.

Broun, K. S., and Krause, H. D. (1982). Paternity blood tests and the courts. *Proceedings of the American Blood Banks International Conference of Inclusion Probabilities in Parentage Testing,* 171–200.

Finkelstein, M. O., and Fairley, W. B. (1970). "A Bayesian Approach to Identification Evidence," *Harvard Law Review,* **83,** 489–517.

Hummel, K. (1982). Paternity case analysis. *Proceedings of the American Blood Banks International Conference of Inclusion Probabilities in Parentage Testing,* 505–506.

Kiefer, J. (1979). "Comments on Taxonomy, Independence and Mathematical Models (With Reference to a Methodology of Machol and Singer)," *Mycologia,* **71,** 343–378.

[6] *Cramer v. Morrison,* 88 Cal.App.3d 873, 881-82; 153 Cal.Rptr. 865 (1979).

Lee, C. L. (1982). Open discussion of logic, assumptions, calculations and methods of expressing results. *Proceedings of the American Blood Banks International Conference of Inclusion Probabilities in Parentage Testing,* 133–147.

Nijenhuis, L. E. (1982). A critical evaluation of various methods of approaching probability of paternity. *Proceedings of the American Blood Banks International Conference of Inclusion Probabilities in Parentage Testing,* 103–112.

Walker, R. H. (1982). Analyses of parentage test case. *Proceedings of the American Blood Banks International Conference of Inclusion Probabilities in Parentage Testing,* 443–488.

The Probability of Reversal in Contested Elections

Dennis C. Gilliland
Department of Statistics and Probability
Michigan State University
East Lansing, Michigan

Paul Meier
Department of Statistics
The University of Chicago
Chicago, Illinois

ABSTRACT

In general, a contested election will not be declared void on the basis of a mere possibility that the result would be reversed in the absence of the irregularities. Courts have sometimes based their decisions on their perceptions of the magnitude of a probability that the result would be reversed in the absence of the irregularities. Sometimes the evidence consists of proof that a certain number k of illegal votes were cast with no evidence of for whom the votes were cast. The (hypergeometric) probability of reversal under random removal of k votes from the total cast has been proposed as the probability on which to base the decision in some such instances. In this chapter we examine implications of this approach and investigate the use of mixture models as alternatives.

1. Introduction

It is not uncommon for election results to include a few illegal votes because of errors and carelessness on the part of the election workers and voters. These and other irregularities will often go unnoticed unless the election results are so close as to cause a candidate to scrutinize carefully the poll books and totals. When a candidate decides that the irregularities were such as to leave the results in doubt, a challenge may be raised that can lead ultimately to a contested election and a court hearing.

The mayoral election in Ann Arbor, Michigan, on 4 April 1977 is a case in point. Albert Wheeler was declared the winner over Louis Belcher on the basis of a one-vote margin, 10,660 to 10,659. Legal action resulted in a case before Visiting Circuit Court Judge James J. Kelly, Jr., in Circuit Court of Washtenaw County, where testimony showed that at least 20 persons were allowed to vote who should not have been allowed. In most of these instances, the person lived outside the boundaries of the city (which are very irregular) and had mistakenly registered to vote in the city. There was testimony that two voters were concurrently registered in different localities and that one vote had been cast in the name of another person. After considerable legal wrangling, plaintiff's attorneys succeeded in calling a few of these voters to the stand and questioning them about how their votes were cast. Some refused to answer and were held in contempt of court. [Their right to refuse was later upheld by the Michigan Supreme Court (see *Belcher v. Mayor of Ann Arbor,* 1978).] In any event, this court's remedy for resolving the uncertainty in the illegal votes was not a cure and, of course, could not be attempted in the usual case of an illegal vote with the voter not identified. The court ultimately voided the election and as it turned out, plaintiff Louis Belcher won the rerun election 14,404 to 14,122 on 3 April 1978.

The decision maker must make a choice, often in a timely fashion, among the options: letting the election results stand, reversing the election, and voiding the election. With some jurisdictions and types of elections, the option to void is not available; the option to reverse is essentially ruled out unless the contestant can prove an altered outcome [see *Harvard Law Review* (*HLR*) 1975, p. 1323, footnote 127]. When the contestant demonstrates only a probability that the results would be reversed but for the irregularities, the viable options are to let the election stand or to void the election. According to *HLR* (1975, pp. 1323–1324),

> The shortcomings of the avoidance remedy, together with the generally applicable reasons for reluctance to provide postelection relief, have led tribunals to base their decisions whether to avoid an election largely upon their perceptions of the probability that the outcome was altered. If that probability is low, relief is held not to be warranted; only if the probability of altered outcome is high will tribunals intervene and provide relief.

In this paper we expand on Gilliland and Meier (1983) and critique the approach proposed by Finkelstein and Robbins (1973) for modeling the uncertainty resulting from illegal votes where there is no evidence concerning the disposition of these votes. Specifically, they arrive at a "mathematical" probability of reversal (altered outcome) using an urn model. Consider an election involving two candidates A and B. Let a and

b denote the numbers of votes cast for A and B, with $a > b$, and suppose that k illegal votes are included in the total $s = a + b$. Based on knowledge of a, b, and $k \geq a - b$ and, with "no evidence" to indicate for whom the illegal votes were cast, should the election of A be declared void? Finkelstein and Robbins (1973, p. 242) propose the following approach to resolving the question:

> Consider all the votes cast in a primary election as balls placed in an urn: black balls which predominate are those votes cast for the winner; white balls are for the loser. A certain number of balls representing the irregular voters are then withdrawn at random from the urn, an operation which corresponds to their invalidation. What is the probability that, after the withdrawal, the number of black balls no longer exceeds the number of white? Note the key assumption that the balls are withdrawn at random, i.e., that each ball has the same probability of being withdrawn. In terms of the real election situation, each voter is deemed to have the same probability of casting an invalid vote. This assumption will of course be untenable if evidence of fraud or patterns of irregular voting indicates that a disproportionate number of improper votes were cast for one candidate. But in the absence of such evidence, the assumption of random distribution of the improper votes is warranted. Whether or not mathematics is used to assess the probabilities, some implicit or explicit view as to the pattern of irregular voting seems inevitable. The assumption that each voter had an equal probability of casting an improper vote is the only neutral and non-arbitrary view that can be taken when there is no evidence to indicate that the probabilities are not equal. Thus in *Ippolito*, *DeMartini*, and other cases, where there was no evidence to disturb the assumption of randomness, the mathematical probability analysis depicted by the urn model is a correct expression of the intuitive probability used by the Court of Appeals in formulating the burden of proof standard for a new election.

Here the unknown number x of illegal votes for A is represented by the number of black balls in the random sample of k taken from the urn. The withdrawal of the k illegal votes reverses or ties the election if and only if $x - (k - x) \geq a - b$ [i.e., $x \geq \frac{1}{2}(k + a - b)$].

Any mathematical model M in which x is represented and given a probability distribution results in a "mathematical" probability of reversal

$$PR_{\mathrm{M}} \stackrel{\mathrm{Def}}{\equiv} P\left[x \geq \frac{1}{2}(k + a - b)\right]. \tag{1}$$

With the Finkelstein and Robbins urn model, x is given a hypergeometric distribution, and we write PR_{U} for expression (1) under their model.

HLR (1975, pp. 1324–1325) is not critical of this approach. Rather, one finds:

> Tribunals have generally reached judgments about the likelihood of altered outcome in uncertain cases on an intuitive basis, by comparing the number of votes affected or the nature of the wrongs involved with the margin of victory. In certain kinds of cases, however, greater precision may be possible. Finkelstein and Robbins have developed a mathematical formula that in some cases would enable tribunals to determine the probability that the number of illegal votes accepted altered the results of an election. By assuming that "each voter [has] the same probability of casting an invalid vote," and by calculating either for all election districts combined or, preferably, for each separate precinct the number of illegal votes cast in the particular race being contested, along with the number of votes each candidate received, probability theory can be used to establish the likelihood that enough of the illegal votes were cast for the winner to alter the result.

In this chapter, we argue against the Finkelstein and Robbins urn-model approach by showing that it induces an unrealistic probabilistic behavior for x. The implicit assumption that all subsets of size k are equally likely to be the set of illegal votes is a strong assumption that does not appear to be justified.

We contrast the urn model with mixture models where a probability distribution F is placed on the elements of a set of models. The set consists of binomial models, including the symmetric binomial model, which gives a probability of reversal PR_{U} very close to that of the urn model's hypergeometric. The probability of reversal PR_F based on the mixture model is simply the F average of the values PR_{M} coming from the models M in the set. The set and F may be chosen so that the mixture model is symmetric with respect to the two candidates, in which case we show that if $k > 1$ and F is nondegenerate, then $PR_F > PR_{\mathrm{U}}$.

Cases used as examples by Finkelstein and Robbins (1973) and Ward (1981) show that tribunals often reach the decision to void when such action is unwarranted on the basis of the urn model. In these cases, it appears that the intuitive assessments of the likelihood of altered outcome exceed the mathematical probabilities of reversal based on the urn model.

We illustrate our ideas with cases that are discussed by Finkelstein and Robbins (1973) and Ward (1981) and with the contested 1977 mayoral election in Ann Arbor, Michigan. In addition, we comment on the contested election discussed by Downs et al. (1978).

In this application, there is a sparcity of data available with which to test and compare various models and against which to compare particular choices of the mixing distribution F in the mixture models. We consider our analysis more as rebuttal to the choice of the urn model rather than as an endorsement for any particular alternative approach. We hope that this investigation will stimulate thought on the question of the appropriateness of models for the purpose of quantifying the probability of an altered outcome in a contested election.

2. The Urn Model

The development that follows and culminates in expression (6) is a brief summary of results from Finkelstein and Robbins (1973) that are relevant to our discussion. Recall that $a > b$, $a + b = s$, that x denotes the number of illegal votes in the total a recorded for candidate A. With the urn model, x has a hypergeometric distribution with probabilities

$$h(x) = \frac{\binom{a}{x}\binom{b}{k-x}}{\binom{s}{k}}, \qquad x = 0, 1, \ldots, k, \tag{2}$$

mean

$$E(x) = ka/s, \tag{3}$$

and variance

$$V(x) = \frac{kab(s-k)}{s^2(s-1)}. \tag{4}$$

In some instances, the binomials $B(k, a/s)$ and $B(k, \frac{1}{2})$ are good approximations to the hypergeometric. For the symmetric binomial,

$$b_k(x) = \binom{k}{x}\left(\frac{1}{2}\right)^k, \qquad x = 0, 1, \ldots, k, \tag{2$'$}$$

$$E_k(x) = \frac{1}{2}k, \tag{3$'$}$$

and

$$V_k(x) = \frac{1}{4}k. \tag{4$'$}$$

For simplicity, we sometimes treat the probability of reversal based on the urn model as the binomial tail probability

$$B_{k;a-b} \overset{\text{Def}}{\equiv} \sum_{x \geq (\frac{1}{2})(k+a-b)}^{k} b_k(x) \tag{5}$$

since the symmetric binomial fits the hypergeometric quite well in the typical application to a disputed election.

Using (3) and (4) to obtain z, the standardized version of x, results in

$$PR_U \doteq P[z \geq z^*] \qquad (6)$$

where $z^* = (a - b)\sqrt{(s - k)/sk}$. Using (3′) and (4′) results in $z^* = (a - b)/\sqrt{k}$, which is a good approximation when k is small compared to s.

Paradigm cases discussed in Finkelstein and Robbins (1973) are *Ippolito v. Power* (1968), *DeMartini v. Power* (1970), and *Santucci v. Power* (1969). Ward (1981) discusses two other contested elections. These data together with that related to the contested 1977 mayoral election in Ann Arbor, Michigan, are found in Table 1.

The hypergeometric probability of reversal is well-approximated by the binomial probability of reversal for these cases as can be seen by reference to Table 2.

According to Finkelstein and Robbins (1973), in the *Ippolito* and *Santucci* cases, courts' decisions to void the elections were affirmed by the Court of Appeals, New York's highest court. In the *DeMartini* case, the Court of Appeals reversed the lower courts' decisions to void the election. According to Ward (1981), the Town Council election was upheld, whereas the Maine Commission on Government Ethics and Election Practices voted 2 : 1 to void the contested Maine House election. In Ann Arbor, the trial court voided the election. In *Santucci* and Maine House, the final decisions were totally inconsistent with the probabilities of reversal based on the urn model.

To appreciate better the implications of the urn model, consider a decision rule based on it that calls for voiding an election if and only if $PR_U > 0.05$. (The critical value 0.05 is used for illustrative purposes only.) For ease of calculation, we use the symmetrical binomial to calculate PR_U. The region in the $(a - b, k)$ plane where $PR_U > 0.05$ is displayed in Figure 1, together with the data points determined by the examples in Table 1.

TABLE 1. Examples

Election	a	b	s	k	$a - b$	$\frac{1}{2}(k + a - b)$	z^*
Ann Arbor	10,660	10,659	21,319[a]	20	1	10.5	.22
Ippolito	1,422	1,405	2,827	101	17	59	1.69
DeMartini	2,656	2,594	5,250	136	62	99	5.25
Santucci	58,076	57,981	116,057	448[a]	95	271.5	4.48
Town Council	2,390	2,383	4,773	16	7	11.5	1.75
Maine House	1,193	1,060	2,253	208	133	170.5	8.79

[a] The number of illegal votes 448 in Santucci was taken from footnote 17 of Finkelstein and Robbins (1973). In Ann Arbor, the votes cast for a third-party candidate and the write-ins, 366 votes in total, are not shown.

TABLE 2. Probabilities of Reversal

Election	Hypergeometric $H(a, b, k)$	Binomial $B(k, a/s)$	Binomial $B(k, \frac{1}{2})$
Ann Arbor	0.4119	0.4120	0.4119
Ippolito	0.0592	0.0626	0.0555
DeMartini	7.5×10^{-8}	1.1×10^{-7}	5.1×10^{-8}
Santucci	3.5×10^{-6}	3.6×10^{-6}	3.3×10^{-6}
Town Council	0.0387	0.0389	0.0384
Maine House	2.3×10^{-20}	8.6×10^{-19}	4.4×10^{-22}

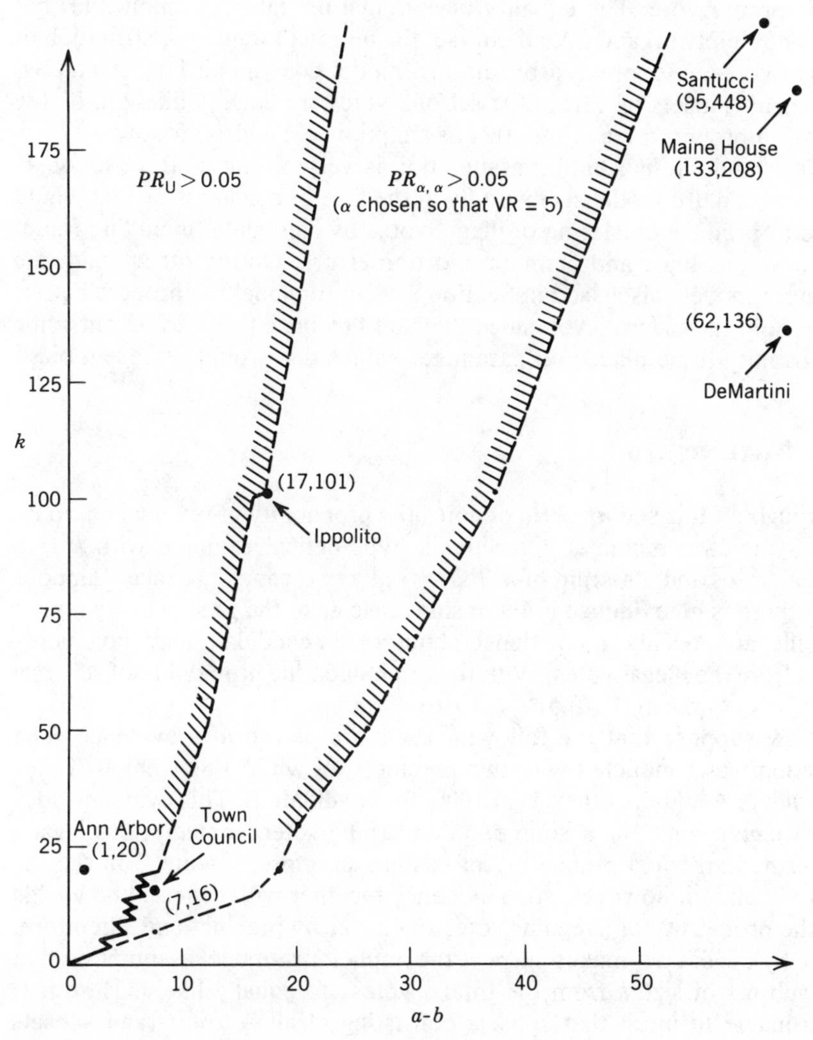

Figure 1. Decision boundary for PR > .05.

For large k, normal approximation to (5) and the use of the continuity correction shows that $PR_U > 0.05$ corresponds to $k > 0.37\ (a - b - 1)^2$. For example, if the margin is $a - b = 25$, then the number of illegal votes k must be at least 213 to cause PR_U to exceed 0.05.

Finkelstein and Robbins (1973, p. 242) state, "The assumption that each voter had an equal probability of casting an improper vote is the only neutral and non-arbitrary view that can be taken when there is no evidence to indicate that the probabilities are not equal." This explicitly stated assumption speaks to the marginal likelihoods of each voter casting an improper vote. The equality of marginal probabilities implies (3) but does not imply (2) and (4). Of course, the notion of random distribution of illegal votes as expressed by the urn model goes much farther and assumes all subsets of k from the set of s votes are equally likely to be the set of improper votes. From this assumption, (2) and (4) follow.

This random distribution assumption is very strong; and, as a consequence, the urn model seems to fly in the face of mechanisms that would tend to produce clustering of illegal votes by candidate, including fraud, unconscious bias, and common election errors. On the other hand, the mixture models discussed in Section 3 seem to model the process better, at least at the macro level, since they are flexible enough to permit some probability to be placed on parameter values that would represent bias.

3. Mixture Models

Throughout this section PR_U denotes the probability of reversal based on the symmetric binomial. Consider a hypothetical example with $k = 6$ illegal votes and a margin of $a - b = 5$ in a two-candidate race. Suppose that there is no evidence in the instant case as to the precinct origin(s) of the illegal votes and no evidence about which candidate may have benefited from the illegal votes. With the urn model, the probability of reversal is $PR_U = P(x = 6) = (0.5)^6 = 0.016$.

Now suppose that the following facts are placed into evidence. The election was conducted with two precincts, of which one went 100% for candidate A and the other went 100% for candidate B. This evidence does not weigh against the assumption concerning voters, namely, that "each voter had an equal probability of casting an improper vote." In the authors' opinion, however, such evidence together with general knowledge of the propensity of irregular votes to cluster by precinct and, therefore, here by candidate, makes suspect the implicit urn-model assumption that all subsets of size k from the total s votes are equally likely. Here it is reasonable to intuit that subsets consisting of all A votes (and subsets consisting of all B votes) are more probable than as determined from the urn model, so that the intuitive likelihood of altered outcome exceeds the urn model's $PR_U = 0.016$.

The urn-model approach does extend to suggest a flexible class of models to use to quantify the probability of reversal. For example, in the hypothetical example given above, the decision maker might assess the likelihood that all six illegal votes came from one precinct as 0.4 (split equally between the two precincts) and place probability 0.6 on the urn model. (This construct can be thought of as three urns with the three different compositions and likelihoods associated with the sample being drawn from each.) The probability of reversal is now the mixture $P(x = 6) = 0.2(1)^6 + 0.6(0.5)^6 + 0.2(0)^6 = 0.209$.

Later we shall discuss more of the rationale for mixture models. In the meantime, we point out certain mathematical properties to demonstrate that the simple urn model is at an extreme position within the class of all symmetric mixture models.

Recall that the urn model for the application at hand essentially induces the binomial distribution $B(k, \frac{1}{2})$ on x. The binomial $B(k, p)$ with $p \neq \frac{1}{2}$ corresponds to a distribution reflecting a greater tendency for illegal votes to have been cast for one candidate or the other. [In HLR (1975, p. 1325, footnote 138) the idea of using such a model is mentioned.] By mixing the distributions $B(k, p)$ with a probability mixing distribution F on p and considering all possible F, a class of models is formed that includes the urn model through specification of F degenerate on $p = \frac{1}{2}$. Furthermore, F that are symmetric in p and $1 - p$ are neutral in regard to the candidates in that the resulting mixture of binomials is symmetric in x and $k - x$.

With a probability distribution F on $0 \leq p \leq 1$, the mixture distribution on x is

$$p_F(x) = \int_0^1 \binom{k}{x} p^x(1 - p)^{k-x}\, dF(p), \qquad x = 0, 1, \ldots, k. \tag{7}$$

The probability of reversal with the mixture model is the tail probability

$$PR_F \overset{\text{Def}}{\equiv} P_{F;a-b} \overset{\text{Def}}{\equiv} \sum_{x \geq (1/2)(k+a-b)}^{k} p_F(x). \tag{8}$$

Some special properties obtain when F is symmetric. From the fact that $E(x) = E(E(x \mid p)) = E(kp) = kE(p)$, it follows that for symmetric F,

$$E_F(x) = \frac{1}{2}k. \tag{9}$$

From the fact that $V(x) = E(V(x \mid p)) + V(E(x \mid p)) = kE(p(1 - p)) + k^2V(p)$, it follows that, for symmetric F,

$$V_F(x) = \frac{1}{4}k + k(k - 1)V(p). \tag{10}$$

Provided $k > 1$ and $V(p) > 0$ (i.e., F is nondegenerate), we see that the variance (10) induced on x by the symmetric mixture model exceeds that (4′) induced by the urn model. Finally, we note that, for symmetric F, (7) and (8) imply

$$PR_F = \frac{1}{2}\int_0^1 \sum_{x \geq (1/2)(k+a-b)}^{k} \binom{k}{x}\{p^x(1-p)^{k-x} + p^{k-x}(1-p)^x\}\, dF(p). \tag{11}$$

By representing the binomial tail probabilities with incomplete beta functions (see Feller, 1957, Chapter VI, (10.9)), it is easily shown that the integrand is minimized at $p = \frac{1}{2}$ (uniquely so unless $k = 1$). Hence, if $k > 1$, then

$$PR_F \geq PR_U \tag{12}$$

for all symmetric F with equality if and only if F is degenerate at $p = \frac{1}{2}$. Henceforth we exclude the case $k = 1$ from consideration.

4. Beta Mixtures

A tractable class of mixing distributions, the conjugate family to the binomials $B(k, p)$, is the family of beta distributions (see Ferguson, 1967, Chapter 3). This is a two-parameter family with parameters denoted α and β. The mixture distribution on x that results is called the beta binomial and is given by

$$p_{k;\alpha,\beta}(x) = \frac{\dfrac{\Gamma(\alpha + x)}{\Gamma(\alpha)x!}\,\dfrac{\Gamma(\beta + k - x)}{\Gamma(\beta)(k - x)!}}{\dfrac{\Gamma(k + \alpha + \beta)}{\Gamma(\alpha + \beta)k!}}, \qquad x = 0, 1, \ldots, k, \tag{13}$$

which is a form convenient for calculation by recursion. The mean and variance of this distribution are

$$E_{k;\alpha,\beta}(x) = \frac{\alpha k}{\alpha + \beta} \tag{14}$$

and

$$V_{k;\alpha,\beta}(x) = \frac{\alpha\beta k(k + \alpha + \beta)}{(\alpha + \beta)^2(\alpha + \beta + 1)}. \tag{15}$$

For the beta binomial distribution on x, the probability of reversal is the tail probability

$$PR_{\alpha,\beta} \stackrel{\text{Def}}{\equiv} P_{k;\alpha,\beta;a-b} \stackrel{\text{Def}}{\equiv} \sum_{x \geq (1/2)(k+a-b)}^{k} p_{k;\alpha,\beta}(x). \tag{16}$$

Consider a beta binomial with the same mean (3) that x has under the urn model. The variance ratio VR of this distribution to that of the hypergeometric (binomial) with this mean is

$$VR = \frac{k + \alpha + \beta}{1 + \alpha + \beta} \cdot f \tag{17}$$

where $f \equiv (s - 1)/(s - k)$ (and f is taken to be 1 for the binomial). For a specified VR with $f < VR < kf$, the α, β values that result in mean (3) and the given VR are

$$\alpha = \frac{a}{s} \frac{(kf - VR)}{(VR - f)}, \qquad \beta = \frac{b}{s} \frac{(kf - VR)}{(VR - f)}. \tag{18}$$

Table 3 gives the probability of reversal based on the hypergeometric and for the beta binomial with α, β parameters chosen to have the same mean as the hypergeometric and with variance VR times that of the hypergeometric, $VR = 5, 10, 20$.

When $\alpha = \beta$, the beta mixing distribution is symmetric in p and $1 - p$, so that the beta binomial $P_{k;\alpha,\alpha}$ is symmetric about its mean

$$E_{k;\alpha,\alpha}(x) = \frac{1}{2} k \tag{19}$$

with variance

$$V_{k;\alpha,\alpha}(x) = \frac{1}{4} k \frac{k + 2\alpha}{1 + 2\alpha}. \tag{20}$$

Note that by (17), the variance ratio of the symmetric beta binomial $P_{k;\alpha,\alpha}$ to the binomial $B(k, \frac{1}{2})$ is

$$VR = \frac{k + 2\alpha}{1 + 2\alpha}. \tag{21}$$

From (21) it follows that the choice $\alpha = (k - 5)/8 > 0$ results in $VR = 5$. For $k > 5$ and this choice of α, the largest $a - b$ that results in $PR_{\alpha,\alpha} >$

TABLE 3. **Probabilities of Reversal**

Election	Hypergeometric $H(a, b, k)$	Beta Binomial $VR = 5$	$VR = 10$	$VR = 20$
Ann Arbor	0.4119	0.4669	0.4833	0.4994
Ippolito	0.0592	0.2483	0.3222	0.3828
DeMartini	7.5×10^{-8}	0.0091	0.0516	0.1391
Santucci	3.5×10^{-6}	0.0223	0.0793	0.1635
Town Council	0.0387	0.2613	0.3938	0.5000[a]
Maine House	2.3×10^{-20}	6.0×10^{-6}	0.0012	0.0189

[a] In Town Council, $VR = 20$ cannot be realized with the beta binomial.

0.05 was calculated. The resulting decision boundary is plotted in Figure 1 along with the decision boundary based on the $B(k, \frac{1}{2})$.

Limiting and special cases of $P_{k;\alpha,\alpha}$ are worth noting. As $\alpha \to 0$, $P_{k;\alpha,\alpha}$ approaches the distribution that puts probability $\frac{1}{2}$ on each point $x = 0$ and $x = k$. This limit is an extreme position where, in the application, all illegal votes are from A (with probability $\frac{1}{2}$) or from B (with probability $\frac{1}{2}$). As $\alpha \to \infty$, $P_{k;\alpha,\alpha}$ approaches the binomial $B(k, \frac{1}{2})$ distribution induced by the urn model.

Rules of thumb such as "$k \geq m(a - b)$ before voiding an election" have been used in some cases according to Finkelstein and Robbins (1973, footnote 20). These rules are not inconsistent with probabilities of reversal based on a mixture model. For example, with $\alpha = \beta = 1$, x is uniformly distributed on $x = 0, 1, \ldots, k$, so that $PR_{1,1} = [k + 1 - \frac{1}{2}(k + a - b)]/(k + 1)$ where the brackets denote the greatest integer function. Here $PR_{1,1} > 0.25$ if and only if $[\frac{1}{2}k - \frac{1}{2}(a - b) + 1] > 0.25(k + 1)$, which for large k corresponds to $k > 2(a - b)$.

We now examine the conditional distribution of the random variable x given certain statistical information y. We demonstrate very different behavior depending on whether the distribution of x is $B(k, \frac{1}{2})$ or is a mixture $P_{k;\alpha,\alpha}$. This demonstration is intended to illustrate the implicit strength of randomness as expressed by the urn model.

Consider a sample space Ω consisting of all 2^k possible sequences of A's and B's of length k, and define the random variable on ω by $x = x(\omega)$, the number of A's in the sample point ω. One may think of ω as the ordered outcome revealing the candidate benefiting from each illegal vote. The urn model essentially assigns uniform probability 2^{-k} to each $\omega \in \Omega$ from which x has the binomial distribution (2′) and the probability of reversal is given by (5). With the symmetric beta binomial mixture model, the distribution is given by (13) with $\alpha = \beta$, and the probability of reversal is given by (16) with $\alpha = \beta$.

Now we consider the thought experiment consisting of generating an ω according to a distribution and then determining y, the number of A's in n of the k components of ω (chosen at random without replacement). Thus, y and $n - y$ are determinations of the numbers of illegal votes for A and B among n votes chosen at random from the pool of k illegal votes. It is straightforward to verify that, in the urn model, the conditional distribution of $x \mid y$ is

$$y + B(k - n, \tfrac{1}{2}); \tag{22}$$

and, in the mixture model, the conditional distribution of $x \mid y$ is

$$y + P_{k-n;\alpha+y,\alpha+n-y}, \tag{23}$$

a translation of a beta binomial distribution. Using the notation from (5) and (16), the conditional probabilities of reversal given y are, respectively, the tail probabilities

$$PR_{\mathrm{U}|y} \overset{\mathrm{Def}}{\equiv} B_{k-n;a-b+n-2y} \tag{24}$$

and

$$PR_{\alpha,\alpha|y} \overset{\mathrm{Def}}{\equiv} P_{k-n;\alpha+y,\alpha+n-y;a-b+n-2y} \, . \tag{25}$$

The contrast between (24) and (25) is striking. In the former, determination of the status of n votes randomly selected from the pool of k illegal votes leaves the symmetric $B(k-n, \frac{1}{2})$ as a measure of the uncertainty of the number of votes for A in the balance. In the latter, $P_{k-n;\alpha+y,\alpha+n-y}$ puts more probability on outcomes consistent with the determination y (e.g., if $y > \frac{1}{2}n$, then $\alpha + y > \alpha + n - y$, so that $P_{k-n;\alpha+y,\alpha+n-y}$ puts more probability on outcomes where A benefits more than B from the $k-n$ illegal votes).

As a numerical example, consider *DeMartini,* where $k = 136$ and $a - b = 62$. Under the urn model, $PR_{\mathrm{U}} \doteq B_{136;62} < 10^{-7}$, and, with $\alpha = 16.375$ so that $P_{136;\alpha,\alpha}$ has variance five times that of the binomial $B(136, \frac{1}{2})$, $PR_{\alpha,\alpha} = 0.008$. Suppose $n = 36$ votes are selected at random from the $k = 136$ and it is determined that $y = 36$ (i.e., that all 36 were cast for A). Under the urn model, and (24), the conditional probability of reversal is $PR_{\mathrm{U}|36} = 0.006$; the statistical information that 36 randomly selected illegal votes were all for candidate A leaves the probability of reversal at a very small level. However, under the mixture model with $\alpha = \beta = 16.375$, $n = y = 36$, (25) and a calculation show that the conditional probability of reversal is the substantial $PR_{\alpha,\alpha|36} = 0.974$.

Good (1950, pp. 35, 70; 1965, pp. 19–20) refers to the technique of using a thought experiment to help assess probabilities as "the device of imaginary results." Hill has used a similar method in many of his articles. Hill (1974, p. 533) writes, "It is particularly useful in this connection to consider the form inference takes when the data is extreme in some fashion or the other. If a model does not respond, in a reasonable way to such extreme data, then it should be modified accordingly." Recently, Diaconis and Freedman (1983) discuss the application of fictitious samples that they regard as a useful way of thinking about priors.

The urn model considers all subsets of k votes from the total cast s to be equally likely to be the set of illegal votes. Finkelstein and Robbins (1973, p. 242) attempt to justify this on the basis of absence of information to the contrary. When there are symmetric and balancing arguments for alternatives, the principle of indifference suggests that the alternatives have equal probabilities; lack of evidence does not justify such a specification (see Black, 1967, p. 474).

Carnap (1945) criticizes the unrestricted use of the principle of indifference on the same grounds illustrated by our thought experiment (i.e., that resulting conditional behavior conflicts with a presystematic concept of probability) (see also Foster and Martin, 1966, p. 25). Carnap proposes an assignment of probability which is uniform on "structures" and then uniform across "states" within structures. In our thought experiment, this is the measure induced on state sequences ω by the $\alpha = \beta = 1$ mixture model.

In the context of his compound decision problem, Robbins (1951) proposed the use of the uniform prior, $\alpha = \beta = 1$, for reasons including a (statistical) consistency property. In the compound application, the statistical information consists of measurements taken on each component with a distribution depending on whether the component is A or B. In our earlier thought experiment the statistical information consists of complete information on a subset of n of the k components chosen at random.

5. Discussion

Are mathematical models relevant or useful in quantifying the probability of reversal (altered outcome) in a contested election where there are known to be k illegal votes, k exceeds the recorded margin $a - b$, and there is "no evidence" to indicate for whom the illegal votes were cast?

There is little question but that the decision problem created by such a contested election forces the decision maker to assess the likelihood of altered outcome, although such an assessment may be intuitive and may not be expressed explicitly. Therefore, it seems natural to advance mathematical models to help meet the need to assess likelihood. Unfortunately, by its very nature a no-evidence contested election does not generate an x value that can be measured. Therefore, no reference set of empirical experiences has been generated against which to judge the appropriateness of various models for the uncertainty in x in a no-evidence contest.

There is, however, some empirical evidence about the values of x for contested elections of the "evidence" type. If such evidence shows that x, the number of illegal votes for the winning candidate, is much more variable than predicted by the urn model, then there is reason to discount PR_U or to judge it to be an underestimate in a no-evidence contest.

The Ann Arbor election provides some data with which to illustrate the type of analysis that can be accomplished to provide information about the behavior of x. The 20 illegal votes appeared in 12 of the 76 voting subdistricts. Random placement of 20 balls into 76 equally likely cells results in 12 or fewer being occupied with probability 1.6×10^{-4}. Since not all subdistricts had the same number of votes, we conducted a Monte Carlo study consisting of 250 repetitions of the experiment of selecting 20 votes at random from the total 21,319 as actually divided among the 76 subdistricts in Ann Arbor. The numbers of occupied cells averaged 17.3 with a standard deviation of 1.3. In no repetition was the number of occupied cells less than 14. Thus, the $k = 20$ illegal votes clustered within subdistricts (precinct) in Ann Arbor to a much greater extent than what is likely from a purely random placement of k votes among the total cast. Furthermore, the proportions P_A cast for candidate A within the 76 subdistricts had a bimodal distribution with mean 0.539 and standard deviation 0.215. An analysis of residuals based on predicting votes for candidate A on a random partitioning of the aggregate into subdistricts rejects the hypothesis that votes for A were randomly distributed among subdistricts. This is also typical of election results. The net effect of the clustering of illegal votes by subdistrict and the clustering of votes for candidate A by subdistrict produces a clustering of illegal votes by candidate even if there is random distribution of illegal votes within subdistrict. The result does not change the conclusion that would be reached by the urn-model approach to this contested election, but it does illustrate why there is a greater variability in x than induced by the urn model applied to the aggregated votes. Models for clustering, such as the mixture model, offer some promise for modeling this behavior of x, at least at the macro level.

Finkelstein and Robbins (1973, footnote 9) suggest that the urn model may be applied within voting subdistricts when the illegal votes are identified on this basis. The resulting distribution on x is that of the sum of hypergeometrically (approximately binomially) distributed summand variables. The resulting distribution on x, however, may be considerably different from $B(k, \frac{1}{2})$. Recognition of this fact may contribute to the decision to discount the urn model's applicability in the aggregated form. Moreover, the hypergeometric distribution may even understate the variability in the distribution of illegal votes within a subdistrict because of clustering by candidate.

A mathematical model advanced to quantify the uncertainty in x for a particular no-evidence contested election will be suspect unless there is empirical evidence to support its reasonableness. The action of tribunals in *Santucci, DeMartini,* and *Maine House,* with the exception of the Court of Appeals in *DeMartini,* suggest that intuitive assessments of the probability of altered outcome greatly exceeded PR_U in each case. According to Ward (1981), the chairman of the Maine Commission on Governmental Ethics and Election Practices, which ruled on the *Maine House* contest after hearing arguments including testimony from a professor of mathematics on the urn model and PR_U, said:

> Acceptance of a purely mathematical approach for predicting voter activity ignores myriad human variables generally perceived to be an inherent and important part of politics and generally perceived as working in no scientifically calculable manner.

Social costs may help partially explain what appears to be the reluctance of tribunals to let contested elections stand in the face of extremely small values of PR_U. Costs may be deemed to be very large for the error of letting an election stand when in fact the illegal votes caused an altered outcome or, for that matter, when the number of illegal votes is large regardless of the (unknown) distribution. According to Ward (1981), another Maine Commission member said:

> Our over-riding concern is reflected in the statement in the legislation which created the Commission: 'It is essential under the American system of representative government that the people have faith and confidence in the integrity of the elective process . . .' . . . The voters could continue to believe that the election outcome might have been different. 'Common sense' indicates that 208 (invalid) votes could affect the outcome of a contest involving a margin of only 133 votes. The lay person does greatly overestimate the probabilities of reversal. As long as this occurs, however, erroneously, faith and confidence in the accuracy of the outcome is undermined. . . . Therefore the decision comes down to a matter of balancing competing weighty considerations. On the one hand, there (is) the extreme improbability that the election would have been reversed. . . . On the other hand, there are more subjective attempts somehow to restore faith and confidence in the electoral process . . .

This member apparently was convinced that $k = 208$ and $a - b = 133$ suggests an "extreme improbability" of altered outcome and may have given some weight to the value PR_U in reaching this conviction. Here the urn model's hypergeometric probability of reversal is less than 10^{-19}.

The urn and mixture models we have discussed specify the number k of illegal votes. When there is proof of the existence of k illegal votes, then k is a lower bound for the actual number of illegal votes. Straf (1983) raises this issue and suggests generalizing the models by regarding k as random.

We have not discussed or suggested a threshold value for PR that, if exceeded, should lead to voiding the contested election. Such a discussion would be useful, but would necessarily involve social policy issues.

Straf (1983) makes an interesting point in regard to threshold value. Assume that there are no ties. If a new election is considered a toss-up and if the new election being lost by the rightful winner of the contested election is regarded as an error, then the action of voiding the contested election results in an error with probability 0.5. Hence, if this error is as serious as failing to void the first election when B is the rightful winner, the threshold value for PR should be near 0.5. By following this in a strict sense, the decision maker will void the contested election only if $PR \geq 0.5$, which is never consistent with the fact that candidate B is behind candidate A in the contested election (i.e., $b < a$).

Since decision makers void contested elections where intuitive and "mathematical" probabilities of reversal are much lower than 0.5, in fact, near 0, the pervading philosophy is that the error created by a second election is not so serious. Renowned practicing attorney Tom Downs describes this philosophy as "election cures errors." (Justice T. M. Kavanaugh used this phrase in a dissenting opinion to describe the doctrine being created by the majority opinion in *Carman v. Secretary of State* (1971).) Straf's comments will help remind the decision makers that "election creates errors" as well.

Some might think that proportional reduction is the appropriate remedy for a contested election of the type we have been discussing. Here the k illegal votes are removed with those taken from the two candidates maintained in the proportions $a:b$. Here the election results stand since a margin of victory is preserved for candidate A. Proportional reduction provides a point estimate for the results after the removal of k illegal votes whereas the methods discussed in this chapter concern the probability distribution of what the results would be. When one of the available options is to void the contested election and to run a new election, it is reasonable for the decision maker to take into account the magnitude of the likelihood of reversal.

There is election law in Michigan that concerns elections, the use of random removal, and the notion of proportional reduction. Consider the analysis of an election subject to an irregularity that was made by Justice Cooley in Michigan over 100 years ago in the case of *People v. Cicott* (1868). The facts were that some paper ballots were in the ballot boxes that should have been withdrawn by election officials on election night but were not. Michigan law called for the random removal of (paper) ballots from the ballot box so as to equalize the number remaining and the number of voters who participated in the particular election. [Current Michigan law calls for the same remedy (see *Michigan Compiled Law,* 1984).] Justice Cooley found that the election should not be declared void, and what he had to say is pertinent.

> Shall the election be declared void for that reason? This would be simply to disfranchise the electors because in some unexplainable manner a fraud, accident or mistake has occurred, for which they are in no way responsible, and which may not at all affect the result.

> The statute says they shall not be thus unjustly disfranchised, but that the votes shall be disposed of by a rule which will operate as equally and justly between the parties as is possible under the circumstances. That rule is based upon the doctrine of probabilities. It directs the drawing from the box of a number equal to the surplus found there, and the equalization of the count of the box and the lists in that mode.
>
> Now, it is apparent that in this drawing the chances are very much greater that ballots duly cast will be drawn and destroyed than that the ballots wrongly in the box will be drawn, for the plain reason that the legal votes will always greatly exceed the surplus in number, unless there has been gross fraud on the part of the inspectors. If the surplus is one in a vote of a hundred, the probabilities are ninety-nine to one that the vote drawn will be one of those which was regularly and legally cast, and that thereby a legal voter will be deprived of his suffrage. Nevertheless, as each ballot is usually one of a number designed to be allowed to particular candidates and counted against others given to other candidates, the drawing may still work no injustice, since each candidate will probably lose by it a number proportioned to the relative number of ballots appearing for him in the box, and thus the relative proportions will be preserved. The actual drawing might-differ from the antecedent probability; but this is the theory which the statute proceeds upon, and the variance can seldom be great.
>
> Now, the relative proportion of votes cast for the two candidates in the two wards in question was such that, had the drawing taken place, Mr. Williams would probably have lost by it one more vote than Mr. Cicott, and no more. The probability is, therefore, that the election, as between these two candidates, has not been affected by the irregular action of the inspectors beyond a single vote.

We see that the Michigan law makes the urn model operational. Justice Cooley remedied a failure to carry out the law by letting the result stand by reasoning that the "expected" result under such action is proportional reduction, which preserves the result of the election.

This chapter has proposed mixture models for modeling the uncertainty in election results that contain a certain number of illegal votes. Mixture models are not limited to this role. Consider the election discussed by Downs et al. (1978). In this case, each of two precincts had a punch card voting device that reversed the votes cast on it. Punch cards from voting devices were comingled within precincts and the evidence consisted of the fact that one out of five voting devices reversed votes in one precinct, and one out of four in the other. The first precinct had a total of 455 votes, the second had 291 votes. No other evidence was available about physical positions of the improper voting devices within the polling places and about the distributions of votes cast among the devices.

The position that one out of five votes was reversed in the first precinct and that one out of four votes was reversed in the second precinct for each candidate is a deterministic but neutral position relative to the evidence. This position enables one to make simple algebraic adjustments to the precinct totals. The election outcome was not reversed by these adjustments.

The model used by Downs et al. (1978) specified the number of votes reversed in the precincts to have independent binomial distributions $B(455, \frac{1}{5})$ and $B(291, \frac{1}{4})$. This model is also neutral with respect to the evidence, and it allows for some variability about the deterministic position. On the basis of this model, the election did not have an altered outcome because of the irregularity. (Here the null hypothesis of altered outcome was rejected with a z-value of 4.5.)

Rather than specifying p to be $\frac{1}{5}$ and $\frac{1}{4}$, a mixture model approach puts distributions on p-values in each of the precincts with means $\frac{1}{5}$ and $\frac{1}{4}$. This maintains a near-neutral position with respect to the evidence. Yet, the uncertainty in regard to p, which arises from the fact that some physical positions for voting devices attract more votes than others within a precinct, is modeled by the distribution on p. This approach allows for a greater degree of clustering by voting device than does the binomial model. Recalculating the significance of the data using mixture models with uncertainty with respect to p modeled by distributions with variances 0.02 about the neutral points (a variability that is conservatively large in this case) changes z from 4.5 to 2.9. The point is that the mixture model approach here, as in situations with k illegal votes, does not require a specification of a parameter value, does require specification of a distribution on parameter values, and results in more probability being associated with reversal.

Our study has concerned the use of the urn and mixture models to assess the likelihood of altered outcome in certain types of contested elections. We have argued that the urn model assessment PR_{U} may be unrealistically small because this expression of random distribution of illegal votes ignores clustering effects. For a mixture model, PR_F is the average probability of reversal across a class of models $B(k, p)$. With $B(k, p)$, $p \neq \frac{1}{2}$, one candidate or the other tends to benefit from the illegal votes. The net effect of a nondegenerate choice of mixing distribution F on p, which is neutral with respect to the evidence, is to produce a model that anticipates clustering by candidate and results in $PR_F > PR_{\mathrm{U}}$.

In the final remarks of his seminal paper, Tribe (1971) summarizes the carefully developed theme that there is considerable danger in the union of mathematics and the trial process. Perhaps this concern should not be so great in this particular, narrow application, where there is an immediate need for some sort of assessment of likelihood. We believe that mathematical models with empirical support for their underlying foundations can best meet this need and hope that this paper will motivate the search for such support.

Acknowledgments

The authors thank Tom Downs, Dorian Feldman, Bruce Hill, Suzanne Krumholz, Miron Straf, and the editors for their constructive suggestions and comments.

References

Belcher v. Mayor of Ann Arbor, **402** Mich. Rep. 132–134 (1978).

Black, M. (1967). Probability. In *The Encyclopedia of Philosophy,* New York: Free Press, Vol. 6, pp. 464–479.

Carman v. Secretary of State, **384** Mich. Rep. 443–460 (1971).

Carnap, R. (1945). "On Inductive Logic," *Philosophy of Science,* **12,** 72–97.

DeMartini v. Power, 262 N.E. 2d 857–858 (1970).

Diaconis, P., and Freedman, D. (1983). Frequency properties of Bayes rules. In G. E. P. Box, T. Leonard, and C. F. Wu (eds.), *Scientific Inference, Data Analysis, and Robustness,* New York: Academic Press, pp 105–115.

Downs, T., Gilliland, D. C., and Katz, L. (1978). "Probability in a Contested Election," *American Statistician,* **32,** 122–125.

Feller, W. (1957). *An Introduction to Probability Theory and Its Application,* New York: Wiley, Vol. I, 2nd ed.

Ferguson, T. S. (1967). *Mathematical Statistics: A Decision Theoretic Approach.* New York: Academic Press.

Finkelstein, M. O., and Robbins, H. E. (1973). "Mathematical Probability in Election Challenges," *Columbia Law Review,* **73,** 241–248.

Foster, M. H., and Martin, M. L. (eds.) (1966). *Probability, Confirmation and Simplicity,* New York: Odyssey Press.

Gilliland, D. C., and Meier, P. (1983). Probability of reversal in contested elections. *Proceedings of the American Statistical Assoc., Social Sciences Section,* Annual Meetings in Toronto, Aug. 15–18, 1983, 145–151.

Good, I. J. (1950). *Probability and the Weighing of Evidence,* New York: Hafner.

Good, I. J. (1965). *The estimation of probabilities: An essay on modern Bayesian methods.* Research Monograph No. 30, Cambridge, Mass.: MIT Press.

Harvard Law Review (1975). "Developments in the Law—Elections," **88,** 1111–1339.

Hill, B. M. (1974). On coherence, inadmissibility, and inference about many parameters in the theory of least squares. In S. E. Fienberg and A. Zellner (eds.), *Studies in Bayesian Econometrics and Statistics,* Amsterdam: North-Holland, pp. 555–584.

Ippolito v. Power, **241** N.E. 2d 232–236 (1968).

Michigan Compiled Laws (1984), 168.802.

People v. Cicott, **16** Mich. Rep. 322–323 (1868).

Robbins, H. (1951). Asymptotically sub-minimax solutions of compound decision problems. *Proc. Second Berkeley Symp. Math. Statist. Prob.* Univ. of California Press, pp. 131–148.

Santucci v. Power, **252** N.E. 2d 128 (1969).

Straf, M. L. (1983). Discussion—statistics in the legal setting. *Proceedings of the American Statistical Assoc., Social Sciences Section,* Annual Meetings in Toronto, Aug. 15–18, 1983, 152–154.

Tribe, L. H. (1971). "Trial by Mathematics: Precision and Ritual in the Legal Process," *Harvard Law Review,* **84,** 1329–1393.

Ward, J. E. III (1981). "The Probability of Election Reversal," *Mathematics Magazine,* **5,** 256–259.

Comment

Herbert Robbins
Columbia University
New York, New York

The Gilliland–Meier (GM) chapter is a critique of the 1973 note by Finkelstein and Robbins (FR) on the use of mathematical probability in election challenges. The issue is, what probability distribution for x, regarded as a random variable, should be used in evaluating the reversal probability $P[x \geq (a - b + k)/2]$? Of course, we are free to posit *any* distribution that we please on the set of values $0, 1, \ldots, k$, but presumably we have to persuade the judge to go along with the one that we favor.

In the single urn model of FR, when only global results are available, x is assigned the hypergeometric distribution in which all the $\binom{s}{k}$ choices of k balls from an urn containing $s = a + b$ balls are regarded as equally likely, and x is the number of white balls chosen. In this model

$$Ex = ka/s, \qquad \operatorname{Var} x = kab(s - k)/s^2(s - 1).$$

In the multiple urn model by subdistricts, x is regarded as the sum of independent hypergeometric variables x_i, one for each subdistrict, and

$$Ex = \Sigma\{k_i a_i/s_i\}, \qquad \operatorname{Var} x = \Sigma\{k_i a_i b_i(s_i - k_i)/s_i^2(s_i - 1)\}$$

In terms of the standardized variable $z = (x - Ex)/\sqrt{\operatorname{Var} x}$, the probability of reversal in the multiple urn model is

$$P\left[z \geq \frac{(a - b + k) - 2\Sigma\{k_i a_i/s_i\}}{2\sqrt{\Sigma\{k_i a_i b_i(s_i - k_i)/s_i^2(s_i - 1)\}}}\right], \tag{1}$$

where z will be approximately $N(0, 1)$ in most cases.

In close elections with a and b approximately equal to $s/2$, the single urn model will have $Ex \simeq k/2$ and $\operatorname{Var} x \simeq k/4$, and x will be approximately

Bin$(k, 1/2)$when $k/s \simeq 0$. GM regards the probability of reversal calculated on this assumption as too small because of "the clustering of illegal votes by subdistrict and the clustering of votes for candidate A by subdistrict" (Sec. 5, paragraph 4). I agree with them. Their proposal symmetric beta-mixed binomial model for the distribution of x increases the probability of reversal by maintaining Ex at $k/2$ but increasing Var x to $(k/4)[(k + 2\alpha)/(1 + 2\alpha)]$ for some $0 < \alpha < \infty$. But I much prefer the multiple urn model by subdistricts put forward in FR (footnote 9, p. 243, and p. 248). (There are several misprints on p. 248; in particular, the denominator in the last formula should be that of (1) above.) The effect of using the multiple urn model is not to increase the value of Var x above $k/4$ *but to increase* Ex *above* $k/2$ *when the data warrant it,* and thereby also to increase the probability of reversal to a value determined by a rational process. GM does mention the multiple urn model, but in the last sentence of paragraph 4 of Section 5 seems to reject it for a reason that remains mysterious to me after several readings; I look forward to a clarification of this point.

The multiple urn approach does not involve an attempt to model by a specific random mechanism the way in which each illegal voter chose to vote, but rather to quantify in the mind of a neutral observer the amount of uncertainty caused by the $k = \Sigma k_i$ illegal votes. In contrast, the GM model of a (symmetric or not) beta-mixed binomial seems to specify such a mechanism: first, a value p is chosen according to some beta (α, β) prior, and then each of the k illegal voters independently tosses a coin with probability p of heads to decide whether to vote for A (heads) or B (tails). The validity of such a model is by no means apparent to me. That by leaving open the values of α, β it can produce such results as "$x = 0$ and $x = k$ each have probability about $\frac{1}{2}(\alpha = \beta \simeq 0)$", "$x = 0, 1, \ldots, k$ are equally likely $(\alpha = \beta = 1)$", and so on, may or may not be regarded as a defect. But, absent fraud, I much prefer the multiple urn model and do not see the need for providing x with any other mean, variance, or distribution. A general argument for urn models in law has recently been made by Levin and myself (Levin and Robbins, 1983, esp. Sec. V).

I like the 1868 quotation in GM from Judge Cooley. It reminds me that I have several times suggested to lawyers the following procedure for settling disputed elections of the "no evidence" type. Let urns be filled by district with black and white balls in the observed numbers, and from urn i let k_i balls be removed at random without replacement; award the victory to whichever color remains in the majority. Alas, the lawyers have taken a dim view of this proposal, preferring an evaluation of the mathematical probability of a reversal with a cutoff value to be set by the judge, as more in keeping with the dignity of the law. Since the exact probability distribution of z in (1) is complicated, a computer simulation to approximate the value of (1) will often be in order. In the Holtzman–Celler case cited in FR (footnote 23) I did this some thousands of times, with no reversals observed.

Elections are also put in doubt when some valid votes are lost or when some qualified voters are prevented from voting by accidental circumstances. Such cases require an entirely different form of analysis from those with illegal votes. Both sources of error may be present simultaneously, and there will be considerable argument about how best to analyze such hybrid cases. An informed discussion of this question might well begin now, since because of our electoral college system it can be anticipated that challenges will occur for election to the highest office in the land.

Additional References

Levin, B., and Robbins, H. (1983). "Urn Models for Regression Analysis, with Applications to Employment Discrimination Studies," *Law and Contemporary Problems,* **46,** 247–264.

Rejoinder

Dennis C. Gilliland / Paul Meier

We agree with Robbins that an analysis by subdistrict is appropriate when the numbers of illegal votes are identified by subdistrict. FR (1973) relegates this recommendation to a footnote (footnote 9), where the impression is created that the analysis by subdistrict may be optional. HLR (1975, p. 1325) finds it preferable. The rationale for assuming the results from separate subdistricts to be mutually independent is not clear.

The basic question concerns the quantification of the uncertainty in an unknown x. Consider a set of votes that includes k illegal votes, of which x are for candidate A. The set may correspond to the votes recorded in a particular subdistrict or an aggregate of subdistricts. (Robbins uses the subscript i when describing the analysis for subdistrict i.) FR asks that the decision maker posit a hypergeometric distribution for x, and Robbins comments that he sees no need for providing x with any other distribution at the subdistrict level of analysis.

Consider a hypothetical example with all $k = 20$ illegal votes in one precinct where $a = 250$ and $b = 250$. There is no evidence of fraud or about which if either candidate benefited from the illegal votes. Suppose that candidate A has a margin of 9 votes overall, so that x must be 15 or more in order for there to be reversal. FR and Robbins suggest that the urn model is the proper means to quantify in the mind of a neutral observer the amount of uncertainty in x. With their approach,

$$p(x) = \binom{250}{x}\binom{250}{20-x}\bigg/\binom{500}{20}, \qquad x = 0, 1, \ldots, 20,$$

and $PR_{\mathrm{U}} = P[x \geq 15] = 0.02$.

The observer may be neutral in regard to the evidence and the candidates in postulating any distribution for x such that $p(x) = p(20 - x)$, $x = 0, 1, \ldots, 9$. FR and Robbins are asking that the observer go beyond this and regard all $\binom{500}{20}$ subsets of size 20 to be equally likely to be the subset of illegal votes. The burden is on them to offer evidence or reasons in support of the urn model as a reasonable model to describe distributions in numbers of illegal votes between candidates.

Consider some examples of the types of actions and errors that result in illegal votes. Fraud is an obvious example. Another is the bias, possibly subconscious bias, of an election official working at the precinct. In Ann Arbor, illegal votes came from persons who resided in certain geographical areas outside precinct boundaries. Each of these factors can work selectively for one candidate in allowing illegal votes for that candidate. This results in a clustering of illegal votes by candidate by precinct to a degree not anticipated by the urn model. That there may be no specific evidence of such actions and errors in a case does not, in our opinion, justify the imposition of a distribution based on the unsupported assumption of a purely random distribution among subsets of voters, even at the precinct level of aggregation. We believe that as data are developed in this area, it will be found that the empirical variance of x will be much larger than that anticipated by the urn model.

Rational decision making requires the careful assessment of the evidence and notice of the lack of evidence in some areas of concern in regard to the case under consideration. It includes the examination of distributions of illegal votes by candidates across other elections where actions and errors have led to illegal votes. Based on this, the decision maker will express his or her uncertainty in x and weigh the consequences related to the various alternative actions that are available.

We suggest that the class of mixture models, which essentially includes the urn model as a boundary case, may be a rich enough class from which to choose a distribution for x for most decision makers and for most applications. We are not suggesting that a mixture model serves to model the processes through which illegal votes are cast. It is a means to model an internal correlation between illegal votes and to produce variances that are larger than those dictated by the urn model. Of course, there are other methods that model clustering as well and achieve these ends.

We conclude with another hypothetical. All $k = 20$ illegal votes are in a precinct where $a = 300$ and $b = 200$. There is no evidence of fraud or about which, if either, candidate benefited from the illegal votes. Suppose that candidate A has a margin of 11 votes overall. Decision maker Smith accepts the FR fiat and calculates $E(x) = 12$, $V(x) = 4.62$, and $PR_U = 0.05$. On the other hand, decision maker Jones considers the possibilities for actions that lead to illegal votes and, perhaps, data from other elections concerning the distributions of illegal votes among candidates. Suppose that Jones finds the beta binomial distribution with $E(x) = 12$ and $V(x) = 20$ expresses his or her uncertainty with respect to x. For Jones, $PR = 0.25$.

Scientific Data and Environmental Regulation

Gordon J. Apple
Maun, Green, Hayes, Simon, Johanneson and Brehl
St. Paul, Minnesota

William G. Hunter and Soren Bisgaard
Department of Statistics
University of Wisconsin
Madison, Wisconsin

ABSTRACT

This chapter addresses the role of scientific data in environmental regulation. The structure of the federal regulatory framework is discussed and the areas where science and law intersect are highlighted. In particular, the interdisciplinary nature of the environmental decision-making process is examined. Through the example of a case study the problems inherent in making environmental decisions with uncertain data are outlined and the process of judicial review of such decisions is analyzed. Approaches for improving the relationships between the actors in the decision-making process are discussed.

1. Introduction

In this chapter on the use of scientific data in environmental regulation, we explain some elements of environmental law, explore key issues at the intersection of science and law, and highlight certain statistical problems of scientific inference that arise in this context. The central issue is the use of scientific data in making environmental decisions under uncertainty. Our purpose is to examine how the interests and responsibilities of scientists, lawyers, and decision makers interact with one another when they are grappling with this issue.

Environmental problems are becoming increasingly important because of two factors: population growth and technological development. More people need more resources—food, energy, space, and raw material for housing, clothing, and other goods. Technological development entails the manufacture and dispersal into the environment of increasing numbers of new chemicals and other products, in ever increasing amounts. The capacity of the environment to safely assimilate the production and waste of our modern industrial society is limited. Levels of hazardous substances in our environment are increasing. Forms of animal and plant life are disappearing. Places of natural beauty are being destroyed. These facts are moving environmental law from the wings toward center stage.

Rogers (1977, p. 1) defined environmental law as "the law of planetary housekeeping." It embraces questions of property rights (e.g., allocation of water rights, access to solar energy, and regulation of land use) and problems concerning public health (e.g., regulation of pollution in liquid, solid, and gaseous forms). This chapter focuses on problems related to public health that pose important legal and statistical difficulties.

In order to make sensible environmental policy, reliable data are needed. Without such data it is difficult to discover that environmental problems exist and to develop public policy to solve these problems. Good data can be hard to find. A 1979 Comptroller General's Report to Congress, in acknowledging that scientific uncertainty was a significant problem in environmental regulation, stated that:

> Major constraints plague the Environmental Protection Agency's ability to set standards and issue regulations. The most important factor is the inconclusive scientific evidence on which it must often base decisions. Numerous court suits result.

Ten years after the Stockholm Conference on the Human Environment, White (1982), in addressing the issue of inadequate data, observed that:

> [I]t is extremely hard to appraise what, in fact, has happened to the principal components of the environment—the atmosphere, marine environment, inland waters, lithosphere, terrestrial biota, and people . . . [As] revealed by the difficulty in measuring changes, a more coherent effort needs to be made to monitor key parameters.

Clearly, there is an urgent need to gain more information about the environment and thereby to reduce present levels of uncertainty. Whatever the present level of knowledge, it should be used wisely in framing environmental regulations. Scientists, lawyers, and decision makers can best approach this task by learning about each other's areas of expertise.

2. The Federal Regulatory Machine

To understand the role of science in environmental regulation, it is necessary to have at least a cursory knowledge of the workings of what Abelson (1978, p. 487) termed "the federal regulatory machine." How environmental policies are formulated and implemented through laws and regulations is a complex process. The uncertainty surrounding relevant scientific facts is a major cause of this complexity. Moreover, the regulatory machine runs not only on scientific facts but also on economic, political, psychological, and sociological factors. Not surprisingly, with such mixed fuel, the machine does not always hum at maximum efficiency. The main actors in the regulatory process are the Congress, the president, the administrative agencies, and the courts. We present a brief description of the part each plays in the design, operation, and maintenance of the regulatory machine.

The Congress

Congress designs the federal regulatory machine. At different times, it enacts legislation for the benefit of constituents, special interest groups, or the nation as a whole. The passage of major environmental laws, such as the Clean Air Act, the National Environmental Policy Act, and the Toxic Substances Control Act, is for the benefit of all. A brief examination of the factors underlying the passage of these laws will illuminate Congress's role in environmental regulation.

Of the many forces that led to the environmental laws of the 1960s and 1970s, the most significant and compelling was the nation's growing awareness that it was wallowing in and, perhaps, dying from the unwanted by-products of its prosperity. National spectacles of environmental pollution were commonplace; oil spilled into the water from blowouts and shipwrecks, smog blanketed Los Angeles and other urban areas, Lake Erie was pronounced dead, and the Cuyahoga River in Ohio caught fire. Environmental groups proliferated and highlighted problems and spoke about dangers that the nation faced. Some of the issues described were international in scope. Many listened, and the response of the general public was to demand action by their elected representatives. The people wanted laws to protect the environment. Congress reacted to these demands. Environmental legislation was drafted, bills were introduced, testimony was presented by scientists and lay people, committee reports were issued, compromises were reached, bills were passed, and eventually new statutes were signed into law by the president.

What we have quickly painted, with a broad brush, is a picture of the reactive legislative process. Environmental laws are almost always reactive in that they are a response to problems that already exist, as Quarles (1976, pp. xv–xvi) has noted:

> Since our government depends on public demand, the government does not begin to attack a problem until that problem has become severe. The government always has to catch up, to find solutions to problems that long before have grown out of control. Although our government may be responsive, it is so only by delayed reaction.

The intensity of the reaction is a reflection of the perceived degree of the public need and demand.

Public policy embodied in environmental legislation forms the framework of the federal regulatory machine. The statutes provide broad (and sometimes vague) guidelines for the administrative agencies. From this statutory basis the agencies derive their power and direction for the promulgation of regulatory programs. The time lag between the emergence of a problem, the recognition by a small number of people, the building of public opinion, the response by legislators, and the enforcement by agencies is often decades.

The President

The president acts as the chief executive officer for the operation of the federal regulatory machine. As head of the executive branch of government he appoints, subject to Senate confirmation, the department heads and administrators who direct the day-to-day activities of the government. These chief operating officers are political appointees who, with rare exceptions, embrace the philosophical viewpoint of the president and follow his directives. Through them, the president prescribes his policies on how the government should be run. For example, the administration of President Ronald Reagan took control of the Executive Branch in 1981 with the aim of dismantling some of the federal regulatory machine. The administration's subsequent effort to control and cut back regulations was an impressive display of presidential power. In a relatively short time the flow of new regulations was drastically curtailed and old regulatory programs were often rescinded or rendered ineffective through nonenforcement, budgetary cuts, or other means. (See, for example, the *New York Times*, 9:1, 1982.) The ability of the president to affect the operation and size of the regulatory machine is substantial.

The Administrative Agencies

The administrative agencies operate the federal regulatory machine. Some of the best known are the Environmental Protection Agency, the Occupational Safety and Health Administration, and the Food and Drug Administration. Through the promulgation and enforcement of environmental standards, guidelines, and regulations, agencies attempt to follow the directives of Congress. Agency administrators are delegated the authority from Congress to make both technical and public policy decisions in implementing environmental law. They are usually granted broad discretion in deciding when and how to regulate.

Although administrators prefer to have a concrete evidentiary basis for their decisions, there are times when one does not exist. Faced with the choice of acting on incomplete data or waiting until better data become available, an administrator must look to the intent of Congress for guidance. If a statute is precautionary in nature and seeks to provide the highest level of protection, administrative action may be compelled despite the uncertainties. If, however, a statute sets forth relatively strict thresholds for action, the agency cannot act under a cloud of uncertainty and must wait for more evidence. In interpreting legislative intent, administrators are also influenced by the policies of the president.

Administrative agencies have several strengths that make them well-suited to deal with uncertainty in the decision-making process. First, they are specialized institutions that have the capability to develop the kind of expertise that comes from continued exposure to the problem area. Second, the agencies are multidisciplinary institutions able to assemble the appropriate mix of professionals needed to accomplish their objectives. Given these strengths, the agencies have the resources to assess and evaluate the data generated in the decisionmaking process. (For a detailed examination of administrative law and procedure, see Davis, 1972).

Notwithstanding these resources, the decision-making process is arduous and filled with conflict. Interested parties to the process (e.g., industry, environmentalists, and state governments) seldom agree on the merits of the outcome. They may believe the decision is arbitrary and capricious, that it exceeds the delegated powers of the agency, that there is inadequate evidence, or that proper procedures were not used. The result can be a petition to the courts seeking review of the decision.

The Federal Courts

The federal courts pass judgment on the agencies' operation of the federal regulatory machine. The branches of the federal judicial system involved in environmental law are the U.S. District Courts, the U.S. Court of

Appeal, and the Supreme Court of the United States. Through the process of judicial review, the courts examine final regulatory decisions and determine if they satisfy varied legal requirements. The courts can affirm the agency's decisions, reverse them, affirm in part and reverse in part, or remand the decisions and require further agency action. They are the final arbiters of conflicts involving the federal regulatory machine. (For detailed analysis of the federal judicial system, see Hart and Wechsler, 1973.)

On the bottom tier of review courts are the federal district courts. The district courts are courts of first impression or initial jurisdiction and as such render decisions based on the evidence introduced and argued before them.

On the intermediate tier, the appellate level, are the U.S. Courts of Appeals. In general, the appellate courts do not hear new evidence. Their purpose is not to retry the case but to review the record of the lower-level proceeding to determine whether a legal error was committed that merits reversal or modification of the original decision. An exception to this defined role of the appeals courts occurs in federal environmental law, where the courts often serve as the court of first impression in the review of federal environmental regulations. Decisions rendered by the federal Courts of Appeal are usually decided by panels of three judges and are accordingly called *panel opinions*. If a matter is of great importance, however, all 4 to 26 justices of the circuit may choose to sit *en banc* and hear the case. A party aggrieved by the panel decision may petition the court for a *rehearing en banc*. Petitions for a rehearing, however, are seldom granted.

The final word on the agencies' operation of the regulatory machine comes from the Supreme Court of the United States. Appeals from decisions by the circuit courts of appeal concerning administrative agencies are heard at the discretion of the Supreme Court by *writ of certiorari*. *Certiorari* is a process by which a party petitions the Supreme Court to review a decision of the Court of Appeals. Should the Supreme Court decide that the issues raised in the case are important enough to merit its attention, it grants *certiorari*. If *certiorari* is denied, the decision of the appeals court stands as the final word on the matter.

3. The Intersection of Science and Environmental Law

This section concerns problems of environmental law in which quantitative data must be assessed. Sometimes satisfactory solutions to problems of data analysis demand consideration of inferential methods and conceptual frameworks that are used in both science and law. We first discuss the importance of statistical methods in environmental law, then identify some troublesome aspects of environmental data, and finally examine the differing methods and concepts used by the scientist and the lawyer in the determination of harm and causation.

The Importance of Statistical Methods

In environmental law, scientific data serve as the basis for decisions made at three extremely important junctures: (a) determining whether a particular situation represents an unacceptable risk to public health and, if so, (b) establishing appropriate regulatory policy that will satisfactorily deal with the problem, and (c) resolving disputes when one or more parties have been accused of violating such regulations.

An important issue in environmental law is the way regulations deal with environmental data. Environmental data are sometimes contaminated with a substantial amount of experimental error (see, e.g., Hunter, 1980). Consequently, to use environmental data correctly, lawyers and others must cope with a formidable array of statistical problems. Unfortunately, the participants in the regulatory process are generally unfamiliar with these problems and the statistical tools available to cope with them. The implications for society can be severe if such data are handled improperly. Although inexact numerical data are often the basis on which environmental policy is established, the stochastic nature of environmental data is often simply ignored. Regulations are sometimes written and interpreted as if such data are exact. Even when standards are couched in statistical terms, these terms are often garbled. If a decision is made incorrectly at the administrative level, it is likely that technical errors will go unrectified, especially if the decision is reviewed by a judiciary that is untrained in science and bound to a minimally intrusive review standard.

Many important environmental concerns that have received wide publicity involve statistical issues. They include the possible dangers of fluorocarbons (it is feared that these chemicals might be depleting the ozone in the stratosphere), 2,4,5-T (a herbicide and plant growth regulator), saccharin, cigarette smoke, benzene, red dyes No. 4 and No. 40, nitrites, asbestos, nuclear radiation, lead in gasoline, water pollutants, air pollutants, pesticides, formaldehyde, and other hazardous substances.

Statistical assessments have been introduced as evidence in a variety of environmental lawsuits. Recent cases have included testimony and exhibits covering quantitative risk assessment and degrees of certainty. Judges, trained in the law, have been presented with evidence on the correlation between air and blood lead levels and with expert testimony on the statistical significance of dose-response curves extrapolated, on the basis of the linear nonthreshold hypothesis, from regions of high exposure levels, where data were available, to regions of low exposure levels, where data were unavailable. Landmark cases involving public health and environmental regulations warrant careful study from both statistical and legal points of view. Cases, for example, in which scientific and statistical data were reviewed include *Ethyl Corporation v. EPA* (1976) regarding regulation of lead additives in gasoline, *Industrial Union Department, AFL-CIO v. American Petroleum Institute* (1980) regarding occupational exposure limits for benzene, and *Gulf South Insulation v. United States Consumer Product Safety Commission* (1983) regarding a ban on the use of urea-formaldehyde foam insulation.

Although in this paper we mainly investigate statistical and scientific aspects of environmental law, we recognize that other aspects—such as political and economic—are crucially important in determining how policy is developed. See, for example, Portney (1978), Scherer (1979), Brickman and Jasanoff (1979), Ashford et al. (1983), and Gillespie et al. (1979). The last reference tells the story of two pesticides, Aldrin and Dieldrin, for which markedly different regulatory action was taken in the United States and Great Britain, even though the same evidence was available in both places.

Troublesome Aspects of Environmental Data

Environmental data often present numerous statistical problems to scientists and engineers trying to measure and analyze them and to the regulators trying to make informed decisions based on them. Some troublesome aspects of typical sets of environmental data include measurement error, serial correlation and seasonal fluctuations, and complex cause-and-effect relationships. Factors such as these need to be considered when decisions are made concerning the collection and analysis of data to answer the questions at issue. In the data collection phase, statistical issues may involve the design of experiments and clinical trials, sample surveys, epidemiological studies, or a census. In the analysis phase, statistical techniques will be most effective when used innovatively as an adjunct to, rather than a replacement for, information and methods available from relevant fields. In particular, since most important environmental problems are multidisciplinary, effective analysis often requires that statisticians work cooperatively with specialists in health, engineering, biology, chemistry, meteorology, economics, and law. [See Pollack et al. (1983) for some amplification of these points.]

Measurement Error and Missing Data

One of the problems in environmental law is that in many instances the entire legislative and regulatory apparatus is constructed on data that are quite shaky. Figure 1 presents data that illustrate this problem. The mea-

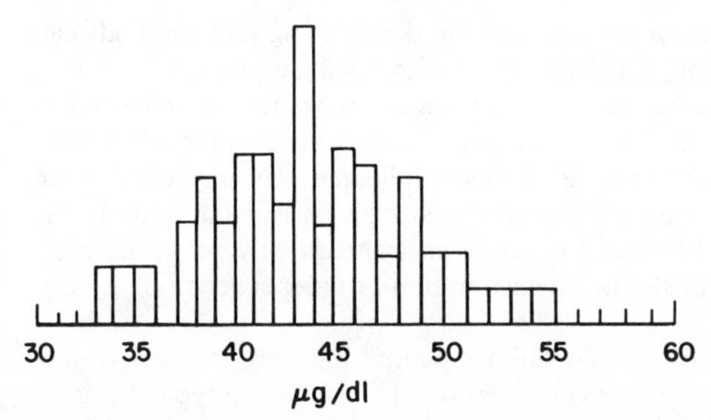

Figure 1. Distribution of blood lead measurements made by 100 laboratories.

sured values of lead in blood made by approximately 100 separate laboratories are displayed. Each laboratory was supplied with a single sample from a standard source. Ideally all measurements should have been identical, all being equal to the best estimate, 41 micrograms per deciliter (μg/dl), which we shall call the true value. Note, however, the wide variability in the reported results. This degree of variability is inherent in such measurements. If it is reported that a certain proportion of the population in some locality has blood lead values below or above a specific level, it is important to know whether the blood lead values being referred to are true or measured quantities. In the illustration here, although the true value is 41 μg/dl, a substantial portion of the measured values is below 40 μg/dl, which is the value most often used as the demarcation level to indicate that a health problem may exist. Conversely, if the true level were below 40 μg/dl, a substantial portion of the measured values may exceed 40 μg/dl. If one is to think carefully about problems in environmental law, one must take measurement errors of this kind into account and be aware of the distinction between true and measured values. One must also recognize that measurement error is only one component of the more general experimental error, which makes reported data even less reliable than they otherwise may appear to be. Systematic errors, in this situation and others, can be more important than random errors. Data resulting from locating, counting, and identifying tumors or determining cause of death in environmental health studies, for example, can be subject to gross error.

Serial Correlation and Seasonal Fluctuations

Environmental data are often collected sequentially (e.g., air and water pollution measurements) and hence are frequently serially correlated, that is, autocorrelated. Most statistical theory, however, is based on the assumption that data are not serially correlated, and using such methods can be seriously misleading. Ordinary regression analysis, which rests in a critical way on this assumption, for instance, is sometimes misused in analyzing environmental data. One reason for high serial correlation is the rapid rate at which data are sometimes collected. Usually data taken closely together in time or space will be positively serially correlated. The important point is that standard procedures are frequently not robust with respect to violations of the usual assumption of independence of experimental errors.

Another complication is that environmental data are often periodic. Ambient temperature at a given location, for example, typically goes through yearly and daily cycles. Certain types of air pollution will have more severe health effects in the summer than in the winter. Biological phenomena, such as migration patterns, exhibit seasonal rhythms. Models used in the analysis of such data usually incorporate appropriate known periodic variations. With many types of biological data (e.g., aquatic life), erratic and quite pronounced random fluctuations may be superimposed on more or less regular seasonal fluctuations.

Complex Cause-and-Effect Relationships

Given the intricate fabric of the ecosystem with its myriad cause-and-effect relationships, including interactions among variables, many of which are unknown, it is not surprising that variables that have an important influence are sometimes overlooked and unrecorded or, even if recorded, are not used. The presence of such lurking variables can confuse investigators who are trying to understand the intricate linkages and delicate balances in the environment. Furthermore, an effect might only manifest itself many years after the initial cause is present (long latency periods).

Complex cause-and-effect relationships present enormous difficulties in the regulation of health hazards. Permissible exposure limits to carcinogens are often based on quantitative risk assessments that extrapolate from animal data, since epidemiologic data is usually lacking or inconclusive. Affected parties (at-risk workers or regulated industries) frequently file lawsuits challenging the existence of a causal relationship, the validity of the data, or the methodology used in quantifying the risk. See *Public Citizen Health Research Group v. Auchter* (1983) regarding regulation of ethylene oxide and *Gulf South Insulation v. United States Consumer Product Safety Commission* (1983) regarding regulation of urea-formaldehyde insulation.

Determination of Harm and Causation

In this subsection we explore how the concepts of harm and causation are treated when disputes are litigated under the recent standards written into major environmental legislation. An environmental dispute involving possible threats to the public health is fundamentally a dispute over whether a harm is occurring, whether a given party is responsible, and, if the answer to both of these questions is yes, whether it should be compelled to cease or modify the activity or compensate the injured parties. These questions all revolve around the concepts of harm and causation.

Harm

The concept of harm in environmental regulation relates to both identifiable damage and harm that is forseeable based on past experience and predictions of risk. For example, groundwater contaminated by a hazardous waste dump is an identifiable harm, whereas the risk of increased incidence of cancer from drinking the water is an intangible harm based on predictions of risk.

Identifying environmental harm is extremely difficult when there exists a long latency period between exposure and harm. The threats posed by substances such as organochlorides, phenoxy herbicides, and polychlorinated biphenols are not immediately apparent. The problem of a long latency period compounded by several features of ecosystems make identification of causal relationships more difficult (Gelpe and Tarlock, 1974, pp. 404–407). First, the immediate or near-term effect of an environmental change often differs appreciably from ultimate long-term effects. Second, intervening or multiple sources of causation are more likely to complicate the relationship when harm is temporally remote from exposure. Finally, in some cases the harm may be geographically remote from the point of production or exposure.

To promulgate regulations, an agency must establish that there is a need to regulate. The issue of how certain an agency must be before it can regulate environmental harm has been characterized as a debate over whether the risk that the harm will occur must be certain (i.e., based on past evidence) or whether the risk can be based on evidence suggesting, but not proving, harm (Brown, 1976, pp. 504–506).

The question of what degree of certainty is required to justify an agency decision to regulate was answered quite differently by the majority and the dissent in *Ethyl Corporation v. Environmental Protection Agency,* 541 F.2d 1 (D.C. Cir., 1976) (*Ethyl*). In that case several manufacturers of leaded "antiknock" gas additives sought review of an order by the Environmental Protection Agency (EPA) requiring a phased reduction in the lead content of leaded gasoline. Regulation was undertaken by the administrator of the EPA on the basis of a standard that required that "a significant risk of harm" to the public health be demonstrated as a consequence of leaded fuel usage. The EPA had promulgated lead phase-down regulations after concluding that environmental lead exposure was a major health problem in the United States (38 Fed.Reg. 33734, 1973). EPA found that a small but significant portion of the urban adult population as well as 25% of children in urban areas were overexposed to lead (38 Fed.Reg. 33734, 1973). Although EPA acknowledged that lead from gasoline was but one source of lead exposure in the environment and therefore only part of the problem, it regulated leaded gasoline since it was a source of air and dust lead that could be readily and significantly reduced in comparison to other sources (38 Fed.Reg. 33734, 1973).

The dissent in *Ethyl* argued that a significant risk of harm could only be established by proof of actual harm:

> If there can be found potential harm from lead in exhaust emissions, the best (and only convincing) proof of such potential harm is what has occurred in the past . . . from which the Administrator can logically deduce that the same factors will produce the same harm in the future. [*Ethyl,* dissent, p. 95]

The majority, on the other hand, concluded that the standard applied by the administrator did not require direct proof that lead emissions themselves had caused actual harm. Rather, the court recognized that the wording of the statute suggested that the statute was "precautionary" in nature and that regulatory action was justified to protect public health. The court found that:

> Sometimes, of course, relatively certain proof of danger or harm from [environmental] modifications can be readily found. But, more commonly, "reasonable medical concerns" and theory long precede certainty. Yet the statutes—and common sense—demand regulatory action to prevent harm, even if the regulator is less than certain that harm is otherwise inevitable. [*Ethyl,* p. 25]

Unfortunately, there is no numerical standard for indicating how much less than certain an administrator who takes regulatory action may be. It is a question that must be addressed on a case-by-case basis.

Causation

Before administrators can regulate, they must also have established that the thing to be regulated is or can be the cause of the actual or potential harm. Contemporary environmental problems typically pose extraordinary difficulties in establishing causation. In many areas of environmental regulation, especially in efforts to regulate carcinogens, there is no direct and simple connection between the harm and the suspected cause. For example, epidemiologic studies may indicate that workers exposed to agent *X* have an increased evidence of lung cancer. It may also be shown that within the group of workers studied, those who smoke account for the majority of lung cancer victims. Questions arise about whether agent *X* is the cause of the cancer, an initiator of carcinogenisis, a cancer promoter that acts in synergism with smoking, or, perhaps, whether it is a human carcinogen at all. It is quite likely that science cannot provide definitive answers, especially if there is a long latency period between exposure and harm. Deciding if a sufficient causal connection exists to justify regulatory action is a difficult task [see *Allen v. U.S.* (1984), for a discussion of these issues in the context of civil litigation].

To justify its regulations of leaded gasoline, EPA had to make an initial showing that there was a significant risk of harm to the population from elevated blood lead levels. EPA did not have to prove beyond a reasonable doubt that a significant risk existed, but merely had to show that, based on the available evidence, a reasonable person *could* agree that it was a plausible conclusion.

Meeting this rational basis standard can be difficult. When the risk is one of a remote harm, and when causation cannot be well established, there may be links missing in the chain of reasoning from known fact to alleged harm. The evidence necessary to bridge these gaps may be unobtainable. If the courts required the agencies to have a complete chain of facts, few environmental regulations would be upheld.

The Environmental Protection Agency's lead phase-down regulations addressed the issue of causation in two parts. First, EPA found that lead from gasoline was the most ubiquitous source of lead found in the air, dirt, and dust in urban areas (38 Fed.Reg. 33734, 1973). EPA also noted that human exposure to this lead takes place by inhalation and by ingestion of dirt and dust contaminated by air lead fallout (38 Fed.Reg. 33734, 5). Second, EPA pointed to scientific studies that indicated that airborne lead contributed significantly to lead exposure in the general population, although other studies did not indicate a strong correlation between air lead levels and blood lead levels (38 Fed.Reg. 33735). EPA then cited additional studies that indicated that contaminated dust and dirt from motor vehicles exhaust were believed to be important exposure routes contributing to lead poisoning in children (38 Fed.Reg. 33734). EPA concluded that the totality of the evidence was sufficient to establish a causal link between lead in exhaust and lead in the human body, and therefore regulation of leaded gasoline was justified.

The issue of causation sharply divided the court in *Ethyl*. The dissent believed that a causal connection had to be established by a chain of scientifically probable assertions:

> [T]he thought process by which an agency reaches its conclusion on informal rulemaking resembles a chain. If there is a link missing, then the agency, to reach the conclusion that it did, was required to take an arbitrary jump in its logic to reach that conclusion. [*Ethyl,* Wilkey, J., dissenting p. 98]

Judge Wright, writing for the majority, strongly disagreed with this position. The regulatory standard involved, he argued, was precautionary in nature:

> Where a statute is precautionary in nature, the evidence difficult to come by, uncertain, or conflicting because it is on the frontiers of scientific knowledge, the regulations designed to protect the public health, and the decision that of an expert administrator, we will not demand rigorous step-by-step proof of cause and effect. . . . The Administrator may apply his expertise to draw inferences from suspected, but not completely substantiated, relationships between facts, from trends among facts, from theoretical projections from imperfect data, from probative preliminary data not yet certifiable as "fact", and the like. [*Ethyl,* p. 28]

The court deferred to the expertise of the agency in its determination that the data showed a sufficient causal connection to justify regulation. More significantly, the court pointed to the legislative intent, which it claimed called for a "precautionary standard," and found that to provide the highest level of protection for public health regulatory action could precede scientific proof.

4. Judicial Review

The decision-making process of an agency is often challenged in the courts. Ideally, the role of the court in reviewing an agency decision is confined to legal analysis. The court ensures that the agency decision is within the scope of its powers granted by Congress, that the procedures followed in reaching the decision satisfy due process requirements and that the decision is rational and well reasoned (Rogers, 1979, p. 699, p. 705; Administrative Procedures Act 706).

When courts are confronted with administrative decisions that are based on complex legal, factual, and policy considerations, however, the judicial review function strays from the ideal. The type of legal analysis necessary in this situation for adequate judicial review is disputed. Some members of the judiciary believe that an agency decision must be upheld if adequate procedures have been followed in the decision-making process. The majority, however, require a "hard look" at both the procedural adequacy of the decision-making process and the sufficiency of the factual "substantive" grounds for the decision.

One of the central issues in this dispute is the scope of the judge's role when confronting decisions made under uncertainty. Some of the special problems involved can be better appreciated by considering the job of the lawyer and the statistician when working with clients on difficult problems. The lawyer who counsels on environmental law questions and the statistician who deals with complicated problems of analysis or design must both decide how deeply they need to immerse themselves in the subject matter field in order to do a competent job. In principle the answer is the deeper the better. For example, the more one knows about the biological, chemical, and physical mechanisms that govern how lead in gasoline is dispersed in the environment, how lead in the environment is transported to the blood, and how lead in the blood affects people's health, the more informed the legal or statistical assistance will be. Judgment determines how much of one's time is allocated to such "background" activities. Good lawyers and good statisticians behave in the same way when handling such complex assignments. They study the background information, they learn something—and often a great deal—about the subject. They read a lot. They also consult knowledgeable experts, and they may even visit the key sites to see for themselves the rivers, the factories, and the laboratories that are involved. They put considerable emphasis on this phase of work knowing that trying to operate in the absence of such understanding and trying to do a "purely" legal or statistical piece of work is foolhardy and is likely to be deficient in some crucial way.

The situation from a judge's point of view in the United States is quite different. Notably, the judge's scope is severely restricted when it comes to consulting experts and filling in apparent gaps in the record. Although the court may appoint its own experts under Federal Rule of Evidence 706, it rarely does so, and even then the experts do not participate in the decision-making process. The lawyer or the statistician can have an expert alongside and ask for clarification of tricky points. "What's going on here?" "How does this work?" "How is the pollutant removed from the effluent?" Not so with the judge. The judge must make a decision solely on the basis of the record. Federal Circuit Judge McGowan, in a recent congressional hearing, graphically explained how one particular administrative law case looked to him:

> I had the Weyerhauser case—the pulp and paper company—under the Clean Water Act. There were 180,000 pages in that record. All I had to do was decide whether it was arbitrary and capricious. What was my job at this point? To be able to understand—and this was essentially chemical engineering—my job was to really master the differences between these differing operations of the different companies so that I could make a reasoned disposition of the case. I would have needed to spend six months on that. With no expert help available to me, no expert that I could call in and say, "What about this?

> Let's put our feet up on the table, and you can explain it to me," I had to do it by myself. ["Hearing," (1981), pp. 930–931, *Weyerhauser v. Costle,* 590 F.2d 1011 (D.C. Cir. 1978)]

A statistician or other outsider to the legal system might imagine that there would be a tendency for a judge to go toward one of two extreme positions when faced with complex environmental cases: either (a) to ignore all the factual issues and simply address legal procedural matters easily within judicial competence or (b) to master all the subject matter knowledge and make a decision in the full light of known scientific and technological facts. The first might be called a purely procedural review, the second an all-encompassing substantive review. Neither is sensible, however, because the first limits the judge's responsibility and the second gives the judge too much responsibility and places excessive power in a judiciary that is not accountable to the electorate. The first could be done by most knowledgeable members of the judiciary. The second requires too much time; in fact, a single case might swallow up a judge for years.

In most cases, the central issue that the judge must address in reviewing an administrator's decision is whether that decision had a rational basis in the administrative record or whether it was arbitrary and capricious. To do this, a judge cannot ignore all substantive facts and fly completely to a purely procedural review; nor can a judge hope to master all the substantive knowledge and fly to the other extreme of an all-encompassing substantive review. The degree to which one immerses oneself in the technical details of the case and seeks to evaluate technical merits determines where one stands on the spectrum between these two unrealistic extremes. The competing practical views are aptly characterized by Judge Bazelon and the late Judge Leventhal in their concurring opinions in *Ethyl.*

Procedural Versus Substantive Review

Judge Bazelon's position of informed procedural review is that the judiciary is limited to reviewing the procedures followed by an agency to determine their compliance with the law and thus their adequacy for reaching a reasoned decision. He notes:

> In cases of great technological complexity, the best way for courts to guard against unreasonable or erroneous administrative decisions is not for the judges themselves to scrutinize the technical merits of each decision. Rather, it is to establish a decision-making process that assures a reasoned decision that can be held up to the scrutiny of the scientific community and the public. [*Ethyl,* p. 66. citing *International Harvestor Co. v. Ruckelshaus,* 478 F.2d 615, 652 (1973), (Bazelon, C.J., concurring)]

Judge Bazelon's reliance on procedural safeguards rests, in part, on his belief in the institutional incompetence of the courts to do otherwise. He illustrates this perceived incompetence and suggests a better approach in stating that:

> Because substantive review of mathematical and scientific evidence by technically illiterate judges is dangerously unreliable, . . . we will do more to improve administrative decision-making by concentrating our efforts on strengthening administrative procedures. [*Ethyl,* p. 67]

Without disputing the need for procedural review, Judge Leventhal harshly criticizes what he sees as Judge Bazelon's abdication of substantive review of complex administrative decisionmaking. Judge Leventhal states that the responsibility of the judiciary is to review agency decisions to

> assure that the agency exercises the delegated [legislative] power within statutory limits, and that it fleshes out objectives within those limits by an administration that is not irrational or discriminatory. [*Ethyl,* p. 68]

In his view, to fulfill this responsibility the court must conduct a substantive review of the record and reasoning that form the foundation of an agency's action.

It is important to recognize that a distinction exists between substantive judicial review and a review that is substantive in the scientific sense. A review that is substantive in the judicial sense does not analyze data in an attempt to reach a conclusion or inference of scientific fact. Rather, what is meant is that the court must review supporting technical reports and studies in its effort to decide whether reasoned decision making has taken place. "The court does not make the ultimate decision, but it insists that the agency take a 'hard look' at all relevant factors" (Leventhal, 1974, p. 514).

Judge Leventhal dismisses Judge Bazelon's fear that judges will have to become scientists. He exhibits faith that judges can learn enough to properly review technical decisions in stating:

> Our present system of review assumes judges will acquire whatever technical knowledge is necessary as background for decision of the legal questions. . . . The aim of judges is not to exercise expertise or decide technical questions, but simply to gain sufficient background orientation. . . . The substantive review of administrative action is modest, but it cannot be carried out in a vacuum of understanding. [*Ethyl,* p. 69]

Judicial Treatment of Statistical Evidence in Ethyl

The relative merits and drawbacks of the procedural and substantive review approaches can be illustrated by a comparison of the treatment of statistical evidence in the majority and dissenting opinions in *Ethyl*. The controversy discussed below illustrates the desirability and concomitant weakness of substantive review of environmental issues by judges with little or no scientific expertise.

As previously discussed, the case concerned regulations promulgated by the Environmental Protection Agency calling for a phased reduction in the lead content of leaded gasoline because of health concerns. These regulations were issued in November 1973, under Section 211 of the Clean Air Act, which authorizes the administrator to regulate gasoline additives whose emission products "will endanger the public health or welfare." Ethyl Corporation, joining other industry petitioners, appealed the promulgation of the regulations to the Court of Appeals for the District of Columbia Circuit. This appeal was heard in September 1974. They claimed that the administrator had misinterpreted the statutory "will endanger" standard, that his decision was without support in the evidence, and that it was arbitrary and capricious.

The appeal was first heard by a three-judge panel of the court. The panel opinion (January 1975), authored by Judge Wilkey, minutely reviewed several of the technical studies relied upon by the administrator. This opinion concluded that, given the administrator's interpretation of the "will endanger" standard, "his analysis reflected a clear error of judgment upon the available evidence," and, therefore, the regulations could not be upheld. (*Ethyl Corporation v. Environmental Protection Agency,* 5 ELR 20096, 20114 (D.C. Cir. 1975)).

The agency immediately petitioned for a rehearing. An order was issued in March 1975 granting a rehearing *en banc* (before the entire circuit court) and vacating the panel opinion. Supplemental briefs were filed, and the case was reheard before the entire court in May 1975. In March 1976 the court issued an opinion authored by Judge Wright, who had also written for the dissent in the panel opinions. It upheld the agency's regulation (*Ethyl Corporation v. Environmental Protection Agency,* 541 F.2d 1 (D.C. Cir. 1976)).

In *Ethyl* the record ran to over 10,000 pages. Many scientific studies were introduced. It seems safe to say that none was perfect for the simple reason that any experiment when examined closely enough will be found to contain at least some blemishes and imperfections; perhaps major flaws will be discovered. It is easier to cope with a smaller rather than larger number of studies. Some studies in *Ethyl* were discarded; they were given a weight of zero. Were too many studies treated in this way? How seriously must a study be flawed to be given a weight of zero? Is it possible, in some quantitative and rational way, to give weights between zero and one to germane studies? Such weighting is done qualitatively by administrators, judges, and others. The opinion discusses this point as follows:

> A word about our approach to the evidence may be in order. Contrary to the apparent suggestion of some of the petitioners, we need not seek a single dispositive study that fully supports the Administrator's determination. Science does not work that way; nor, for that matter, does adjudicatory fact-finding. Rather, the Administrator's decision may be fully supportable if it is based, as it is, on the inconclusive but suggestive results of numerous studies. By its nature, scientific evidence is cumulative: the more supporting, albeit inconclusive evidence available, the more likely the accuracy of the conclusion. If, as petitioners suggest, one single study or bit of evidence were sufficient independently to mandate a conclusion, there would, of course, be no need for any other studies. Only rarely, however, is such limited study sufficient. Thus, after considering the inferences that can be drawn from the studies supporting the Administrator, and those opposing him, we must decide whether the cumulative effect of all this evidence, and not the effect of any single bit of it, presents a rational basis for the low-lead regulations. [*Ethyl,* at 75–76]

In complicated cases, it must be tempting for administrators, judges, and others to assign zero or very small weights to some studies mainly to make their job easier. Do they do this? For whatever reason, in *Ethyl,* for example, the so-called Azar study was given very little weight (Azar et al., 1973). We are puzzled by this action. Acting as careful scientists, the authors of that study indicated various shortcomings in their investigation. They are not fatal flaws. Nevertheless, these points were seized upon by the court as reasons to disregard entirely the information gathered by these scientists.

Two central questions faced by the administrator, and hence by the court, were these: (a) Did a causal relationship exist between air lead levels and blood lead levels? and (b) If such a relationship exists, what is the percentage of air lead that is absorbed into the blood through the lungs? The court had before it a variety of scientific studies that addressed these questions. Some of them supported EPA's position, and some of them did not. The relevant scientific evidence can be divided into three categories: (a) theoretical calculations, (b) epidemiological studies, and (c) clinical studies.

The theoretical calculations were a series of logical deductions based on a set of assumptions, such as the total amount of air inhaled per day by an average person and the percentage of lead in the air that was absorbed into the blood. The purpose of the calculations was to find the critical air lead level that would cause the bloodlead level to exceed 40 micrograms per 100 milliliters. This value, 40 μg/100 ml, was regarded at the time by the administrator as a level that it would be prudent not to exceed for health reasons. (The administrator indicated that uncertainty existed about the exact value that should be used for this purpose. In the majority opinion, the court favorably commented on the administrator's candor in addressing this issue). The theoretical calculations proceeded as follows. It was assumed that a "standard man" breathes 20 m^3 of air each day and absorbs 30% of the lead it contains. Calculations showed that 30 μg of lead would be absorbed daily if this "standard man" was exposed to a concentration of air lead of 5.0 $\mu g/m^3$, which is a level found in many larger cities. A subject absorbing 30 μg of lead daily would reach a blood lead level of 40 μg/100 ml within a year. This conclusion rests on the assumption that 30% of the lead in inhaled air will be absorbed into the blood. Data are needed to substantiate this assumption.

Two controlled experiments demonstrated that lead in air finds its way into the blood of human beings who breathe that air. A lead-isotope study involved the exposure of a single individual for 160 days to air with an elevated concentration of lead (Rabinowitz et al., 1973). After this person had reached equilibrium, it was determined that 28% of the blood lead had come from the air. The remainder came from other sources, for example, from food. The second experiment, the Albany Chamber Study (1973), used volunteers who were prisoners, and exposed them to three different levels of air lead. Higher levels of air lead were correlated with higher levels of blood lead.

As usual, the epidemiological studies were most vulnerable to criticism because of the large number of uncontrolled lurking variables that could influence the results. Much importance was attached to the Daines study (1972), which covered a two-year period, that showed that, among a sample of black women, there was a correlation between air lead and blood lead, and that the average value of both these two measured quantities decreased as the distance to a major road increased. The Seven Cities Study investigated the relationship between air and blood lead in seven U.S. cities. Sources of lead other than the air were neither controlled nor measured. Overall, the results did not show that higher air lead levels were clearly correlated with higher blood lead values. When comparisons were made between metropolitan and suburban areas of the same city, however, a fairly consistent pattern was revealed: higher air lead and blood lead levels were found in the urban populations. It was argued that climate and diet differences were less pronounced within a city as compared to differences among cities. Even so, these urban–suburban comparisons did not give, individually, statistically significant results. Collectively, however, it was argued by the EPA that the relationship was established. Petitioners disputed this point.

Scientific data did not supply unequivocal answers to the key questions at issue. Many individuals concerned with this case had to come to grips with the problem of determining what information could be properly extracted from the data surrounded, as they were, with a cloud of uncertainty. These individuals included the administrator, the scientists, the lawyers, and the judges. The method of analysis and the language used in the analysis of the data were not common among all individuals. There were, therefore, serious problems of communication. Not surprisingly, lawyers and judges, when dealing with statistical and scientific ideas, made statements that were unclear and sometimes confusing when judged from the point of view of a professional statistician or scientist.

Two ideas that seemed to be difficult for nonstatisticians to understand were the nature of a hypothesis test and the meaning of the concept of statistical significance. In the opinions the word *significant* appears many times—sometimes in its limited technical statistical sense, sometimes in its everyday sense as a synonym for *important,* and sometimes in a way that makes it difficult to know which meaning was intended. That a result could be statistically significant and scientifically or practically unimportant did not seem to be well understood. Nor did the following fact: the statement that a set of data does not allow one to reach the conclusion that a tested hypothesis is discredited is not equivalent to the statement that the set of data allows one to conclude the tested hypothesis is correct. This fact is sometimes expressed this way: failing to reject a hypothesis is not the same as accepting it.

There is a short discourse in note 112 on the meanings of *significance* in statistics and everyday use. It is actually a minilecture from the majority to the dissenting judges, who are scolded for misunderstanding this term. In this discourse a statistical text is quoted. But the understanding of terminology that is revealed in that note seems, from a statistical point of view, to be fleeting glimpses rather than something that pervades the court's majority and dissent opinions. Indeed, immediately following the citation from the statistical text, we find this statement:

> Typically, scientists refuse to certify an observed relationship as "significant" unless they are 95% certain that the data could not have been generated randomly. [*Ethyl,* note 112]

From a statistical point of view, this language is at least unusual; some statisticians would claim that it is meaningless.

The majority and dissenting opinions in both the panel and *en banc* decisions engaged in a "hard look" substantive review of the evidentiary support for the administrator's decision. In engaging in thorough and searching substantive review, the majority and the dissent held widely contrasting views regarding the statistical certainty to be required of intermediate conclusions in the reasoning process from observed facts to ultimate conclusion. Judge Wilkey, writing for the majority in the panel opinion, believed that the administrator's action was infirm in part because:

> [The Administrator] appears to have reached a conclusion as an independent deduction but without any chain of scientific facts or reasoning leading him ineluctably to this conclusion. [*Ethyl,* pp. 2011–2012].

Judge Wright, writing for the dissent in the panel opinion, rebuked the majority for their inflexible insistence on scientific "fact" apart from probabilities:

> Contrary to the majority's suggestion that [a]ll true risk assessment is based on facts and nothing else," one must recognize that, at least in a metaphysical sense, all facts are themselves nothing more than risks, or *statistical probabilities*. This is certainly true in a legal sense as well. . . . As the fact-risk distinction has evolved in this quite technical branch of the case law . . . "fact" refers to that which is scientifically proved (although, again, science only "proves" to within a statistical- and not a factual-certainty), and "assessment of risk" refers to all other permissible inferences that may be drawn from the proved facts and other data. [*Ethyl,* note 32 at p. 20128, citations omitted, emphasis added]

The controversy over "scientific facts" versus "statistical probabilities" was developed more fully in the *en banc* opinion in the context of a debate between the majority and dissent over whether the Environmental Protection Agency had acted unfairly with regard to an "intermediate conclusion" of an epidemiological study that showed a relationship between the levels of lead in the air and in the blood in urban and suburban areas of several cities (Tepper and Levin, 1972). This Seven Cities Study, first looked at air lead and blood lead levels between the several cities and concluded that there was no discernible relationship between these two variables on an intercity basis. (For example, although air sampled in Greenwich Village had more lead than that sampled in Washington, D.C., blood lead levels were lower in Greenwich Village (Tepper and Levin, 1972, p. 58, Table 14).

When the authors of the study looked at the relationship between air lead levels and blood lead levels *within* a particular metropolitan area, however, they reached entirely different results. In three metropolitan areas data were collected from two population groups (smokers and non-smokers). In each of the six samples (three cities and two population groups) both the air lead levels and the blood lead levels were higher at the urban site sampled than at the suburban site sampled. The language of the study is worth quoting at length:

> The probability of obtaining six relationships in this direction (urban > suburban) out of six comparisons is approximately equal to $1/2^6 = 1/64 = 0.015$, if one assumes either direction is to be equally likely. *Consequently it is concluded that urban mean blood levels are significantly higher than these from related suburban areas (at the 1.5% level), based on data from three paired locations in Philadelphia, New York, and Chicago. . . .*
>
> The observations that urban levels of blood lead are significantly higher than suburban levels, but that air concentration of lead are [sic] not clearly reflected in blood lead levels generally, suggest that other variables are more important than ambient air lead levels in determining concentrations of lead in the blood. . . .
>
> In a given metropolitan area, urban–suburban comparisons tend to minimize the influence of diet and climate. *That airborne lead contributes to the relatively higher blood lead concentrations in inner-city populations would seem to be the most probable interpretation of these consistent observations.* [Tepper and Levin, 1972, pp. 32, 49, emphasis added]

This statistical analysis was the target of "critical comments" and the data was consequently reanalyzed by the Environmental Protection Agency. The reanalysis ("Additional Analysis of the Seven City-Lead Study") demonstrated, in the words of the majority, that "this 'consistent observation' had less than a 1 in 1000 likelihood of being the product of chance" (majority, *Ethyl,* note 112 at p. 51).

When the case was reheard *en banc* the majority and the dissent took sharply divergent views of what constituted "statistical certainty." What gave impetus to this debate was the fact that the reanalysis of the Seven City Study had been released in such a way that only one of the industry petitioners had had an adequate opportunity to comment.

Judge Wilkey, now writing for the dissent in the *en banc* opinion, objected that administrative due process had been violated since "it was the additional analysis *only* which gave any statistically significant support to the Administrator's conclusions" and the other industry parties were not given an opportunity to comment on the reanalysis (dissent, *Ethyl,* p. 83, emphasis in original). The original analysis, said the dissent, yielded "statistically insignificant results" (dissent, *Ethyl,* note 62 at p. 83). Just why the dissent felt that the original analysis yielded "statistically insignificant results" is not made clear. Writing for the majority,

Judge Wright's characterization of the dissent's criticism is, however, plausible:

> The dissent's emphasis on the importance of the reanalysis of the Seven Cities Study appears to be based on a misunderstanding of the meaning of statistical significance. Statistical significance simply expresses the level of assurance we can have that certain data was not the product of random relationships. . . . There is, of course, no reason why the Administrator cannot rely on observed relationships among data which he is less than 95% certain reflect a true underlying relationship between the phenomena that the data measure. . . . The dissent's assertion that only the reanalysis provides support for the Administrator must be 95% certain before he credits any evidence. [majority *Ethyl,* note 112 at p. 51, citations omitted]

Judge Wright thus makes it clear that, at law, "statistical significance" cannot simply be equated with the achievement of a 5% significance level. In the following passage he argues that, just as "95% certainty" is not required of intermediate conclusions, neither is it required of the ultimate findings of fact in the context of regulatory proceedings.

> Petitioners demand sole reliance on *scientific* facts, on evidence that reputable scientific techniques certify as certain. Typically, a scientist will not so certify evidence unless the probability of error, by standard statistical measurement, is less than 5%. That is, scientific fact is at least 95% certain.
>
> Such certainty has never characterized the judicial or the administrative process. It may be that the "beyond a reasonable doubt" standard of criminal law demands 95% certainty. But the standard of ordinary civil litigation, a preponderance of the evidence, demands only 51% certainty. A jury may weigh conflicting evidence and certify as adjudicative (although not scientific) fact that which it believes is more likely than not. . . . The standard before administrative agencies is no less flexible.
>
> Agencies are not limited to scientific fact, to 95% certainties. Rather they have at least the same fact-finding powers as a jury, particularly when, as here, they are engaged in rulemaking. [majority, *Ethyl,* note 58 at p. 28, emphasis in original, citations omitted]

It could be argued that the nonstatisticians were unknowingly using a Bayesian mode of analysis and that their language was closer to that of the Bayesian rather than that of the frequentist. This may be so. What is clear, from a professional statistician's point of view, is that their explanation, as it stands, is garbled. [For discussions of the presentation of Bayesian statistical analyses in legal proceedings, see Fienberg and Kadane (1983) and Lindley (1977).]

In effect, the court in *Ethyl* deferred to the expertise of the EPA in its interpretation of the data. Industry petitioners sought to have the court reject the logic of the agency's decision-making process and to make its own analysis. The court, although engaging in "hard look" substantive review, refused to second-guess the agency in its analysis of the scientific evidence.

Present Standards of Judicial Review

The conflict between the use of either procedural or substantive review of environmental regulations has, to a great extent, been resolved. Recent decisions by the Supreme Court and Federal Circuit Courts have found that the "hard look" judicial review championed by Judge Leventhal is the standard by which substantive regulatory actions should be judged. Courts are also required to review the procedures used in agency decision making. However, Judge Bazelon's idea that the courts should enhance the regulatory process by "concentrating our efforts on strengthening administrative procedures" has been rejected (*Ethyl,* p. 67).

In *Vermont Yankee Nuclear Corp. v. Natural Resources Defense Council,* 435 U.S. 519 (1978) (*Vermont Yankee*), the Supreme Court reversed two lower court opinions authored by Judge Bazelon that implicitly sought to impose procedures on the Nuclear Regulatory Commission beyond the statutory minima of the Administrative Procedure Act. The Circuit Court decisions had found that through inadequate procedures the commission had failed to weigh and document the environmental effects of two nuclear power plants and, therefore, invalidated the grant of licenses. In rejecting Judge Bazelon's position that imposition of better procedures would improve the decision-making process, the Supreme Court stated that there was no basis that "permitted the court to review and overturn the rulemaking proceedings on procedural devices employed (or not employed) by the Commission so long as the Commission employed the statutory minima . . ." (*Vermont Yankee,* p. 548). The opinion serves as a warning to lower courts about how far they can go in the procedural review of agency decision making.

In *Vermont Yankee,* the Supreme Court also gave a warning as to the limits of substantive review. In harsh language the Court noted that "[t]he fundamental policy questions (regarding nuclear energy) appropriately resolved in Congress and in the state legislatures are *not* subject to reexamination in the federal courts under the guise of judicial review of agency action" (*Vermont Yankee,* p. 558, emphasis in original). Although this language suggests that the courts are not to engage in the "hard look" of substantive review, the Supreme Court in a subsequent decision dispelled any such notion. In *Industrial Union Department, AFL-CIO v. American Petroleum Institute,* 448 U.S. 607 (1980) (*Industrial Union*), a plurality of the Supreme Court upheld a lower court decision that had rejected the substantive basis for an Occupational Safety and Health Administration standard for exposure to benzene. The Supreme Court upheld the decision, but only after making an admittedly "detailed examination of the record" (*Industrial Union,* p. 2872). The review was so thorough that the dissent accused the plurality of making a *de novo* examination of the factual evidence (*Industrial Union,* note at pp. 2890–2891). Whether the Court engaged in *de novo* review is questionable; what is certain, however, is that the Court took a "hard look" at the agency decision rather than defer to its expertise as it had done in *Vermont Yankee*. [For discussion of "hard look" review, see Marcel (1983).]

Finally, it should be noted that labeling "hard look" judicial review as substantive may be a misnomer.

Under the "hard look" doctrine of judicial review, judges do not engage in fact-finding. Instead, they review the scientific validity of procedures used in the fact-finding process. Judges are confined to reviewing the record in light of accepted procedures in the scientific community. Scientists might claim that to label such review as substantive rather than procedural is misleading.

Summary

In analyzing judicial review of agency decision making, one must remember that Congress, in enacting environmental legislation, has generally sketched only the broad features and basic outlines. Congress has done so under the assumption that the expert agencies would fill in the details and that the courts would assure that congressional intent was followed. The net result is that both the agencies and the courts are placed in the position of making quasi-legislative decisions. If the level of judicial review increases, courts become further entangled in legislative decision making. If the level of judicial review decreases, the agencies are given the freedom to legislate. The problem, then, is the proper division of essentially legislative responsibilities between the agencies and the courts. Judge McGowan provides guidance in noting:

> "[T]he question of the scope of judicial review is essentially a problem of allocating Congress' unfinished lawmaking tasks among two sets of unelected officials/judges and administrators. This problem arguably should be left to a variable process that fluctuates according, for example, to the relative expertise of the two sets of officials.

> . . . [A] gross allocation of duties to the courts [too much review] would undo a relatively finely tuned system of shared judicial and administrative responsibility for interpreting and enforcing broadly delegative legislation.'' [McGowan, 1977, note 5 at p. 1166]

5. Conclusions and Recommendations

Issues of environmental health are difficult to understand because of the interplay of many factors. Rachel Carson described the essence of the problem as follows:

> When one is concerned with the mysterious and wonderful functioning of the human body, cause and effect are seldom simple and easily demonstrated relationships. They may be widely separated both in space and time. To discover the agent of disease and death depends on a patient piecing together of many seemingly distinct and unrelated facts developed through a vast amount of research in widely separated fields. [Carson, 1962]

In the face of these scientific uncertainties, in the presence of so many lurking variables, it is challenging indeed for society to promulgate laws that will protect its well being and ensure that the environment will be satisfactory to future generations. It is understandable that actors in this drama sometimes appear diffident. As Judge Leventhal once observed:

> [C]ounsel are often helpful in elucidating a technically complex record, and I have suggested ways in which an appellate court could be more flexible in making use of their knowledge. But surprisingly often in argument before a court that enjoys the services of an excellent bar, the judges' questions on technical matters are fended off rather than answered, even after whispered conferences with the assistant nearby. [Leventhal, 1974, at 551, citation omitted]

Environmental lawyers cannot be expected to be trained as scientists, engineers, or statisticians. They can, and should, however, have enough background in the sciences that they understand the scientific method so they will be able to work intelligently when setting environmental standards and settling disputes. What Frank (1957) has written about the goal of educating citizens in general about science applies in particular to lawyers and judges:

> Many people have believed that this goal could be achieved by popularizing the results of science, by adult education courses in which intelligent and interested men and women could absorb the ''facts''

> discovered by scientists in a digestive way. Conant, however, made the point that by absorbing "results" and "facts" laymen could not acquire any judgment about reports of scientists. What the citizen needs rather is an understanding of how the mind of the scientist works in getting results, and along with it, in what sense those results are "valid" or "reliable" and can be used as a basis of judgment.

Statisticians, who have a special interest in seeing that the scientific method is carefully applied, can help in educating receptive attorneys in the language and logic of science.

Conversely, environmental lawyers should take it upon themselves to educate the statisticians and scientists in the language and thought pattern of law. The role of the courts in matters of scientific controversy remains poorly understood in parts of the scientific community. Scientists and statisticians who have been through the experience of testifying as an expert witness in a legal proceeding are often heard to complain about the manner in which lawyers have seemingly twisted their testimony. Sometimes there is real merit to such complaints, but more often what is perceived as a failure of justice is rather a lack of congruence between two dissimilar sets of goals and ways of reaching them. Scientists and statisticians are concerned with the search for truth. They are most comfortable when they are allowed to qualify a statement. The scientific process is envisioned as an unending one in which it is assumed that further studies may be done to delve deeper into the questions presently under investigation. The focus is on learning, not on making decisions. The focus is on continual improvement of hypotheses and models. It is on the future. The legal system, on the other hand, is concerned with justice. Decisions have to be made in a short time. The focus is on the present. The legal system usually cannot afford the luxury of extended analysis to resolve ambiguities and uncertainties.

Perhaps the discomfort of some statisticians with working within the legal system is also a function of the mental pictures they carry of themselves as professionals. Statisticians can view themselves either narrowly as "producers" or broadly as "consultants." Producers collect, summarize, and interpret data. Such work is often required when environmental matters are at issue. It is our belief that statisticians called on to participate in a legal proceeding should consider to what extent their contributions could be made more valuable by assuming the role of consultants to a given side of a dispute as well as producers of data. This is particularly true in the environmental field, where the key questions of fact are usually not narrow statistical ones. In such cases the statistician's role is more akin to that of coach than to that of player. The statistician, who is concerned with how we learn from data, can help coordinate testimony of the individual scientists, engineers, epidemiologists, and others to ensure that the statistical analyses used are proper and, as far as possible, that the conclusions are mutually compatible. Since statistics is the science of science, the most valuable contribution that a statistician may make in a particular legal proceeding is to be sure that scientific method is used, not abused.

For statisticians to function most effectively in the environmental area, they, even more so than lawyers, must learn something of the specialized fields (biology, chemistry, epidemiology, toxicology, etc.) often involved in environmental controversies. This holds true whether the statistician is involved in litigation, a rule-making proceeding, or is merely an interested observer of environmental law and policy.

If the administrative agencies, upon which Congress so much depends to formulate environmental policy, are to discharge their responsibilities in a manner in keeping with the seriousness of the underlying problems, then it is extremely important that they have competent scientists and statisticians. Given the high levels of uncertainty commonly confronted, it is imperative that these professionals play a part—wisely and competently—in making environmental decisions, especially the difficult ones.

Acknowledgment

The authors wish to thank Peter L. Brightbill for his assistance on early drafts of this chapter. This work was supported in part by the National Science Foundation under Grant SES-9018418.

References

Abelson, P. H. (1978), "The Federal Regulatory Machine," *Science,* **200,** 487.

Administrative Procedure Act of 1946, 5 U.S.C. Section 551 et seq.

Albany Chamber Study (1983). Knelson, J. H., R. J. Johnson, F. Coulston, L. Goldberg, and T. Griffin, "Kinetics of Respiratory Lead Intake in Humans," in *Proceedings of the International Symposium on Environmental Health Aspects of Lead;* CEC, CID, Luxembourg, pp. 391–401.

Allen v. U.S. 588 F.Supp. 247 (D. Utah 1984).

Ashford, N. A., Ryan, C. W., and Caldant, C. C. (1983), "Law and Science Policy in Federal Regulation of Formaldehyde," *Science,* **222,** 894.

Azar, A., Habibi, K., and Snee, R. (1973), "Relationship of Community Levels of Air Lead and Indicies of Lead Absorbtion," in *Proceedings of An International*

Symposium on Environmental Health Aspects of Lead; CEC, CID, Luxembourg, pp. 581–594.

Brickman, R., and Jasanoff, S. (1979). "Concepts of Risk and Safety in Toxic Substances Regulation," unpublished manuscript, Cornell University, Program on Science, Technology, and Society.

Brown, B. V. (1976). "Projected Environmental Harm: Judicial Acceptance of a Concept of Uncertain Risk," *Journal of Urban Law,* **53,** 497–531.

Carson, R. (1962). *Silent Spring,* New York: Houghton-Mifflin.

Clean Air Act, Pub. L. No. 88-206, 77 Stat, 392 (1963) as amended.

Comptroller General, Report to the Congress of the United States (1979). "Improving the Scientific and Technical Information Available to the Environmental Protection Agency in Its Decisionmaking Process," U.S. General Accounting Office, September 21, 1979, Washington, D.C.

Daines, R. H., et al. (1972). "All Levels of Lead Inside and Outside of Homes," *Industrial Medicine,* **41,** 26.

Davis, K. (1972). *Administrative Law Text.* St. Paul: West.

Ethyl Corporation v. Environmental Protection Agency, 541 F.2d 1 (D.C. Cir. 1976), Cert. Denied, 426 U.S. 941 (1976).

Federal Register 38: 33734 (1973). Environmental Protection Agency: Regulation of Fuels and Fuel Additives—Control of Lead Additives in Gasoline.

Fienberg, S. E., and Kadane, J. B. (1983). "The Presentation of Baysian Statistical Analyses in Legal Proceedings," *The Statistician,* **32,** 88.

Frank, Philip (1957). *Philosophy of Science,* Englewood Cliffs, N.J.: Prentice-Hall.

Gelpe, M. R., and Tarlock, A. D. (1974). "The Uses of Scientific Information in Environmental Decisionmaking," *Southern California Law Review,* **48,** 371–427.

Gillespie, B., Eva, D., and Johnston, R. (1979). "Carcinogenic Risk Assessment in the United States and Great Britain: The Case of Aldrin/Dieldrin," *Social Studies of Science,* **9,** 265–301.

Gulf South Insulation et al. v. United States Consumer Product Safety Commission, 701 F.2d 1137 (5th Cir. 1983).

Hart, H. M., Wechsler, H. (1973). *Hart and Wechsler's The Federal Courts and the Federal System,* 2d ed., by Paul M. Bator, Paul J. Mishkin, David L. Shapiro and Herbert Wechsler, Mineola, NY: Foundation Press.

"Hearings": *Regulatory Procedures Act of 1981: Hearings on H.R. 746 Before the Subcommittee on Administrative Law and Governmental Regulations of the House Committee on the Judiciary,* 97th Congress, 1st Session (1981).

Hunter, J. S. (1980). "The National System of Scientific Measurement," *Science,* **210,** 869.

Industrial Union Department AFL-CIO v. American Petroleum Institute, 448 U.S. 607 (1980).

Leventhal, H. (1974). "Environmental Decisionmaking and the Role of the Courts," *University of Pennsylvania Law Review,* **122,** 509–555.

Lindley, D. V. (1977). "Probability and the Law," *The Statistician,* **26**(3), 203–220.

McGowan, C. (1977). "Congress, Court, and Control of Delegated Power," *Columbia University Law Review,* **77,** 1166.

Marcel, K. S. (1983). "The Role of the Courts in a Legislative and Administrative Legal System—The Use of Hard Look Review in Federal Environmental Litigation," *Oregon Law Review,* **62,** 403.

National Environmental Policy Act, Pub. L. No. 91-190, 42 U.S.C. 4331 et seq. (1969).

Pierce, R. J., and Shapiro, S. A. (1981). "Political and Judicial Review of Agency Action, *Texas Law Review,* **59,** 1175.

Pollack, A. K., Hunter, W. G., and Apple, G. (eds.) (1983). "Proceedings of the Seventh Symposium on Statistics and the Environment," *The American Statistician,* Part II, **37.**

Portney, P.R. (ed.) (1978). *Current Issues in U.S. Environmental Policy,* Baltimore: The Johns Hopkins University Press.

Public Citizens Health Research Group v. Auchter, 554 F.Supp. 242 (D.D.C. 1983), Rev'd, 702 F.2d 1150 (D.C. Cir. 1983).

Quarles, J. (1976). *Cleaning Up America An Insiders' View of the Environmental Protection Agency,* Boston: Houghton-Mifflin.

Rabinowitz, M., Wetherill, G. H., and Kopple, J. D. (1973). "Lead Metabolism in the Normal Human: Stable Isotope Studies," *Science,* **182,** 725.

Rogers, W. H. (1977). *Handbook on Environmental Law,* St. Paul.: West.

Rogers, W. H. (1979). "A Hard Look at Vermont Yankee: Environmental Law Under Close Scrutiny," *Georgetown Law Review,* **67,** 699.

Scherer, F. M. (1979). "Statistics for Government Regulation," *The American Statistician,* **33,** 1–5.

Tepper, L. V., and Levin, L. S. (1972). "Seven Cities Study—A Survey of Air and Population Lead Levels in Selected American Communities." Department of Environmental Health, College of Medicine, University of Cincinnati, Cincinnati, Ohio. Final Report (Contract Ph-22-68-28) 73 pp.

Toxic Substances Control Act, Pub. L. No. 94-469 (1976) 15 U.S.C. 2601 et seq.

Vermont Yankee Nuclear Corp. v. National Resources Defense Council, 435 U.S. 519 (1978).

White, G. F. (1982). "Ten Years After Stockholm," *Science,* **216,** 569.

Comment

D. E. Splitstone
IT Corporation
Pittsburgh, Pennsylvania

More than any other area of the law, environmental law focuses on the intersection of public policy and the expanding frontiers of science. The wave of environmental regulation during the 1970s reflects the governmental response to the perceived public concern over the environment. In this very brief time frame, advances in technology have enabled the measurement of alleged environmental contaminants to parts per trillion where measurement to only parts per million was possible before. However, the development of methodology to assess risk to human health and environmental welfare in the light of this newfound ability of measurement has significantly lagged behind. A reluctant judiciary has been asked to fill this void. Apple, Hunter, and Bisgaard provide an excellent discussion of the enormous size of this task and the question of its appropriate delegation to the judiciary.

The sentiments of the judiciary were concisely expressed by Chief Judge Markey (1982) at the Sixth Symposium on Statistics and the Environment when he said:

> I am not sure that federal judges, immune from the political process, should ever be involved, under any circumstances, as arbiters of the degree of risk acceptable to the public. That task is much better performed by the people's representatives in the Congress, whether directly or through delegation to an agency. Unlike the courts, Congress and the agencies have investigative staffs and the opportunity for input from numerous segments of the public affected or likely to be affected by the degree of risk.

The Supreme Court, ruling in *Chevron, U.S.A. Inc., v. Natural Resources Defense Council* (1984) involving the Environmental Protection Agency's (EPA) air-emissions trading policy, expressed much the same sentiment:

> [P]olicy judgments are more properly addressed to legislators or administrators not to judges; judges are not experts in the field and are not part of either political branch of the government. Courts must, in some cases, reconcile competing political interests, but not on the basis of the judges' personal policy preferences. In contrast, an agency to which Congress has delegated policymaking responsibilities may, within the limits of that delegation, properly rely upon the incumbent Administration's views of wise policy to inform its judgments. While agencies are not directly accountable to the people, the Chief Executive is, and it is entirely appropriate for this political branch of the government to make such policy choices—resolving the competing interests which Congress itself either inadvertently did not resolve, or intentionally left to be resolved by the agency charged with the administration of the statute in light of everyday realities.

A major cause, perhaps the root cause of much of the environmental litigation, is that Congress is unwilling to set a firm policy regarding the level of risk that is acceptable. For example, the Clean Air Act contains words such as "promote public health and welfare" with an "adequate margin of safety," the Toxic Substance Control Act specifies only "unreasonable risk," and the Federal Insecticide, Fungicide and Rodenticide Act is concerned with "unreasonable adverse effects on the environment." The EPA is then left to determine what is "unreasonable" and what is an "adequate margin of safety."

The courts have on several occasions affirmed that this was in fact the intent of Congress. As to the powers and duties of the EPA Administrator under the Clean Air Act, the court in *Ethyl Corporation v. Environmental Protection Agency* (1976) indicated that the salient technical expertise is within the agency. Thus statistical as well as other technical disagreements should not be argued in court, unless they reveal a glaring deficiency in the agency's reasoning on a particular issue. How far the courts should go to investigate a potential deficiency may be the subject of debate, as indicated by the authors. It is clear, however, that it would be of benefit to have a more technically informed judiciary and Chief Judge Markey (1982) has made some suggestions about how that may come about.

The statistical issues yet to be resolved in environmental law and regulation are legion. The resolution of some, such as extrapolation of dose-response relationships to low dose levels (see Park and Snee, 1983) may add significantly to the body of statistical knowledge. Other issues of equal importance concern the statistical properties of environmental measurements. Hunter (1980) has provided a very fine discussion of these measurement problems. Although in general the statistical problems of environmental measurement are not very difficult, they are very interesting when their potential to impact policy, standard setting, and enforcement is considered.

Brown et al. (1976) investigated the error associated with the electron microscopic counting of asbestos fibers in water. This work was performed while the investigators served as court-appointed consultants in the *Reserve Mining* case (1974) (*United States v. Reserve Mining Co.*). The investigators conclude:

> [I]t is apparent that we have no measure of the "true" concentration of amphibole fibers. Laboratories differ and we have no way of knowing which may be nearest the truth. If, however, we wish to compare concentrations at two sites or under two different sets of circumstances, we can do so, in a relative sense, if at least two provisions are fulfilled: (1) the concentration measures must all be done in the same laboratory using the same equipment and the same technician, and (2) several aliquots of each sample of water must be examined independently.

The results of this study have implications beyond the *Reserve Mining* case, as these investigators concluded:

> The large variability of the counting technique must be considered when planning experiments in which measured amounts of asbestos are given to experimental animals. It must also be assumed that large numbers of fibers that cannot be resolved with the light microscope are present in most asbestos fiber preparations. Thus, any study purporting to relate fiber size to biological effects should state fiber size, as well as number, in terms of both light and electron microscopic determinations.

Thomas (1977) provides an excellent discussion of the Reverse Mining Case.

Sophisticated measurement techniques are not the only ones that require statistical attention. The regulation of the emission of air pollutants on the basis of the visual perception of smoke has a history as long as the history of air-pollution regulation. Prior to 1881 and the enactment of municipal legislation in Chicago, Illinois, and Cincinnati, Ohio, it had to be proved in each individual case that smoke was injurious or offensive to the senses. With the enactment of this initial legislation, smoke became a nuisance per se. This early legislation prohibited smoke of a degree defined variously as "dense," "black," or "gray." Problems associated with the definitions of "dense," "black," and "gray" led to the adoption, in 1919, of a standardized method for defining smoke density—the Ringelmann chart—by its inclusion in a smoke ordinance for the City of Boston passed by the Massachusetts legislature. Court challenges to the use of the Ringelmann chart in Los Angeles County Rule 50 and Section 24252 of the California Health and Safety Code during the 1950s in *People v. International Steel Corporation* and *People v. Plywood Manufacturers of California* resulted in a ruling that a trained inspector need not have a Ringelmann Chart in his possession while observing smoke plumes (U.S. Environmental Protection Agency, 1978).

The certified observer's perception of emission opacity is a major enforcement tool used by the various regulatory agencies. Few studies of the precision and accuracy of this measurement have been performed, yet millions of dollars for control systems have been spent to achieve compliance with regulations based on this measurement (see Splitstone and Lin, 1983). Currently, continuous emission monitors, which measure light transmittance and allegedly translate transmittance into units of fume opacity, are being employed. It is, however, not at all certain whether these continuous monitors track the visual emission opacity measurement (see Splitstone and Erickson, 1983). Hunter (1980) points out that

> [M]andated measurement methods are required by regulatory agencies and other government groups. These methods exist for measuring almost all physical, chemical, and biological phenomena. The methods have been culled from the literature, from the organizations that write voluntary standards, and some have been developed by the agencies. Few provide adequate estimates of precision, and fewer still provide an evaluation of interlaboratory bias. The societal costs of these poor measurements are large. Much needs to be done to meet the physical and statistical requirements for establishing and maintaining dependable measurements. Excepting those directly supported by the National Bureau of Standards, most of the nation's measurement systems are uncontrolled.

If risks are to be assessed, informed and intelligent regulations written, and compliance to be determined, the precision and accuracy of the measurements on which these functions depend must be investigated and the measurement system controlled. That it is possible is demonstrated by the system of weights and measures. To bring it about requires more attention by the statistically informed.

Neyman (1977) indicated the need to consider competing risks in the formulation of environmental policy. It is the responsibility of those familiar with the art and practice of statistics to make sure this message is delivered, not only to our elected representatives but to the public as well. However, as practitioners of statistics, we must also be careful not to infringe on the right of the appropriate policymaker. It is appropriate for us to assess risk. It is not appropriate for us to say how much risk is acceptable. Sometimes there is a very fine line between these two concepts. The National Research Council (1983) has done a fine job of addressing this issue in its report *Risk Assessment in the Federal Government: Managing the Process*.

The authors give us the message that there is a crying need for statisticians to be involved in environmental decision making. Involved, not just as "producers," but as "consultants" in the broad sense of the word. Consultation must be provided outside the formal confines of the court, to our elected and appointed officials, who in many cases do not realize they are in need of a statistical consultant. Anderson (1983a, b, c) has reviewed these needs and some attempts to satisfy them. More, however, is left to be accomplished.

Additional References

Anderson, R. L. (1983a). "Statistics and the Environment," *Amstat News*, American Statistical Association, April.

Anderson, R. L. (1983b). "Statistics and the Environment (II)," *Amstat News*, American Statistical Association, September–October.

Anderson, R. L. (1983c). "Statistics and the Environment (III)," *Amstat News*, American Statistical Association, November.

Brown, A. L., Taylor, W. F., and Carter, R. E. (1976). "The Reliability of Measures and Amphibole Fiber Concentration in Water," *Environmental Research*, **12**, 150–160.

Chevron, U.S.A., Inc. v. Natural Resources Defense Council, Inc., et al., No. 82-1005 (U.S. Supt. Ct. 6-25-84), *21 ERC 1049*, July 13, 1984.

Clean Air Act, 42 U.S.C. §7401 *et seq.*

Federal Insecticide, Fungicide, and Rodenticide Act, 7 U.S.C. §136 *et seq.*

Markey, H. T. (1982). "The Legal Implications of Risk," *The American Statistician*, **36**(3), Part 2, August.

National Research Council (1983). *Risk Assessment in the Federal Government: Managing the Process*, National Academy Press, Washington, D.C.

Neyman, J. (1977). "Public Health Hazards from Electricity—Producing Plants," *Science*, **195**, 754–758.

Park, C. N., and Snee, R. D. (1983). "Quantitative Risk Assessment: State-of-the-Art for Carcinogenesis," *The American Statistician*, **37**(4), 427–441.

Splitstone, D. E., and Erickson, B. C. (1983). "The Relationship Between Visual

Emission Observations and Continuous-Emission Monitor Measurements of Coke-Battery Stacks," *Proceedings: Symposium on Iron and Steel Pollution Abatement Technology,* USEPA EPA-600/9-83-016, April 1983, pp. 201–217.

Splitstone, D. E., and Lin, J. Y. (1983). "The Precision of Visual Emission Observations Within an Integrated Iron and Steel Works," *Proceedings: Symposium on Iron and Steel Pollution Abatement Technology,* USEPA, EPA-600/9-83-016, April, pp. 190–200.

Thomas, W. A. (1977). "Judicial Treatment of Scientific Uncertainty in the *Reserve Mining* Case," *Proceedings of the Fourth Symposium on Statistics and the Environment,* American Statistical Association, pp 1–13.

Toxic Substance Control Act, 15 U.S.C. §2601 *et seq.*

U.S. Environmental Protection Agency (1978). *APTI Course 439 Visible Emissions Evaluation—Student Manual,* National Technical Information Center, PB291861, September.

United States v. Reserve Mining Co., 380 F.Supp. 11 (D. Minn. 1974).

Confidence Intervals in Legal Settings

Herbert Solomon
Department of Statistics
Stanford University
Stanford, California

ABSTRACT

In a number of cases in a legal or regulatory setting, the analysis of sample data is central to the issue at hand, for example, estimating a population mean or some other parameter. The confidence interval constructed and the size of the confidence coefficient are important evidentiary items for the judge, jury, or hearing examiner. Several cases are reported and discussed in which confidence interval estimation bears a heavy burden in the resolution of disputes in a legal setting.

1. Introduction

On a number of occasions in the judicial arena or in governmental agency decision making based on hearings before regulatory bodies, estimates from data collected in surveys, or data resulting from observations, or data produced by clinical trials or laboratory experimentation are evidentiary items to be considered by judge, jury, hearing examiner, or other decision maker. In some situations they are central to resolving issues. How estimates are produced and what one may say of their operational meaning have become the subjects to which adversaries address themselves in direct examination and cross-examination. In this chapter, we discuss confidence interval estimation and its role in legal or quasi-legal settings. In this way we find statisticians assuming an important role in litigation in courtrooms or hearings before administrative judges and examiners.

In addition to the analyses of the data, the statistician is faced not only with presenting the results but also with addressing the court about the meaning of the results obtained through his or her techniques. The notion of variability is well understood in law, but the very specific methods we use to quantify variability are difficult to grasp—although with time this situation is improving. Actually, a number of judges have become conversant in recent years with confidence interval arguments and their meaning. This is especially so on the federal bench. I have tried a number of different ways to explain in court what confidence intervals are telling us. The implications of confidence intervals in litigation are best put forward by discussing actual cases.

In what follows we shall observe the application of the confidence interval notion in legal and prelegal settings. This will be accomplished by reporting on specific cases in which I have appeared or served as consultant and other cases reported in the literature by statisticians or from decisions rendered by courts usually at the appellate level. First, a brief word about the confidence interval notion, which we understand but must make intelligible to the court. We recognize that the data before us do not exhaust all the information in a universe; namely, any estimate we obtain from them is subject to error since it is based on incomplete information. The point estimate, such as the arithmetic mean, does not incorporate any aspect of the variability attached to estimation procedures. It is more meaningful to include variability information provided by the standard deviation. One way of accomplishing this is to construct the interval estimate known as a confidence interval, which is measured by a confidence coefficient expressed as a percentage.

A court usually does not grasp the ad infinitum meaning implicit in a confidence statement, but the width of the interval and the degree of confidence we assert can be digested. Thus, as we know, we should like the range of the interval to be quite narrow and, of course, the confidence coefficient to be high. To achieve this, the sample size would have to be large. Small sample sizes can work to the decided advantage of one of the contesting parties for whom a wide interval can damage the other side's case. Another way of shortening the interval is to reduce the size of the confidence coefficient. This would only upset the judge and is not a desirable action. Thus, sample size and size of confidence coefficient are controlling factors, and we are usually captives of the former.

We have essentially been talking about two-sided confidence intervals. There are occasions where upper one-sided or lower one-sided confidence intervals are more appropriate for issues at hand. Conceptually, the confidence coefficient has the same meaning but these intervals are a bit more difficult for the judge to assimilate.

In the cases to be discussed shortly, the unknown parameter value is estimated from the sample data on hand. The confidence coefficients and confidence interval widths derive from probabilistic considerations applied to the estimators rather than the estimated parameters. There may be situations in which the parameter also has a probability distribution, which if known, can help provide better confidence intervals. There can be some limitations to this approach in legal and regulatory matters. Assumptions about the probability distribution of the parameters or assumptions in general about possible values of the parameters can be fervently contested and thus reduce the credibility or validity of the results. On some occasions, this information is available only from other expert witnesses appearing along with you for an attorney, and this may weaken the thrust of the statistician's study; or more important, the attorney does not wish to make this expert information available to you since nonstatistical testimony and yours, arrived at independently, can certainly strengthen the case. While it would be foolish to rule out what may be termed a Bayesian analysis, the data-dependent confidence interval to estimate a parameter has values that can be replicated by any statistician using the same data, and it has an operational meaning that can serve as an evidentiary item.

2. Estimating Retail Sales Subject to Sales Tax

Let us begin with a case, reported in the literature by Sprowls (1957), that brings out other salient points when a court is presented with confidence interval statements. A large retail chain (Sears Roebuck) sought a sales tax refund from the city of Inglewood, California. Sales made by stores in the city to residents living outside the city limits were not subject to a $\frac{1}{2}$% sales tax. The store in question did a substantial volume of business with out-of-city residents. Taxes were paid on a quarterly basis. An error in measuring the extent of out-of-city sales resulted in an overpayment to the city, which was discovered by the accounting department of the retail chain.

A pilot study based on some sample days indicated an overpayment of $27,000, and a court suit was instituted for its recovery. The store presented as supporting evidence results from a probability sample that independently estimated the refund at approximately the same amount. This probability sample based on a random selection of 33 out of 826 working days yielded a ratio of out-of-city to total sales of 36.69%. The sampling error in this case turned out to be approximately 1.5%. A 95% confidence interval for the value that would result from a complete audit is therefore 33.7–39.7% (36.69% ± 3%). Translated into dollar amounts, these ratios become $28,250 ± $4,200, or approximately $24,000 to $32,500, and thus tend to support the claim of a tax refund of $27,000.

The judge ruled against the introduction of these sampling results as evidence. It is significant but typical that the counsel for the city asked the statistician appearing as an expert witness for the retail chain only one question: "Do you know exactly that the ratio of out-of-city to total sales is 36.69%?" Obviously, no one could make that statement from the sample data. However, the judge permitted a complete audit before he ruled on the case, if the store was willing to undergo this expense. The complete audit yielded a figure of $26,750, just a bit short of the claim of $27,000. In this situation the court was in effect giving zero weight to a random variable (the sample estimate of 36.69%) as a piece of evidence. This seems harsh, and yet may make some sense in that an affluent company was required to spend its own funds to provide a value with zero variability. It would be a different matter if funds were not available to the complainant to accomplish this or if, as is frequently the case, it is impossible to reconstitute all the original observations. Also, note that this took place 30 years ago and courts have become more tolerant and sophisticated in admitting statistical testimony.

3. Estimating Error Rates in Welfare Eligibility

Quality-control programs depend on sampling techniques. It is usually impossible to examine every item in a process for error or flaws. In connection with a number of welfare programs in which federal funds are employed, there is an auditing process to monitor how well eligibility requirements are met and whether proper amounts are paid. It is obvious that errors will come about, and so feasible error rates are determined; these vary from program to program. If error rates are exceeded, a financial sanction can be applied that depends on the amount of the exceedance.

For the Aid to Families with Dependent Children (AFDC) program the error rate is established at 4%. To monitor this exactly would require an audit of every case in the AFDC program and, in addition, to do this in a reliable manner. Obviously, some sampling is in order for a decision to be made on how well eligibility and payout are going. The use of sampling implies immediately there will be some uncertainty attached to any estimate of the error rate.

In a California county (Alameda) in 1981, a sample size of 152 cases was employed to obtain an estimate of error rate. The error rate estimated by the state for this county was 5.74%, and this was translated into a penalty of $944,597. The estimated error rate may be useful as a guide for management decisions, but it does not seem relevant for the application of financial sanctions without taking into account the variability attached to the estimate on the basis of 152 cases.

A more realistic view of the uncertainty can be obtained by computing a confidence interval estimate. When applied to the county data, the 95% confidence interval for the actual error rate is 2.54% to 8.94%. The lower boundary of this interval for the true error rate is well below the designated error rate. It suggests that to be fair in assessing financial penalties, a much larger sample, if not the whole population of cases, would have to be examined to see what the true error rate is. In short, not enough evidence is at hand for an action to be taken either way. If the lower boundary of the confidence interval goes above 4%, this might suggest the notion of a financial sanction, but even here the exact amount would still be subject to argument.

The sample size used to estimate the AFDC error rate in this county is too small to provide an accuracy that should be required if large financial penalties are to be imposed. There is too much arbitrariness in the evidence at hand for a decision about a financial sanction to be made and imposed. What is called for is more discussion between the state and the counties about how this important problem can be handled now and in the future. Actually, a number of California counties received financial sanctions, but the confidence interval approach applied in each county produced lower endpoints smaller than 4%. The state government dismissed the sanctions against the counties and is now looking into new procedures that take confidence interval statements into account for monitoring county performance on eligibility and payout and minimize anticipated litigation. In fact, in 1982 legislation (Assembly Bill No. 1456), the State of California mandated the 95% confidence interval as an index for amount of financial sanction.

4. Estimating Underpayment of Taxes

The Internal Revenue Service reconstructed the cash drop at craps tables at a Nevada casino for three fiscal years beginning 1 April 1961 and ending 31 March 1964. The basis for the reconstruction is the cash count obtained by teams of IRS agents at each of 87 craps table shifts at sampling intervals over this period of time. On the basis of these counts, the IRS alleged that shortages in reported income occurred according to some design by the casino. On the basis of this sampling of a population of over 19,000 craps table shifts for the three-year period, the IRS asked for several million dollars in delinquent income taxes, plus a 50% penalty that alleged fraud. Despite the pejorative connotation of that term, the action was a civil procedure rather than a criminal action that had been dispensed with earlier for other reasons.

It appears upon examination that the IRS was a bit unrealistic in the way it handled the statistical assessment of unreported income. It claimed at first that a team of three agents spelled each other over an eight-hour shift and counted the cash presented to the table for either gaming or for chips. All cash is immediately put into a drop box at the table, and at that time was then counted after the eight-hour shift by the casino and posted in their ledgers. The totals provided by each of the three agents were then accumulated and matched with the house count. The comparison was done at a much later date—for one thing, an income tax return is filed only once a year, and the IRS auditing occurs afterwards. Since the agents were informed to count all bills as $1 bills if they could not read the denomination, the assumption by the IRS was that, at best, their count should be under what the house reported. When comparisons were made over the 87 shifts, the house count was, at times, under the amounts recorded by the IRS agents. This suggested underreporting of incomes ("skimming") to the IRS.

When the issues began to receive scrutiny by both sides, and depositions were taken, the IRS stated that two teams of three IRS agents made the counts rather than one, and that, moreover, that team reporting the larger income provided the value used by the IRS. The IRS gave no thought to individual variability in counts made under casino gambling conditions. This explicit lack of attention to counting errors made by IRS agents is a fundamental mistake in the basis for their claims. It stands out sharply because the IRS itself assumes the possibility of such counting errors by requiring the use of two teams of agents to estimate cash transactions at each craps table. The use of the maximum value in this situation is an especially poor decision. It can be demonstrated by a very simple model that the employment of this rule can lead to a probability as high as 0.75 that the casino can be accused of understating its income when in actuality it is not.

The central issue here, however, and it is a statistical one, is how to measure the influence of the counting error on estimating the true income. It was possible, from all the data available, to arrive at an estimate of the standard deviation of counting error made by the IRS agents. This turned out to be approximately $690. A 95% confidence interval or 99% confidence interval would give us two or three times this amount divided by $\sqrt{2}$ on either side of the sample mean (based on a sample of size two) as a range in which, with the stated confidence, the true value (actual cash drop per eight-hour shift) could fall. That is, swings of $976 or $1464 on each side of the mean value obtained from the two counts of the IRS teams. When this confidence interval procedure was applied over the 87 shifts, it was found that with a lower one-sided 95% confidence interval, 75 times out of 87, the house count fell inside the interval, and for a lower one-sided 99% confidence interval, 82 out of 87 values fell inside the interval. This is also accomplished for two-sided 95% confidence intervals, and it was found that 17 out of the 87 shifts fall outside the interval, and for two-sided 99% confidence intervals, that 9 out of the 87 shifts fall outside the interval. In the first case, there were 9 cases in which the house count is too large and 8 cases in which the house count is too small, and for the second situation, there were 4 cases in which the house count is too large and 5 cases in which the house count is too small.

Arguments can be presented for either one-sided or two-sided confidence intervals in this analysis. The two-sided confidence interval makes sense if one is looking for deviations from house count in either direction—namely, human counting errors are just as likely in one direction as the other around a true value. The one-sided confidence interval could be based on the mechanism in the IRS instructions by which some cash drop may be missed altogether, or reported as a lower value whenever there is doubt about actual magnitude. Under these alternatives, the true cash drop is likely to be larger than IRS reconstructed cash drop, provided there is no IRS fallibility in counting.

One could go on with more statistical tests that help resolve the situation. What is important, from the point of view of measurement evidence, is that statistical analysis suggests that differences between IRS counts and house reported income could fall within an interval accounted for by human counting errors experienced by IRS agents under casino gambling conditions. This does not mean that the casino was reporting income correctly. It demonstrates only that the evidence provided does not eliminate human counting error as the reason for the differences.

The case offered provides evidence based wholly on statistical analysis, a method central to the government's approach, which when applied correctly could have suggested to the government that its argument did not have a good evidentiary foundation. As is possible in these cases, the motivation for the proceedings contained factors that might not be directly related to a simple income tax discrepancy. This discussion is reported in Solomon (1982).

5. Estimating Number of Uninsured Automobiles

In an article by Katz (1975) another instance is given. In that article Katz discusses his analysis of a very specific case in the State of Michigan in which he was an expert witness and some of the problems faced by a statistician in a courtroom. The issue at hand was to provide an estimate of the number of uninsured automobiles in the population of passenger cars. It was illegal to drive a car without insurance in Michigan, and so this would obviously complicate the response to be received from the owner of a car that fell into the sample. These individuals had to be assured that the survey was for informational purposes only, and that no action would be taken against them if they responded. Katz traces how in the sample of 249 vehicles out of a total population of over 4,000,000 he arrives at a confidence interval estimate of roughly 2–6%. How to present this analysis and its results are the interesting part of the paper since in effect the statistical analysis of the data obtained by systematic survey sampling is essentially a textbook operation.

6. Estimate of a Coefficient of Variation for Tire Wear Ratios

This is an account of the statistical analysis of National Highway Traffic Safety Administration (NHTSA) test data on tread wear of tires over a standard course. The central issue is whether a criterion—5% coefficient of variation for tread depth in experimental data—is tenable in the light of NHTSA test data. The Rubber Manufacturers Association (RMA) claimed it was not. A theme proposed by an expert for the RMA, that the upper endpoint of the confidence interval estimate for the coefficient of variation should not exceed 5%, ignores the effect of sampling variability on interval estimation of a population value. In this way it seems too demanding and unrealistic.

The development that leads to the computation of 95% confidence intervals for the coefficients of variation (CV) for each auto in each of three tire categories is given in Solomon (1977). These intervals are given in the last column of Table 1 and have a width that is strikingly small. Moreover, the upper end is either under 5% or slightly above it. For radials, all eight intervals do not reach 5% at the upper end; for belted/bias tires, three out of six do not reach 5% and the highest upper end is 6.08%; for bias tires, all nine intervals have an upper value slightly exceeding 5% and go only as high as 7.03%. For the sample sizes involved, these upper endpoints are quite small for a population CV value of 5% and provide adequate justification for NHTSA's assertions about meeting a CV of 5% or less. Most important, all lower endpoints are below 5%.

The coefficient of variation under study here is the ratio of the standard deviation of regression slopes to the mean of regression slopes. The regression line is simple linear regression between tread depth and miles traveled.

From the data it appeared that the variability in regression slopes over all trials was small. The possibility of homogeneity of variances suggests an analysis by which some pooling of data can be accomplished to provide more informative confidence intervals by which a 5% coefficient of varia-

TABLE 1. 95% Confidence Intervals for Each CV for Three Types of Tires[a]

Tire Type and Test Number	95% Confidence Interval	
	RMA	NHTSA
Radial		
A	2.71–17.98	2.47–4.39
B	2.70–17.90	2.30–4.09
C	0.98– 6.46	2.26–4.03
D	1.26– 8.31	2.72–4.84
E	1.52–10.03	2.46–4.38
F	1.20– 7.90	2.58–3.30
G	0.83– 5.46	2.15–3.83
H	1.01– 6.64	2.24–3.99
Belted/Bias		
A	1.46– 9.64	3.08–6.02
B	2.39–15.84	3.11–6.08
C	3.08–20.50	2.62–5.13
D	1.97–13.03	2.12–4.16
E	0.57– 3.77	2.35–4.60
F	0.84– 5.55	2.17–4.25
Bias		
A	2.23–14.78	3.75–6.68
B	2.56–16.95	3.83–6.83
C	5.29–35.82	3.93–7.01
D	1.74–11.47	3.94–7.03
E	1.22– 8.01	3.61–6.44
F	1.15– 7.59	3.64–6.48
G	0.94– 6.18	2.95–5.26
H	5.77–39.35	3.43–6.11
I	3.24–21.56	3.00–5.34

[a] All entries are given in percentages

tion criterion can be assessed. The basic data before us are slopes of lines calculated by regression analysis. These estimated slopes would be expected to be normally distributed, and there was no reason in this issue that one should not suppose this to be the case. It would seem quite feasible that one can add greatly to the precision of interval estimation by pooling the variances.

To see whether this could be accomplished, the variances of the data for each auto trial were calculated and Bartlett's test for variances was applied to the trials for each type of tire to see if it might be reasonable to pool the variances by tire category. For the radial tires and the belted/bias tires, Bartlett's test indicated equality of variances at the 5% significance level. Therefore one can safely pool the variances and thereby increase the information available to check on a 5% coefficient of variation.

For the bias tires, there is a large variance for one set of tires in that category, which, if included, would cause rejection of the hypothesis of equality of variances. This estimated variance may be considered an outlier. When it is omitted, the remainder of the eight estimated variances do not provide for rejection of equality of variances by Bartlett's test at the 5% significance level, and therefore they can be pooled. One variance outlier among 23 estimated variances is something that could be expected by chance, and its omission in no way should affect the analysis. For each tire category, the variances were pooled. This leads to the entries in the NHTSA confidence interval column. The RMA analysts did not pool to achieve estimates of variance, and the RMA confidence interval column reflects this. Note that except for two of the bias tires, the lower 95% confidence limit is less than 5% even for the RMA column, but the large width of these intervals can make it difficult to judge the 5% criterion.

There is a fairly dramatic improvement in the confidence intervals obtained by the analysis using pooling which one can see by comparing the two columns in Table 1. One could probably give an argument for pooling the variances over all tires. This would give even smaller width confidence intervals, but pooling by tire category seems more appropriate.

From this analysis one can strongly assert that the population coefficients of variation appear to be less than 5%. The RMA expert was quite demanding and unrealistic in asking that all confidence intervals have an upper value less than 5%. A statistician trying to ascertain that a coefficient of variation has a population or true value of 5% would never insist that the upper values of confidence intervals resulting from test data be 5% or less. One would expect an upper confidence limit to be larger than 5% when the true value is 5%, but would hope that the 95% confidence intervals would be somewhat narrow in order to make the results meaningful.

This analysis is one component of the expert documentation presented by NHTSA. That agency won approval for tire grading at that time (1978), but the issue is still somewhere in the regulatory process.

7. Estimation When Original Data Cannot Be Recovered

In 1961 and 1962, a dispute between a large producer of electrical power and a municipal utility district consumer in California engaged the efforts of some statisticians. These efforts continued intermittently over a period of one year in connection with hearings held before the Public Utilities

Commission of the State of California. An examiner, or at times one of the five commissioners, presided over the hearings. In such cases, after all testimony is heard and all exhibits are accepted into the record, a finding is made by the examiner and is sent to the commission, which makes the final determination. The hearings in this particular matter are identified as Case 7127 before the commission; they fill 25 volumes.

Before we discuss the substantive statistical issues, note that the technical competences called on by both sides initially were wholly in the engineering sciences and that only gradually did the talents of statisticians seep into the hearings. Although the major point of controversy was statistical in nature, this occurred only because one of the expert engineer witnesses employed statistics and was knowledgeable in the subject. The case in question centered on the implications for billing purposes of metering error in the measurement of electrical power demand. A description of the problem follows.

An agreement made in 1955 listed a monthly rate schedule under which the Pacific Gas and Electric Company (PG&E) would furnish electric power to the Sacramento Municipal Utility District (SMUD). The monthly charge was based on the maximum half-hour load placed by SMUD on PG&E facilities. According to the schedule, the billing for electrical power demand each month was obtained by applying the rate of $1.64 per kilowatt (kW) to the demand on PG&E, provided the demand in any half-hour period did not reach or exceed 50 megawatts (mW). If the maximum demand in any such period in one month reached or exceeded 50 mW, the same rate applied except that thereafter the minimum billing was for 50 mW, and demands in excess of 50 mW were payable at the same basic rate. SMUD also purchased power from the U.S. Bureau of Reclamation, and it was the joint demand by SMUD on both agencies that determined the billing. Since the Bureau of Reclamation could supply at most 290 mW, it was deemed that a total demand of 340 mW or more by SMUD on the two agencies was equivalent to a demand on PG&E of 50 mW or more. Thus a decision on whether a critical PG&E billing value was reached was inextricably intertwined with total demand supplied to SMUD by PG&E and the Bureau of Reclamation.

The agreement left metering this demand to the joint approval of PG&E, SMUD, and the Bureau of Reclamation. Several types of power demand (PD) meters are standard for billing purposes; in this network, PD-5 and PD-7 meters were employed. We find the following metering structure on the day for which a registered electrical power demand in excess of 340 mW was contested by SMUD: two PD-5 meters at Hedge, owned by PG&E; two PD-7 meters, one at Folsom and one at Elverta, both owned by the Bureau of Recalmation. The total demand in this network is measured by the algebraic sum of the four readings: specifically, the readings at Folsom and at Elverta are added to the difference of the two readings at Hedge. On the day in question, namely, 21 July 1960, in the half-hour period between 2:30 and 3:00 P.M., the algebraic sum of the meter readings at these three locations depicted a demand of 342.274 mW, above the critical billing value. Thus, beginning 1 August 1960, a minimum charge for 50 mW was in order and was billed monthly by PG&E. SMUD paid this minimum charge for several months, and then instituted litigation on the basis that one of the meters, the PD-7 located at Elverta under the control of the Bureau of Reclamation, was in error on 21 July 1960. In addition, SMUD claimed that the magnitude of the error was such that the reading indicated a demand above the true situation by an amount that caused the total algebraic sum to exceed the critical value of 340 mW. The registered value contested at Elverta was 223.2 mW. It was SMUD's allegation that this meter was reading high by about 2 mW on the average, and this could suggest a total value less than 340 mW.

It should be pointed out that many engineering studies were necessary for this analysis, and these were accomplished by scores of others. It is my conclusion, however, that the statistical flavoring was the most important consideration, and the events in this case should give impetus to the California State Public Utilities Commission and other agencies to develop billing rules that account more concretely for random error in metering devices in order to prevent or to minimize such litigation.

An investigation of the magnitude of the error and the determination of its implications for the billing agreement between SMUD and PG&E need not have arisen had the agreement incorporated some rules for handling expected measurement error. For example, these rules could be based on a range of readings centered on 340 mW. Since a specific analysis to prepare such a range of error and accompanying penalties was not attempted in the preagreement stage, the implicit interpretation is that rules derived from past history or tradition exist for settling controversies initiated by errors in metering. For example, a 2% error rule was referred to at times in the proceedings; this could give one measure of the error tolerable to both sides. Under such a rule, risks are assumed by both parties in exchange for administrative ease and avoidance of meter disputes.

If, however, a 2% rule or some other rule of tolerance is lacking, a determination can be made of the permissible range of error for the specific metering situation presented so that a foundation can be laid and a structure developed in which to examine the issue of the significance of the claimed metering error.

Suppose this is now attempted. We start with the fact that "true demand" is a variable whose measurement is directly sought. This variable is innately continuous and cannot be found directly except through measuring instruments that are engineered to provide a specific discretization.

Thus, we must recognize at the outset that compromises have been made by the parties in the sense that the measurements that *can* be obtained are substituted for the true demand; in this case the number of impulses registered by PD meters becomes the substitute observation on which billing is based.

By the nature of the instrument, two PD meters indicating the same reading can be triggered by two varying true demands; this is because, at best, the PD meter measures by "quantizing." In other words, each PD meter is designed to register an impulse after a certain demand is placed on the system: at Hedge, the impulse is registered after 2.88 mW; at Folsom, after 1.0 mW; and at Elverta, after 1.2 mW. Thus the PD-5 and PD-7 readings give us a basis for finding tolerable error.

The PD meter system introduces measurement error because of the coarseness arising from quantizing. Such error is typical of measurement in general; the quantizing in this case, however, is quite great. Because of it, deviation from total true demand can vary as much as 7.96 mW (the sum of 2.88, 2.88, 1.0, and 1.2 mW) in either direction. The relative frequency of errors in this range of ± 7.96 mW will provide a useful measuring framework for deciding the zone of tolerable error.

Let us now attempt an evaluation employing quantizing error only. The purpose here, of course, is to develop the magnitude of the units in a yardstick that can then be applied along an error axis to see if an observed deviation is tolerable.

We are interested in how the total demand reading will vary due to the quantizing feature of the PD-7 or PD-5. This reading is the algebraic sum of four independent meter readings. Now we obtain the total demand variability, and the resulting standard deviation then gives us the precision for total demand reading. Thus let us develop the variance for one meter and assume the meter has an impulse saver so that variability is due only to quantizing. Note that here we are ignoring possible biases due to lost impulses, which were quite possible at the time in question. Thus we may be underestimating total error.

Let D symbolize true (unknown) half-hour demand on the part of the system measured by a single PD meter and m the quantizing constant; specifically,

$$\begin{aligned} \text{Hedge:} \quad & m = 2.88 \text{ mW} \\ \text{Folsom:} \quad & m = 1.00 \text{ mW} \\ \text{Elverta:} \quad & m = 1.20 \text{ mW}. \end{aligned}$$

If N is a random variable denoting the number of impulses the PD meter actually registers, then the respective probabilities that $N = n$, or $N = n + 1$ (the only possibilities when impulses cannot be lost) are

$$P[N = n] = 1 - r, \qquad P[N = n + 1] = r$$

where

$$r = \frac{D}{m} - n, \quad 0 \leq r < 1,$$

that is, r is the excess true load still available after n stampings but not quite large enough to produce the $(n + 1)$th stamping.

Thus we have a Bernoulli random variable N for each meter, given that r is known; but since r is the remainder obtained by dividing true load by m and since true load is not known, some assumptions must be made to obtain the variance. If r is known, then the variance of the estimate mN is

$$\mathrm{Var}(mN) = m^2 r(1 - r).$$

Unfortunately, the true loads are not known, but an upper bound for the above variance is obtained by setting $r = \frac{1}{2}$. In a similar analysis, SMUD assumed that the unknown r was uniformly distributed (a typical assumption for roundoff errors) over the quantizing interval, and they arrived at a variance of $\frac{1}{6}$ when $m = 1$. This can be compared with the maximum value of $\frac{1}{4}$ when $m = 1$.

Now total demand is

$$m_1 N_1 - m_2 N_2 + m_3 N_3 + m_4 N_4$$

where m_i is the quantizing constant for each meter and N_i is the number of impulses at each meter. Thus the variance of total demand is

$$m_1{}^2 r_1(1 - r_1) + m_2{}^2 r_2(1 - r_2) + m_3{}^2 r_3(1 - r_3) + m_4{}^2 r_4(1 - r_4)$$

or in our specific situation

$$(2.88)^2 r_1(1 - r_1) + (2.88)^2 r_2(1 - r_2) + (1.0)^2 r_3(1 - r_3) + (1.2)^2 r_4(1 - r_4)$$

which has a maximum value of 4.75 and corresponding standard deviation of 2.18 mW. Using the result proposed by SMUD, wherein each $r_i(1 - r_i)$ is replaced by $\frac{1}{6}$, we get $(\frac{2}{3})(4.75) = 3.16$ and a corresponding standard deviation of 1.78 mW.

Note that variability increases with the size of quantizing constant. Note also that, while a standard deviation of 2.18 mW for total demand reading may be a little on the high side, it indicates that we should not be surprised to find an estimated total demand that differs from true total demand by 4.4 mW ($2 \times 2.18 = 4.36$).

The observed total demand value is used for billing purposes, and now we have provided a yardstick for examining and deciding whether an observation of total demand which falls above 340 mW is aberrant or could be produced by random error alone. Moreover, this yardstick depends only on quantizing error. If we apply the 2% rule, an error of 6.8 mW above and below a true load of 340 mW is allowable. This is very close to a three standard deviation range on either side of true demand—6.54 mW for our maximum standard deviation of 2.18 mW and 5.34 for the SMUD approximation leading to a standard deviation of 1.78 mW. These lead to 1.92% and 1.57% error rules, respectively.

Even though each party can incur a loss from quantizing, the fact that quantizing error alone could account for a specific discrepancy in the acceptable band of error dictates that any such discrepancy should be accepted. Doing so is merely acceptance of the risk created by the parties when they agreed to accept quantized estimates of true demand for billing purposes. Any other use of quantized measurement can only lead to uninhibited billing negotiation and administrative anarchy so far as certainty of metering is concerned. On the other hand, if a reading were outside the band of acceptable error, for example as high as 350 mW or as low as 330 mW, and if some assignable causes for such error were immediately apparent, both sides would no doubt agree that quantizing error alone could not possibly account for such large error and that thus the reading of total demand from the revenue meters should not be directly translated into a biling charge, as in the first illustrative situation.

Thus either the 5.34 mW or 6.54 mW error (or error range of 10.68 mW and 13.08 mW, respectively) could serve as a statistically relevant yardstick of tolerable error for determining whether a billing adjustment should be made under the metering terms of the agreement. From the details of the controversy between SMUD and PG&E, this basic yardstick could be enough to resolve the central issue of whether the estimate of total demand from the four meters, namely, 342.274 mW, should be used directly for billing.

The confidence interval argument developed here could be used to resolve the dispute. Note that this case differs from the Sears, Roebuck situation since the conditions on the day in question cannot be retrieved. Additional discussion of this case, especially taking into account the model where only one of the four meters is in error, appears in Solomon (1966).

8. Estimating County Utility Revenues

Payment to a California county (Sacramento) by a public utility (Pacific Gas and Electric) depended on the ratio of customers' gas pipe lengths that were laid on public property to the lengths on customer property. Typically the gas pipe length to a customer was composed of a length in a public area (franchised area) and a length on the consumer's property (nonfranchised area). The ratio of these two lengths entered into a formula that led to payment to the county by the utility—the larger the public segment, the larger the payment.

Approximately 2.75 million gas meters constituted the population of interest in 1983. For each, there is a ratio of franchised gas pipe lengths to nonfranchised gas pipe lengths. In order to estimate the ratio, 540 customers were obtained by systematic sampling of the utility files, roughly a 1-in-every-5000 rule. Of the 540 selected, complete information on 102 could not be obtained, leaving 438 cases for the study. Naturally, at this point, a study of the possible causes for the missing information on 102 records should be initiated. For the purpose of this analysis and to achieve a bound, let us assume that this nonrespondent group did not arise from a different universe. Now we can assume the sample contains 438 observations. Naturally the nonresponse issue has to be examined and can loom large in litigation. The reason for this study was a contention by the county that the ratio of franchised to nonfranchised gas pipe lengths over the county was not as indicated by the utility, and thus additional revenue was owed the county. The survey was accomplished by the utility.

For each of these 438 records, the files contained gas pipe lengths for franchised and nonfranchised areas. The mean length for franchised areas was 21.71, and for nonfranchised areas it was 42.14 feet, yielding a sample ratio of 0.515. The utility had been paying on a basis of a ratio equal to one-half. The precision of this estimate was viewed through confidence intervals.

The utility developed confidence intervals for the true ratio by recourse to the formulas for ratio estimates given in Cochran (1977). These are pretty much the same as confidence intervals proposed by Paulson (1942) for the ratio of correlated normal variables, except that the latter have fewer restrictions. We are in a fortunate situation here since the sample means in the numerator and denominator are each based on large sample sizes (i.e., 438 observations), and so we have the sample ratio of two correlated normal random variables. I applied the "bootstrap" technique to produce confidence intervals based on these 438 values for franchised and nonfranchised areas (Table 2). The first entry in each cell is the value obtained from the formula in Cochran, the second entry in each cell is the bootstrap estimate based on 1000 samplings where the pairs (franchised and nonfranchised lengths) are replicated.

The bootstrap estimates that are not based on distribution assumptions are very close to the formula estimates based on the ratio of two correlated normal variables. This is not too surprising, because of the robustness of the bootstrap procedure and the large sample sizes. At any rate, the confidence interval analysis demonstrates the tenability of a ratio

TABLE 2. Confidence Intervals for the Ratio of Lengths

Confidence Coefficient	Confidence Interval	
99%	0.435	0.612
	0.430	0.605
98%	0.442	0.602
	0.437	0.595
95%	0.453	0.588
	0.451	0.583
90%	0.462	0.575
	0.461	0.571
80%	0.474	0.561
	0.471	0.559

between franchised gas pipe lengths and nonfranchised gas pipe lengths equal to one-half.

Because of this analysis, the public utility commissioned another sample survey, paying heed to problems arising in the initial survey, such as the nonresponse problem, whose results were in line with their claims. I have no knowledge of any final outcome.

9. Estimating Gross Revenue Damage Claims

A corporation, service company, or retail store can sometimes suffer lost sales as a result of unfair or negligent practices by others. One way of checking this is to investigate actual weekly or monthly revenues over a period in contention and compare these with forecasts of what might have been had the affecting variables not occurred. We shall not concern ourselves with liability determination and just look into damage claims.

Typically, a time series analysis will be performed to forecast revenues that will be compared with actual revenues. To do this, a base or historical period will be determined to provide the data to be employed in estimating the parameters of the time series model. Choosing the time series model will also be extremely important. The statistician will be peppered with questions about the choice of the base period and the choice of the time series model. One big issue will be whether the revenue trend is linear or quadratic (the latter can provide big swings, up or down, in forecasts). After the trend is taken out, the question of seasonality will have to be studied, and if it exists, what correlational model best fits the situation.

Let us assume all this has been done, and it can be a rather large and onerous task if the data base is large and the modeling is subtle. We can then find estimated or forecasted revenues (weekly or monthly) and compare them with actual revenues. The differences will be positive or negative, and the algebraic sum over all weeks or months will tell us on the average whether or not business has been lost in terms of gross revenues. This, of course, will be a point estimate, and will not be as informative as a confidence interval estimate for the true shortfall. Naturally, if the confidence interval contains zero shortfall, then claims for damages are defeated. Realistically, a lower one-sided confidence interval is called for in these situations, and typically the lower endpoint is above the zero dollar value. The lower bound serves as a conservative value on which a dollar damage decision can be made. Thus, concluding that a lower 99% confidence bound is $1,000,000 permits one to tell the court that with 99% confidence the true shortfall is above $1,000,000.

We have implicitly applied a standard deviation value in dollars to achieve what has just been stated. Computing the standard deviation dollar value may not be easy in some situations because of correlation in the difference values (actual minus estimated). In such situations it may be possible to bound the standard deviation value and thus get two lower confidence bounds, proposing one or the other depending on the degree of conservatism desired.

I am involved in two such ongoing studies and do not yet have permission to be more specific. Because of this and other reasons, there are other cases in which confidence intervals are presented in litigation but have not yet reached the printed record. On the other hand, a number of cases in which confidence interval arguments have been employed have reached the appellate level and appellate decisions on the federal and state levels are published.

A number of appellate decisions result from litigation under the Equal Employment Opportunity legislation and various state laws on discrimination in employment. The higher courts are sensitive to sample size issues in employment discrimination cases. Baldus and Cole (1982) discuss and analyze the use of statistics in discrimination suits. Other recent appellate cases in which confidence interval arguments are employed are a suit by the owner of a professional baseball team against the Internal Revenue Service on the depreciation of baseball player contracts (565 F.Supp. 524 (1983)); evaluating a risk of abortion under certain medical conditions in a case by doctors against the State of Florida (550 F.Supp. 1112 (1982)); a suit brought by the National Highway Traffic and Safety Administration against the Ford Motor Company in connection with the evaluation of test track data (453 F.Supp. 1240 (1978)). This indicates the variety of situations in which confidence interval arguments are relevant in legal proceedings.

References

Baldus, D., and Cole, J. (1980). *Statistical Proof of Discrimination,* Colorado Springs, Colo.: Shepard's.

Cochran, W. G. (1977). *Sampling Techniques,* 3rd ed., New York: John Wiley & Sons.

Katz, L. (1975). "Presentation of a Confidence Interval Estimate as Evidence in a Legal Proceeding," *American Statistician,* **29,** 138–142.

Paulson, E. (1942). "A Note on the Estimation of Some Mean Values for a Bivariate Distribution," *Annals of Mathematical Statistics,* **13,** 440.

Solomon, H. (1966). Jurimetrics. In F. N. David (ed.), *Research Papers in Statistics,* New York: John Wiley & Sons, pp. 319–350.

Solomon, H. (1977). "Assessment of Analysis Critical of NHTSA Assertions of 5% Coefficient of Variation for True Mean Ratios." National Highway Traffic Safety Administration, Report #NHTSA-7-3474.

Solomon, H. (1982). Measurement and burden of evidence. In J. Tiago de Oliveira and B. Epstein (eds.), *Some Recent Advances in Statistics,* London: Academic Press, pp. 1–27.

Sprowls, R. C. (1957). "The Admissibility of Sample Data into a Court of Law," *UCLA Law Review,* **4,** 233–250.

Index of Cases

Author Index

Subject Index